Advanced Pharmaceutical Analytical Technology

Advanced Pharmaceutical Analytical Technology

Guest Editor

Alexis Oliva

MDPI

Basel • Beijing • Wuhan • Barcelona • Belgrade • Novi Sad • Cluj • Manchester

Guest Editor
Alexis Oliva
Dpto. Ingeniería Química y
Tecnología Farmacéutica
Universidad de La Laguna
Tenerife
Spain

Editorial Office
MDPI AG
Grosspeteranlage 5
4052 Basel, Switzerland

This is a reprint of the Special Issue, published open access by the journal *Molecules* (ISSN 1420-3049), freely accessible at: https://www.mdpi.com/journal/molecules/special_issues/RZ61127PIB.

For citation purposes, cite each article independently as indicated on the article page online and as indicated below:

Lastname, A.A.; Lastname, B.B. Article Title. *Journal Name* **Year**, *Volume Number*, Page Range.

ISBN 978-3-7258-4865-2 (Hbk)
ISBN 978-3-7258-4866-9 (PDF)
https://doi.org/10.3390/books978-3-7258-4866-9

Contents

About the Editor

Alexis Oliva

Alexis Oliva graduated from the University of La Laguna with a degree in Pharmacy in 1990, going on to receive his PhD in Pharmacy and Pharmaceutical Technology in 1995. He is an Associate Professor in the Department of Chemical Engineering and Pharmaceutical Technology at the Faculty of Pharmacy since 2002. He has participated in twelve national and regional scientific research projects.

To date, he has co-authored over 75 scientific publications, 60 of which have been published in international journals ranked in the first and second quartile according to the Journal Citation Reports. He is also the co-author of three book chapters, various theses and has presented over 60 papers at international and national conferences. He is also the Guest Editor in three Special Issues of MDPI-refereed journals and a reviewer for different journals within the field of pharmaceutical analysis.

His current research focuses on biosimilars and interchangeability studies, conducted in collaboration with Professor SC Chow of Duncan University in North Carolina, USA, who collaborates with the FDA on biosimilar interchangeability issues, and Professor João Concálves of the Research Institute of Medicines at the University of Lisbon, who is a European expert in the immunogenicity and pharmacokinetics of biosimilars.

molecules

MDPI

Editorial

Advances in Pharmaceutical Analytical Technology

Alexis Oliva

Departamento de Ingeniería Química y Tecnología Farmacéutica, Facultad de Farmacia, Universidad de La Laguna, 38200 Tenerife, Spain; amoliva@ull.edu.es

The rapid development of new analytical methods has led to advances in many areas, from the formation of new pharmaceutical analytical technologies to the quality control of new drugs at different stages of enhancement and the search for new dosage forms [1–3].

In essence, quality by design (QbD) can be interpreted as a strategy for the maximization of time and cost savings. Optimizing the safety, efficacy, and quality of a drug requires a thorough understanding of the formulation and manufacturing process at every stage of progression. The full implementation of the QbD approach in the pharmaceutical field has still not been realized, despite the undeniable benefits of the QbD approach and the widespread information on the regulatory expectations of QbD.

This Special Issue contains thirteen manuscripts covering the latest advances in pharmaceutical analytical techniques, focusing on applying QbD principles to the development of analytical methods [4–8]. These principles promote the use of DoE and MLR strategies [9,10]. All the articles in this Special Issue perfectly capture these principles.

The papers collected here represent contributions from leading research groups around the world. They provide insights into the current state of and future directions in the field of analytical methods. For example, Klawitter et al. [Contribution 3], developed a new analytical platform for the assessment of the metabolism and the pharmacokinetics of the antineoplastic drug gemcitabine, which is currently under investigation. In Wang et al. [Contribution 2], the advancement of a cell-based in vitro assay for the biological activity of thyroid-stimulating hormone (TSH) was achieved. Some examples of practical applications also stand out, such as a new, environmentally friendly, high-throughput microwell spectrophotometric assay (MW-SPA) for the determination of three selective serotonin reuptake inhibitors (SSRIs) in their pharmaceutical forms [Contribution 11]. Stimac et al. [Contribution 10] analyzed the suitability of multi-detection SEC as a tool for monitoring molecular processes during antibody (IgG)–horse radish peroxidase (HRP) conjugation reactions and analyzed its plausibility for final product quality control. Peña et al. (Contribution 9) designed the first gas chromatography–mass spectrometry (GC-MS) method to simultaneously quantify 17α-ethinylestradiol (EE) and drospirenone (DP) in contraceptive formulations according to the International Council for Harmonization (ICH) Q2(R1) guidelines.

All research articles in this Special Issue demonstrate significant progress in the development and application of analytical methods for quantifying drugs in complex matrices [Contributions 1,3,7], pharmaceutical preparations [Contributions 2,7,9,11], or biological samples [Contributions 5,6,8,10], with applications in different analytical fields. After the satisfactory advancement and optimization of the analytical method, a comprehensive validation method according to the ICH-Q2 (R1) guidelines and QbD principles is performed to verify that the method is suitable for the intended purpose [Contributions 1–11].

This Special Issue also contains two review articles, where valuable insights into recent improvements in detection methods are provided. Lis-Cieplak et al. [Contribution 12], conducted a comprehensive discussion of the main strategies to improve the analytical efficiency of pyrrolizidine alkaloid (PA) determination via mass spectrometry (MS). PAs are toxic compounds that occur naturally in certain plants, but there are many secondary pathways that lead to the PA contamination of other plants, including medicinal herbs and herbal foods, which poses a risk of intoxication in humans. This review examines various detection means, focusing on liquid chromatography (LC) methods combined with mass spectrometry (MS) detection and using different types of analyzers characterized by varying sensitivities. The authors provide valuable guidance for future developments in MS detection by critically analyzing the advantages and limitations of each method.

The second review by Chen et al. [Contribution 13] explains the state of research on the chemical composition, pharmacology, and quality control of F. ferulaeoides, a traditional indigenous Chinese medicine with significant pharmacological activity, including insecticidal, bactericidal, and anti-tumor properties. This work aims to provide a reference for quality assessment and resource use in anticipation of the application of this product in the industry.

Both research papers and reviews are valuable references for researchers involved in the progression and application of analytical techniques, with a particular focus on the pharmaceutical field, and they demonstrate excellent scholarship and provide comprehensive overviews of their respective fields.

Acknowledgments: The Guest Editors wish to thank all the authors for their contributions to this Special Issue, all the reviewers for their work in evaluating the submitted articles, and the editorial staff of *Molecules* for their kind assistance.

Conflicts of Interest: The author declares no conflicts of interest.

List of Contributions

1. Cossu, M.; Sanna, A.; Mangano, G.; Ledda, G.; Chessa, G.; Gallo, P.; Vella, A.; Pecorelli, I.; Sdogati, S.; Gili, M.; et al. Emerging Mycotoxins in Cheese: Simultaneous Analysis of Aflatoxin M_1, Aflatoxicol, and Sterigmatocystin by LC-MS/MS. *Molecules* **2025**, *30*, 1774. https://doi.org/10.3390/molecules30081774.
2. Wang, L.; Gao, J.; Xu, K.; Li, J.; Liang, C. Development and Validation of Two Cell-Based Reporter-Gene Assays for Determining the Bioactivity of Recombinant Human Thyroid-Stimulating Hormone Pharmaceutical Products. *Molecules* **2025**, *30*, 1037. https://doi.org/10.3390/molecules30051037.
3. Klawitter, J.; Easton, M.; Karpeisky, A.; Farrell, K.B.; Thamm, D.H.; Shokati, T.; Christians, U.; Zinnen, S.P. Novel Approaches to Monitor Pharmacokinetics and Metabolism of Gemcitabine-Ibandronate Conjugate in Mice and Dogs. *Molecules* **2025**, *30*, 354. https://doi.org/10.3390/molecules30020354.
4. Ledeţi, A.; Baul, B.; Ridichie, A.; Ivan, D.; Vlase, T.; Tomoroga, C.; Dragomirescu, A.; Vlase, G.; Bertici, R.A.; Man, D.E.; et al. Thermooxidation of Four Sartans: Kinetic Analysis Based on Thermo-Gravimetric Data. *Molecules* **2024**, *29*, 5527. https://doi.org/10.3390/molecules29235527.
5. Zhang, J.; Yu, H.; Shen, Y.; Yang, X.; Wang, Y. Rapid Liquid Chromatography–Tandem Mass Spectrometry Method for Determination of Total and Free Testosterone in Human Serum and Its Application to Monitoring Biomarker Response of Elite Athletes. *Molecules* **2024**, *29*, 5007. https://doi.org/10.3390/molecules29215007.

6. Szumska, M.; Mroczek, P.; Tyrpień-Golder, K.; Pastuszka, B.; Janoszka, B. Determination of Cotinine, 3′-Hydroxycotinine and Nicotine 1′-Oxide in Urine of Passive and Active Young Smokers by LC-Orbitrap-MS/MS Technique. *Molecules* **2024**, *29*, 3643. https://doi.org/10.3390/molecules29153643.
7. Zhang, Y.; Liu, H.; Lv, T.; Xiao, M.; Gao, G. Protein Tyrosine Phosphatase 1B Inhibitors of *Pueraria lobata* Based on the Spectrum–Effect Relationship by Q-Marker Selection. *Molecules* **2024**, *29*, 2731. https://doi.org/10.3390/molecules29122731.
8. Zhang, Y.; Cao, J.; Su, J.; He, T.; Wang, Q.; Wei, F.; Guo, X.; Mei, Q.; Zeng, J. Study of Bitespiramycin Distribution in Rats and Cerebrospinal Fluid of Patients by a Sensitive LC-MS/MS Method with Rapid Sample Preparation. *Molecules* **2024**, *29*, 1037. https://doi.org/10.3390/molecules29051037.
9. Peña, J.; González-Mariño, I.; Pavón, J.L.P. Ultrasound-Assisted Extraction, Followed by Gas Chromatography–Mass Spectrometry for the Simultaneous Quantification of Ethinyl Estradiol and Drospirenone in Contraceptive Formulations. *Molecules* **2023**, *28*, 4978. https://doi.org/10.3390/molecules28134978.
10. Štimac, A.; Kurtović, T.; Halassy, B. Multi-Detection Size Exclusion Chromatography as an Advanced Tool for Monitoring Enzyme–Antibody Conjugation Reaction and Quality Control of a Final Product. *Molecules* **2023**, *28*, 4567. https://doi.org/10.3390/molecules28114567.
11. Darwish, I.A.; Alzoman, N.Z. Development and Validation of Green and High-Throughput Microwell Spectrophotometric Assay for the Determination of Selective Serotonin Reuptake Inhibitors in Their Pharmaceutical Dosage Forms. *Molecules* **2023**, *28*, 4221. https://doi.org/10.3390/molecules28104221.
12. Lis-Cieplak, A.; Trześniowska, K.; Stolarczyk, K.; Stolarczyk, E.U. Pyrrolizidine Alkaloids as Hazardous Toxins in Natural Products: Current Analytical Methods and Latest Legal Regulations. *Molecules* **2024**, *29*, 3269. https://doi.org/10.3390/molecules29143269.
13. Chen, Z.; Zhou, G.; Ma, S. Research Progress of *Ferula ferulaeoides*: A Review. *Molecules* **2023**, *28*, 3579. https://doi.org/10.3390/molecules28083579.

References

1. Gavilano Fajardo, F.; Maggi Tavares, M.; Rashid, A.; Aurora Prado, M. Novel Eco-Friendly Stability Indicating Capillary Zone Electrophoresis Method for Determination of Aripiprazole in Tablet Dosage form: DoE Directed Optimization, Development and Method Validation. *J. Pharm. Sci.* **2022**, *111*, 3340–3351. [CrossRef]
2. Brunelli, C.; Osborne, R.; Yule, G.; Dixon, T.; Bruce, I.; Taylor, M. Automated multifactorial design of experiment and Bayesian optimisation algorithm approaches to method development for the green analysis by supercritical fluid chromatography of a pharmaceutical ingredient. *J. Chromatogr. A* **2024**, *1732*, 465214. [CrossRef] [PubMed]
3. Haile Kassahun, H.; Schepdael, A.; Ketema, G.; Adams, E. Development and validation of a simple and affordable LC-UV method for identification and assay of selected antimicrobial medicines. *J. Pharm. Biomed. Anal.* **2024**, *244*, 116127. [CrossRef]
4. Divya Zambre, D.; Hussain, U.; Sheikh, S.; Jaiswal, S.; Belgamwar, V. Stability-indicating HPLC analysis of Azilsartan Medoxomil potassium: A QbD-based method development and validation. *J. Chromatogr. B* **2025**, *1259*, 124599. [CrossRef] [PubMed]
5. Muchakayala, S.K.; Katari, N.K.; Saripella, K.K.; Schaaf, H.; Marisetti, V.M.; Ettaboina, S.K.; Rekulapally, V.K. Implementation of analytical quality by design and green chemistry principles to develop an ultra-high performance liquid chromatography method for the determination of Fluocinolone Acetonide impurities from its drug substance and topical oil formulations. *J. Chromatogr. A* **2022**, *1679*, 463380. [CrossRef] [PubMed]
6. Hnin, H.M.; Tun, T.; Jansook, P. Development and validation of high-performance liquid chromatography method for the simultaneous quantification of rivastigmine hydrogen tartrate and asiaticoside co-loaded in niosomes: A Box–Behnken design approach. *J. Chromatogr. B* **2024**, *1241*, 124170. [CrossRef]
7. Li, X.; Wang, Y.; Hu, W.; Qong, Q.; Ding, L. Development and validation of pharmacokinetics assays for a novel HER2-targeting antibody-drug conjugate (SHR-A1201): Application to its dose-escalation pharmacokinetic study. *J. Pharm. Biomed. Anal.* **2024**, *240*, 115964. [CrossRef] [PubMed]
8. Adamiszak, A.; Czyrski, A.; Sznek, B.; Grześkowiak, E.; Bienert, A. The Application of the Design of Experiments and Artificial Neural Networks in the Development of a Fast and Straightforward HPLC-UV Method for Fluconazole Determination in Hemato-Oncologic Pediatric Patients and Its Adaptation to Therapeutic Drug Monitoring. *Pharmaceuticals* **2024**, *17*, 1679. [CrossRef]

9. Simões, A.; Veiga, F.; Vitorino, C. Question-based review for pharmaceutical development: An enhanced quality approach. *Eur. J. Pharm. Biopharm.* **2024**, *195*, 114174. [CrossRef] [PubMed]
10. Zhang, J.; Raghavachari, R.; Kirkpatrick, D.; Keire, D.; Xu, X.; Faustino, P. Analytical Procedure Development and Proposed Established Conditions: A Case Study of a Mass Spectrometry Based NDSRI Analytical Procedure. *J. Pharm. Sci.* **2024**, *113*, 3028–3033. [CrossRef] [PubMed]

molecules

MDPI

Article

Emerging Mycotoxins in Cheese: Simultaneous Analysis of Aflatoxin M_1, Aflatoxicol, and Sterigmatocystin by LC-MS/MS

Maurizio Cossu [1], Andrea Sanna [1], Giuseppe Mangano [1], Giuseppe Ledda [1], Giannina Chessa [1], Pasquale Gallo [2], Antonio Vella [3], Ivan Pecorelli [4], Stefano Sdogati [4,*], Marilena Gili [5] and Carlo Boselli [6]

1 Istituto Zooprofilattico Sperimentale della Sardegna "G.Pegreffi", Via Vienna, n.2, 07100 Sassari, Italy; maurizio.cossu@izs-sardegna.it (M.C.); andrea.sanna@izs-sardegna.it (A.S.); giuseppe.ledda@izs-sardegna.it (G.L.);
2 Istituto Zooprofilattico Sperimentale del Mezzogiorno, Via Salute, n.2, 80055 Portici, Italy; pasquale.gallo@izsmportici.it
3 Istituto Zooprofilattico Sperimentale della Sicilia, Via Gino Marinuzzi, 3, 90129 Palermo, Italy; antonio.vella@izssicilia.it
4 Istituto Zooprofilattico Sperimentale dell'Umbria e delle Marche, Via G. Salvemini, 1, 06126 Perugia, Italy; i.pecorelli@izsum.it
5 Istituto Zooprofilattico Sperimentale del Piemonte, Liguria e Valle d'Aosta, Via Bologna, 148, 10154 Torino, Italy; marilena.gili@izsto.it
6 Istituto Zooprofilattico Sperimentale del Lazio E Della Toscana "M. Aleandri"—National Reference Centre for Ovine and Caprine Milk and Dairy Products Quality (C.Re.L.D.O.C.), 00178 Rome, Italy; carlo.boselli@izslt.it
* Correspondence: stefano.sdogati@izsum.it

Abstract: The presence of mycotoxins in cheese is a significant concern due to their potential health risks. Mycotoxins can contaminate cheese through two main routes: indirectly via contaminated animal feed, and/or directly, because of mold growth on dairy products. It has been reported that cheese may contain metabolites of aflatoxin B1 such as aflatoxin M1 (AFM1), aflatoxicol (AFL), and, its precursor, sterigmatocystin (STC). This study presents a reliable method for the simultaneous determination of AFM1, AFL, and STC in cheeses made from ovine, goat, or buffalo milk. The method was developed using single liquid extraction, clean-up by an immunoaffinity column (IAC), and liquid chromatography-tandem mass spectrometry (LC-MS/MS) determination. The method was subjected to initial validation according to EU regulations, which outline the required performance parameters and criteria of analytical methods for official food control. The limits of quantification (LOQs) of the method for AFM1, AFL, and STC are 2.0 ng/kg, 5.0 ng/kg, and 1.0 ng/kg, respectively. The method was applied in a study for the assessment of mycotoxin transfer from milk to cheeses and also their growth.

Keywords: emerging mycotoxins; cheese; aflatoxin M_1; aflatoxicol; sterigmatocystin; LC-MS/MS

1. Introduction

Cheese has been reported to contain mycotoxins, derived from naturally contaminated milk (as raw material) or due to mold growth during cheese spoilage [1,2]. Mycotoxins are toxic secondary metabolites produced by various fungi genera, including *Penicillium*, *Aspergillus*, and *Fusarium* spp. [3]. The presence of these substances in dairy products can have two origins: (a) indirect contamination, resulting from lactating animals ingesting contaminated feed, and (b) direct contamination, occurring due to accidental mold growth on dairy products [4]. Of major concern is the presence of three toxic mycotoxins: aflatoxin

M_1 (AFM1) and aflatoxicol (AFL), which are metabolites of aflatoxin B_1 (AFB1), and sterigmatocystin (STC) [3,5]; AFB1 can be usually found at high concentration in contaminated feed materials and, particularly, in maize-based feeds [6,7].

AFM1 is a monohydroxylated derivative of AFB1, formed and excreted in the milk of lactating animals (cattle, buffalo, goat, and sheep) that have consumed AFB1-contaminated products. AFM1 is relatively stable in both raw and processed milk products. Therefore, if raw milk containing AFM1 is used to make cheese, the toxin will be transferred almost quantitatively to cheese [8].

AFL is a reduced metabolite of AFB1, is considered less toxic than AFB1 but easily reconverted into the parent compound by liver dehydrogenases, prolonging the persistence of the mycotoxin in the body [9]. While AFL is soluble in water and easy to dispose of, its formation does not appear to be a significant detoxification pathway for AFB1 [10]. AFL has been found in both artisanal and industrialized cheese from cow milk, at concentrations which are higher if compared to AFM1 found in the same cheese [11].

STC is a fungal secondary metabolite and intermediate in the synthesis of aflatoxins produced by various *Aspergillus* species, with *A. versicolor* being the main producer. *Aspergillus* species that produce STC are generally xerophilic and can grow in environments with low water activity. Optimal environmental conditions for STC production are temperatures between 23 and 29 °C, water activity >0.76, and moisture content above 15%. STC is a biogenic precursor of aflatoxin B1. Despite its potential carcinogenic, mutagenic, and teratogenic properties, STC has been found in different cheeses contaminated with *A. versicolor*. Since 1987, it has been classified by the International Agency for Research on Cancer (IARC) as group 2B, but there is limited information about its occurrence and analysis in food [12,13]. AFM1, AFL and STC structures are detailed in Figure S1.

The European Union (EU) set a maximum concentration level at 0.050 μg/kg for AFM1 in milk while for AFL and STC, no maximum levels are currently in force.

The aim of this study was to develop a suitable and reliable analytical method to support regulatory authorities for future risk-based legislation concerning the maximum levels for mycotoxins in dairy products. The Italian Ministry of Health recently published a document demanding further evaluations for AFL and STC in cheese to fill the knowledge gap about their occurrence in dairy products and also, possibly, determine their transformation factors (TFs) [14].

Analytical procedures based on liquid chromatography coupled to fluorescence detection or mass spectrometry are valuable tools for assessing mycotoxin concentrations in dairy products, notably cheese. In standard analytical methods, an extraction phase utilizes water and organic solvents like acetonitrile and methanol [15,16], or employs innovative systems based on enzyme-assisted extraction with a cocktail of different enzymes, as proposed by Pietri et al. [17]. The subsequent step usually involves a purification phase. Common purification techniques encompass immunoaffinity columns (IACs) [13], Solid-Phase Extraction (SPE) methods [18,19], Molecularly Imprinted Polymers (MIPs) [20,21], and one-step multifunctional clean-up columns (IA Mycosep®) [22]. The QuEChERS (Quick, Easy, Cheap, Effective, Rugged, and Safe) method also provides an alternative, covering both extraction and purification steps [23].

For instrumental determination, the standard approach includes reversed-phase high-performance liquid chromatography (HPLC) coupled to a fluorescence detector (FLD) for both AFM1 and AFL [5,24] while UV absorption is typically employed for STC [25]. Liquid chromatography coupled to mass spectrometry (LC-MS) or tandem mass spectrometry (LC-MS/MS) can also be utilized for the determination of mycotoxins in cheese [8]. The analytical multi-mycotoxins methods available in the literature are limited and typically focus on analysis of bovine milk cheese [26].

In this study, we have developed a simple and sensitive method for the simultaneous analysis of AFM1, AFL, and STC, that may occur in different kind of cheeses, made from ovine, caprine and buffalo milk. The analytical approach involved a simple liquid extraction, followed by a clean-up with an immunoaffinity column (IAC); then, the determination was carried out using LC-MS/MS. This method enables the analysis of different cheeses at various stages of ripening, allowing for the quantification of mycotoxins studied at very low levels of contamination; at the present time, only for AFM1 has the Commission Regulation (EU) n. 2023/915 set a maximum limit (ML) in milk and milk-derived products, whereas, regarding AFL and STC, ML are not set yet. The European Food Safety Agency (EFSA) is interested in monitoring AFL and STC levels to evaluate possible health risks for consumers [27].

2. Results and Discussion

Firstly, the AFLATEST® WB SR+ immunoaffinity column performances were tested by spiking blank soft, semi-hard, and ripened-hard cheese samples with AFM1, AFL, and STC at 1 ng/g. Recoveries were in the range of 80–100% for AFM1 and AFL, thus proving the suitability of AFLATEST® WB SR+ for AFL determination and overall good preliminary performances for the method. Recoveries for STC were low (<10%), probably due to an unsatisfactory sample clean-up. A freezing step was introduced to precipitate and separate fat and proteins from the liquid extraction phase. Cooling down the sample at −18 °C for 3 h improved the STC recoveries (>70%). A figure describing the improvement in sample clean-up obtained with the freezing step is reported in Supplementary Materials (Figure S2).

The method developed is suitable for AFM1, STC, and AFL analysis in cheese matrixes at very low LOQ values. A comparison between the results obtained in this work and the existing methods in the literature is presented in Table 1. Dairy products are complex matrixes due to their high fat (up to 40%) and protein content, making the analysis of mycotoxins a challenging task. Analytical methods available in the literature are limited and typically focus only on AFM1 analysis in bovine cheese. This study presents an innovative multi-analyte method for the analysis of AFM1, STC, and AFL in cheese obtained from goat, sheep, and buffalo milk (Table 2). The introduction of a purification step using immunoaffinity columns suitable for all three mycotoxins enabled the development of a single-step purification process, effectively eliminating interferences and producing a cleaner final extract. Furthermore, the addition of a freezing step for further purification improved the extract cleanliness and enhanced the efficiency of the immunoaffinity columns. As can be seen, in Table 1, in this work, we obtained comparable recoveries to that reported in the literature by Veršilovskis et al. [19], which appears to have been the best to date. However, examining the precision (RSD_r), LOD, and LOQ, the present method demonstrates better precision and higher sensitivity compared to what has been reported in the literature so far. These improvements can be addressed by the use of ILIS, compensating for bias and instrumental variability, coupled to IAC purification, which allows for the analysis of only selected compounds, removing all matrix-related interferences from the finial extract. The experiments conducted on cheeses at different ripening stages allowed us to determine the LOQ values, ranging from 1.0 to 5.0 ng/kg (Table 3). The LOQ values comply with the criteria established by Regulation 2782/2023 [28] (0.5 × ML), where ML stands for the "maximum level" reported in Regulation (EU) 915/2023 [27].

Table 1. Comparison between the present and previously published methods.

		Recovery %	RSD_r %	LOD (ng/kg)	LOQ (ng/kg)	Approximate RT (min)
Pietri et al., 2016 [17]	AFM1	87.1–96.3	1.4–2.3	5–20	15–60	4.9
	STC	-	-	-	-	-
	AFL	-	-	-	-	-
Iha et al., 2011 [15]	AFM1	80	5.9	3	10	7.0
	STC	-	-	-	-	-
	AFL	-	-	-	-	-
Rodríguez-Cañàs et al., 2023 [29]	AFM1	79.6–101.1	6.63–13.20	6–11	20–36 *	5.9
	STC	73.9–104.3	1.03–5.25	178–1005	588–3316 *	8.4
	AFL	-	-	-	-	-
Carvajal-Moreno et al., 2019 [5]	AFM1	95	-	10	-	8.5–9.5
	STC	-	-	-	-	-
	AFL	98	-	10	-	3.0–5.6
Jakšić et al., 2021 [30]	AFM1	84–112	8–18	10	30	5.7
	STC	-	-	-	-	-
	AFL	-	-	-	-	-
Veršilovskis et al., 2009 [19]	AFM1	-	-	-	-	-
	STC	96–104	17–27	30	100	7.7
	AFL	-	-	-	-	-
Current Study(This work)	AFM1	95	2.0–3.2	0.6	2.0	4.7
	STC	98–103	2.1–2.7	0.2	1.0	6.4
	AFL	96–97	2.7–4.9	1.0	5.0	5.4

*: in concentrated extracts.

Table 2. Quality parameters of cheeses used for study method.

Composition Parameter	Soft (MFFB% $\geq$ 62)	Semi-Hard (62 > MFFB% > 55)	Ripened-Hard (MFFB% $\leq$ 55)
Protein %	15–19	29	29–34
Fat %	20–40	40	31–34
Dry matter %	>30	>70	>75

Table 3. Linearity range, LOD, and LOQ.

Mycotoxin	Linearity Range (ng/kg)	Calibration Curve Equation	R^2	LOD (ng/kg)	LOQ (ng/kg)
AFM_1	2.0–200	$y = 2.05x - 0.38$	0.9986	0.6	2.0
STC	1.0–100	$y = 1.26x - 0.15$	0.9999	0.2	1.0
AFL	5.0–2000	$y = 136.4x - 135.9$ *	0.9982	1.0	5.0

* External standard calibration.

The RSD_r ranged from 2.0 and 4.9% with mean recoveries between 95% and 103%, proving that the developed method complies with performance criteria established by Regulation 2782/2023 for recovery (70–120%) and precision under repeatability conditions ($RSD_r \leq 20\%$). The analytical performances were consistent across different degrees of cheese ripening. Figure 1 illustrates an LC-MS/MS chromatogram of a blank ripened-hard sheep cheese sample; Figure 2 shows the same cheese sample spiked with 25 ng/kg of AFM1 and STC, and 50 ng/kg of AFL; in Figure 3, a chromatogram of a commercial sample of sheep hard cheese, naturally contaminated with AFM1 and STC, is displayed. Supplementary data on precision and accuracy for each cheese type are reported in Table S2.

The high sensitivity and specificity of this method make it particularly useful for food plan control, exposure studies, and risk assessment evaluation, especially for mycotoxins that only recently were put under the magnifying glass of food safety scientists (AFL and STC). The importance of monitoring AFL and STC was recently highlighted by Braun et al., who reported STC in 25% of analyzed processed cereal foods for infants and, for the first time, AFL at a relatively high concentration (1.1 μg/kg) in a milk cereal sample [31]. Cheese is consumed daily by roughly 75% of people between 2 and 19 years in the USA; moreover, cheese consumption is increasing worldwide, with the EU (20.75 kg/year per capita), the USA (17.91 kg/year per capita), and Canada (14.45 kg/year per capita) leading the rankings [32,33]. On these bases, it is evident that, to ensure food salubrity and adequately determine the exposure of populations to contaminants (in this case, mycotoxins), deploying the most sensitive analytical methods available is required.

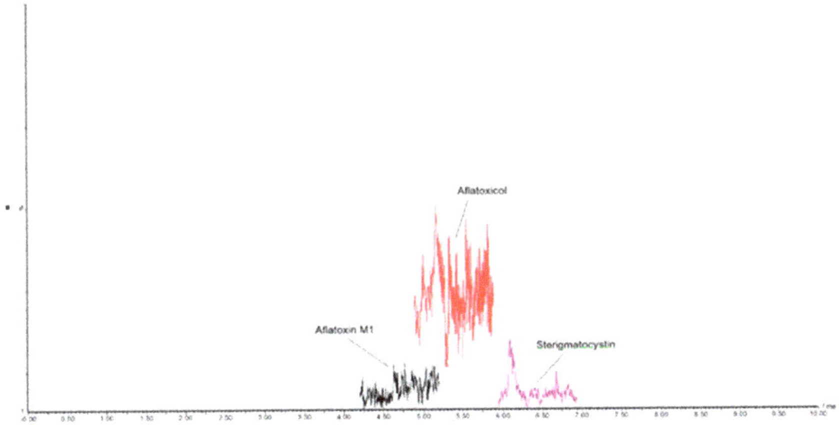

Figure 1. MRM chromatogram of a blank sheep ripened-hard cheese sample (AFM1 < 2.0 ng/kg, STC < 1.0 ng/kg, and AFL < 5.0 ng/kg).

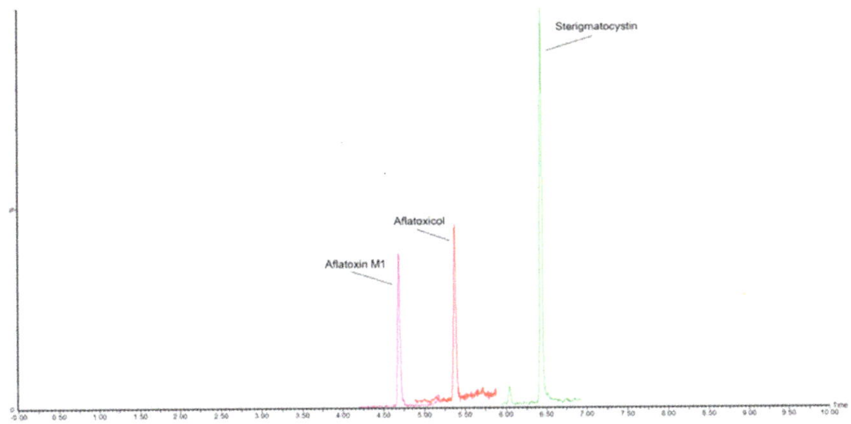

Figure 2. MRM chromatogram of a sheep ripened-hard cheese sample spiked with 25 ng/kg of Aflatoxin M1, 25 ng/kg of Sterigmatocystin, and 50 ng/kg of Aflatoxicol.

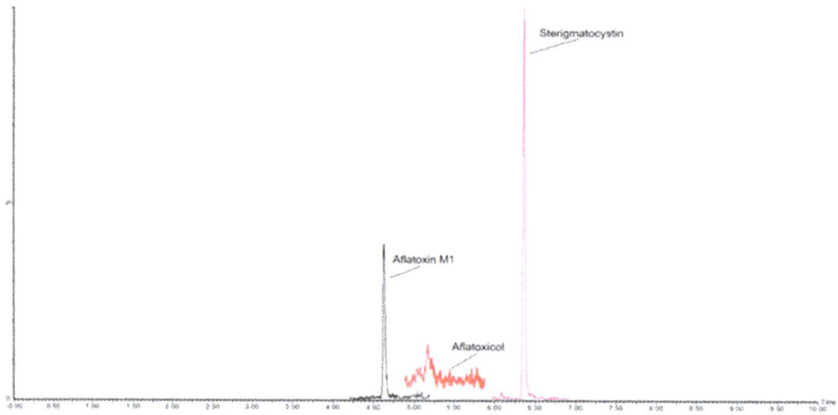

Figure 3. MRM chromatogram of a sheep ripened-hard cheese commercial sample, naturally contaminated with Aflatoxin M1 and Sterigmatocystin at 17 ng/kg and 21 ng/kg, respectively. Aflatoxicol was ≤LOQ.

Preliminary Results of Cheese Samples Analyzed

This method was applied to the analysis of 55 cheese samples (obtained from sheep, goat, and buffalo milk) with different ripening levels (soft, semi-hard, and ripened-hard cheeses) sampled from local market stores (Sassari, Italy). The method's high sensitivity allowed for the detection of STC and AFM1 in 95% (52 out of 55) of the cheeses tested, while AFL was not detected in any of the samples (≤5.0 ng/kg). AFM1 was determined at an average level of 11 ng/kg, with concentrations ranging from ≤2.0 to 43 ng/kg, whereas STC was detected at a mean concentration of 5 ng/kg, in a range from ≤1.0 to 24 ng/kg. More precisely, AFM1 was detected in the range 3–37 ng/kg in buffalo cheese, 4–18 ng/kg in goat cheese, and 3–43 ng/kg in sheep cheese; STC was found between 2 and 4.3 ng/kg in buffalo cheese, 2 and 24 ng/kg in goat cheese, and 2 and 17 ng/kg in sheep cheese. Based on cheese ripening, AFM1 was detected in the range 3–37 ng/kg in soft cheeses, 3–18 ng/kg in semi-hard cheese, and 6–43 ng/kg in ripened-hard cheese, while STC was detected at between 2 and 5 ng/kg (soft), 2 and 21 ng/kg (semi-hard), and 3 and 24 ng/kg (ripened-hard). The detected values for each of the 55 analyzed samples are detailed in the Supplementary Materials (Table S3) together with the average levels (Table S4) and were generally lower than the data obtained by the authors, reported in Table 1. From this preliminary market survey, it can be observed how AFM1 cheese contamination is around 10 ng/kg, with no great difference in terms of lactating species or cheese ripening. In contrast, although detected STC levels are generally lower than those of AFM1, they seem to be, on average, higher in ripened than in softer cheese [13]. Possibly, STC and other mycotoxins may be produced by *Aspergillus* spp. and other fungi on the cheese surface during ripening. STC may diffuse under the crust as ripening progresses and thus be detected in these products in higher amounts with respect to soft cheeses, where the ripening stage is reduced or absent.

Referring to AFM1, the detected levels, considering the transformation factors for cheeses, were well below the maximum limit established by the Commission Regulation (EU) n. 2023/915. These preliminary results, obtained for an ongoing study aiming to evaluate the contamination levels of these mycotoxins in cheeses, suggest that both STC and AFM1 can be present while being within safe limits, at least concerning AFM1.

3. Materials and Methods

3.1. Samples

A total of 11 cheese samples (5 soft, 3 semi-hard and 3 ripened hard) were purchased from local supermarkets. The sampled products included 3 buffalo mozzarella cheeses, 2 buffalo ricotta cheeses, 3 sheep semi-hard cheeses, and 3 sheep hard cheeses. Quality parameters of the cheeses used are reported in Table 2. Cheese hardness parameters were determined using "moisture content on a fat-free basis" (MFFB%) parameter. MFFB% is calculated as [moisture content/(total weight − fat content)] × 100. MFFB was detailed in Council Directive 96/16/EC of 19 March 1996 on statistical surveys of milk and milk products, now repealed by a more recent regulation [34].

3.2. Chemicals and Reagents

Reference Material (RM) of Aflatoxin M_1 was purchased as a solution from Sigma-Aldrich (St. Louis, MO, USA). RMs of Sterigmatocystin, [$^{13}C_{17}$]-Aflatoxin M1 and [$^{13}C_{18}$]-Sterigmatocystin were purchased as solutions from Romer Labs (Tulln, Austria). RM of Aflatoxicol was obtained as a neat substance from LGC Standards (Wesel, Germany).

Certified Reference Material (CRM) ERM-BD284 was obtained from the EC Joint Research Center (Geel, Belgium).

Acetonitrile (LC-MS-grade) for UHPLC mobile phase and methanol (RPE-grade) for sample extraction were purchased from Carlo Erba (Val-de-Reuil, France). Acetonitrile (RPE-grade) for sample extraction, ammonium acetate (LC-MS-grade) as mobile phase additive, and sodium chloride (RPE-grade) were purchased from VWR (Radnor, PA, USA). Dulbecco's Phosphate-Buffered Saline (PBS) was obtained from Sigma-Aldrich (St. Louis, MO, USA).

Ultra-pure water used in the preparation of solutions and UHPLC mobile phases was purified by a Milli-Q® Advantage A10 Water Purification System (Merck KGaA, Darmstadt, Germany).

AFLATEST® WB SR+ immunoaffinity columns were obtained from Vicam (Milford, MA, USA). Columns have a specification >90% recovery in complex samples and a capacity as high as 1000 ng for aflatoxins and their metabolites.

3.3. Working Standard Solutions

Standards of Aflatoxin M1, Sterigmatocystin, [$^{13}C_{17}$]-Aflatoxin M1, and U-[$^{13}C_{18}$]-Sterigmatocystin were available as stock solutions. Aflatoxicol stock solution was prepared by dissolution of the neat substance (purity 98%) in methanol at 1000 µg/mL. Aflatoxin M1, Sterigmatocystin, and Aflatoxicol working solutions were prepared at 2.5–2.5–25 ng/mL, respectively, in acetonitrile, starting from stock solutions by successive dilutions. Mixed working solution of labeled Aflatoxin M1 and Sterigmatocystin was prepared at 2.5 ng/mL.

All the solutions were stored in parafilm-coated amber glass bottles in the dark at −18 °C. Appropriate dilutions with water–acetonitrile 75:25 (v/v) mixture were made to provide standards to be used for calibration purposes.

3.4. Sample Preparation and Analysis

An amount of 2.00 ± 0.05 g of homogenized sample were weighed into a centrifuge polypropylene tube and 80 µL of labeled internal standard solution at 2.5 ng/mL was added. The sample was then mixed by an Ultra-Turrax™ (Staufen im Breisgau, Germany) with 2 mL of water and 5 mL of acetonitrile–methanol 85:15 (v/v) until a milky phase was formed. The extract was centrifuged for 10 min at 9000 rpm and the supernatant was transferred into a 15 mL tube. The remaining residue was extracted again by the above mentioned procedure and the supernatants were combined. The extract was added with

1.0 g of NaCl and stirred vigorously for 1 min. The solubilization of the salt, up to saturation, allows for the separation of the aqueous phase from the organic one. The extract was then placed in a freezer at −18 °C for at least three hours to allow the precipitation of the interfering substances. At the end of this step, 2 mL of organic extract was withdrawn and diluted with 25 mL of PBS. This operation must be carried out immediately after removing the extract from the freezer to avoid re-solubilization of the interferents. The diluted extract was purified through an immunoaffinity column at a flow of about 1 drop/second, avoiding drying out the column. Then, the column was washed by passing 10 mL of water at a rate of about 2 drops/s. At the end of this step, columns were dried by a gentle stream of air for a few seconds.

A glass test tube was placed under the immunoaffinity column and 0.75 mL of the eluting solution of acetronitrile–methanol 1:2 (*v*/*v*) was added and analytes were eluted by gravity. After 3 min, an additional 0.75 mL of the eluting solution was added and eluted by gravity. The column was dried by passing air to quantitatively collect all the eluate. The eluate was dried by a stream of nitrogen at room temperature, taken up again with 0.4 mL of a water–acetonitrile 75:25 (*v*/*v*) mixture with 0.1% formic acid, filtered by 0.2 μm syringe filter, and analyzed by LC-MS/MS.

AFM1 and STC quantification was performed by an isotopically labeled internal standard (ILIS) calibration curve while AFL was quantified by an external standard calibration curve. Concentration levels of the standard calibration curves were between 2.0 and 200 ng/L for AFM1, 1.0–100 ng/L for STC, and 5.0–200 ng/L for AFL. ILIS of AFM1 and STC were at 100 ng/L for each calibration level.

3.5. Optimization of the Extraction and Clean-Up Steps

The optimization step that significantly improved the method's performances primarily targeted the solvent extraction and IA purification. The immunoaffinity columns (IACs) used are certified for isolating aflatoxins B1, B2, G1, G2, M1, M2, and sterigmatocystin, but not aflatoxicol. However, aflatoxicol has a structure very similar to AFM1, allowing it to form an antigen–antibody bond, as noted by the column manufacturers. Preliminary tests were conducted to verify this hypothesis, and experiments at different concentration levels of aflatoxicol confirmed this assumption. The clean-up, particularly the IA purification step, was identified as the most critical phase of the method due to the complexity of the cheese matrix, which contains a wide variety of interferents, including fats and proteins, particularly casein. These components can obstruct the purification columns, thereby limiting the extraction efficiency. To study this effect, recovery tests were carried out using soft, semi-hard, and ripened-hard cheeses, detailed in Table 2, spiked at 1 ng/g. Sample freezing was tested to evaluate extraction efficiency and interference reductions. The cooling process facilitates the mechanical separation of lipid components and proteins, resulting in a cleaner extract and more efficient immunoaffinity purification.

3.6. Instrumentation LC-MS/MS

LC-MS/MS analysis was carried out using a Waters XEVO TQ-S™ triple quadrupole instrument equipped with a Waters Acquity UPLC® system with H-Class pumps (Waters Corp., Wilmslow, UK). An electrospray ionization (ESI) probe in positive mode was used. A Waters Acquity UPLC® HSS T3, 1.8 μm (2.1 × 100 mm) column (Waters Corp., Wilmslow, UK) was used. Mobile phases consist of (A) ammonium acetate 1 mM in water and (B) acetonitrile. The gradient was 1 min isocratic at 90% A, then 5 min with a gradient from 90% A to 100% B. Then, 3 min isocratic at 100% B was completed to wash the column and 3 min at 90% A for column re-conditioning and equilibration.

The flow rate was 0.3 mL/min, column temperature was set at 40 °C, and the injection volume was 10 μL.

Under these UHPLC conditions, the AFM1, AFL, and STC retention times were 4.7, 5.4, and 6.4 min, respectively. The parameters of the mass spectrometer were optimized using the standard solution of each analyte by combined infusion.

The best response was recorded with the following parameters: source temperature 150 °C, capillary voltage 2.0 kV, nitrogen desolvation gas temperature 600 °C with a flow rate of 1000 L/h, cone gas flow 150 L/h, and nebulizer 5 bar. Each analyte of interest was surveyed for optimal conditions and monitored by multiple reaction monitoring (MRM). MS transitions and cone and collision parameters are listed in Table S1 (Supplementary Materials).

3.7. Method Performance Parameters

The study of method performances was conducted in accordance with Regulation (UE) 2017/625 [35] and Regulation (EU) n. 2023/2782 [28], which outlines the required parameters and criteria for analytical methods used in the official control of mycotoxins in food. Specifically, the study included the following parameters: linearity and working range, limit of detection (LOD), limit of quantitation (LOQ), accuracy (in terms of extraction recovery), precision (in terms of relative standard deviation in repeatability conditions – RSD_r), and specificity. Data for all these parameters were obtained using spiked samples or Certified Reference Materials when available.

As linearity criteria were not pointed out by [28,35] this parameter was evaluated based on Guidance SANTE 11312/2021v2 [36]. Linearity was evaluated in solvent using calibration standard curves at different concentration levels for each mycotoxin. The acceptability criteria is a ± 20% tolerance window of back-calculated concentrations (BCCs) of each calibration point from the true value. The concentration ranges studied for the different mycotoxins are presented in Table 3. The instrumental LOD (limit of detection) was evaluated as the concentration showing a signal-to-noise (S/N) ratio ≥ 3 using solvent solutions, while the instrumental LOQ (limit of quantitation) was estimated as the lowest concentration of the analyte in the sample that could be measured with reasonable statistical certainty (S/N ≥ 10). The LOQ values obtained for the three mycotoxins are reported in Table 3.

Accuracy and precision were evaluated using spiked blank cheese samples at 25 ng/kg for AFM1 and STC and 50 ng/kg for AFL (Level 1); 50 ng/kg for AFM1 and STC, 500 ng/kg for AFL (Level 2) and reported in Table 4.. Additionally, accuracy for AFM1 was assessed analyzing a Certified Reference Material milk powder matrix for AFM1 (ERM-BD 284) (Table 5).

Table 4. Method performance evaluation.

Mycotoxin	Replicates (*n*)	Concentration Range (ng/kg)	Performance Evaluation Parameters	
			Recovery %	RSD_r %
AFM_1	6	25	95	2.0
	5	50	95	3.2
STC	6	25	103	2.1
	5	50	98	2.7
AFL	5	50	97	2.7
	5	500	96	4.9

RSDr: relative standard deviation under repeatability conditions.

Table 5. Method performance evaluation on CRM ERM-BD 284.

Certified Reference Material	AFM1 Concentration (ng/kg)	Replicates (*n*)	Recovery %	RSD_r %
ERM BD 284—Milk powder	440	10	101	3.6

The specificity of the method was evaluated through an analysis of different kinds of blank and spiked samples. Specificity criteria were met if the spiked sample signal was at least three times higher than interferences, if any, present in a blank sample [28]. The applicability was estimated through the comparison of retention times (RTs) and MS/MS parameters such as the ion ratio% (IR), which is the ratio between the two MRM transitions of each molecule (qualifier/quantifier q/Q × 100). Mycotoxins were identified in an unknown sample if the IR was between ±30% of that of the standard. Regarding the RT, for AFM1 and STC, the relative retention time (RRT) was used, while for AFL, the absolute retention time was considered. In the first case, an identification range window of ±0.02 min was used, while for AFL, a range of ±0.1 min around the RT of the reference standard was considered. Furthermore, samples of uncontaminated cheeses were used to verify the absence of matrix interference in the MS/MS chromatograms.

3.8. Statistical Analysis

The descriptive parameters of the calibration curve (equation, correlation coefficient, and back-calculated concentrations) were calculated using the instrumental software TargetLynx V4.1 (Waters™, Milford, MA, USA) with the linear regression model. Extraction recovery was calculated as the ratio% between experimental sample concentration (calculated via calibration curve equation) and spiked concentration.

Precision was assessed as the relative standard deviation in repeatability conditions (RSD_r). Both precision and accuracy were calculated using Microsoft Excel (Microsoft Excel 2007, Microsoft Corporation, version 2501, Redmond, WA, USA).

4. Conclusions

A method combining Ultra-Turrax™ sample grinding in the presence of a solvent, followed by clean-up with an immunoaffinity column and LC-MS/MS determination in positive MRM mode, has been developed and optimized. The introduction of a freezing step facilitated the separation of lipid components and proteins that could clog the immunoaffinity column. This additional step increased its efficiency in terms of recovery.

The analytical performance parameters were evaluated for the quantification of Aflatoxin M1, Aflatoxicol, and Sterigmatocystin in goat, sheep, and buffalo cheeses, at level concentrations fitting for the purposes of food control. The optimization approach allowed for the development of a quantitative method, with remarkable sensitivity, excellent linearity response, and adequate precision for the mycotoxins studied, which can also be applicable to exposure evaluations. Only a few methods in the literature are available to quantify AFL and STC and, to the best of our knowledge, the proposed method is the most sensitive. In comparison, this method offers several advantages, including applicability to a wide range of cheeses and suitability for routine analysis. The preliminary results obtained for an ongoing study aimed at evaluating the contamination levels of these mycotoxins in cheeses, demonstrating the applicability of this method.

Supplementary Materials: The following supporting information can be downloaded at https://www.mdpi.com/article/10.3390/molecules30081774/s1: Figure S1: Structures of mycotoxins examined in cheese samples. Figure S2: Sample clean-up without freezing step (left) and with freezing step (−18 °C, 3 h) (right). Table S1. MS transitions and cone and collision parameters. Table S2. Recovery and precision performances at two concentration levels for the different kinds of cheese. Table S3. AFM1, AFL, and STC contamination of 55 cheese samples. Table S4. Average AFM1 and STC concentrations in the analyzed samples.

Author Contributions: Conceptualization, M.C., G.C. and C.B.; methodology, M.C. and S.S.; validation, M.C.; formal analysis, M.C. and G.M.; investigation, A.S. and G.L.; resources, G.C.; writing—original draft preparation, M.C. and G.M.; writing—review and editing, M.C., G.C., P.G., I.P., S.S., M.G., C.B. and A.V.; supervision, G.C.; project administration, G.C.; funding acquisition, C.B. All authors have read and agreed to the published version of the manuscript.

Funding: This research was funded by the Italian Ministry of Health (MOH). Ricerca Strategica grant number IZSLT 01/21 RC "STRATEGICA".

Institutional Review Board Statement: Not applicable.

Informed Consent Statement: Not applicable.

Data Availability Statement: The original contributions presented in this study are included in this article and the Supplementary Materials. Further inquiries can be directed to the corresponding authors.

Conflicts of Interest: The authors declare no conflicts of interest.

Abbreviations

The following abbreviations are used in this manuscript:

AFL	Aflatoxicol
AFB1	Aflatoxin B1
AFM1	Aflatoxin M1
BCC	Back-calculated concentrations
CRM	Certified Reference Material
EFSA	European Food Safety Agency
EU	European Union
ILIS	Isotopically labeled internal standard
HPLC	High-performance liquid chromatography
IAC	Immunoaffinity column
IR	Ion ratio
LC-MS/MS	Liquid chromatography-tandem mass spectrometry
LOD	Limit of detection
LOQ	Limit of quantitation
MIP	Molecularly Imprinted Polymers
MRM	Multiple reaction monitoring
PBS	Phosphate-Buffered Saline
RM	Reference Material
RSDr	Relative standard deviation under repeatability conditions
RT	Retention time
SPE	Solid-Phase Extraction
STC	Sterigmatocystin

References

1. Sengun, I.; Yaman, D.; Gonul, S. Mycotoxins and Mould Contamination in Cheese: A Review. *World Mycotoxin J.* **2008**, *1*, 291–298. [CrossRef]
2. Hymery, N.; Vasseur, V.; Coton, M.; Mounier, J.; Jany, J.L.; Barbier, G.; Coton, E. Filamentous fungi and mycotoxins in cheese: A review. *Compr. Rev. Food Sci. Food Saf.* **2014**, *13*, 437–456. [CrossRef] [PubMed]
3. Sweeney, M.J.; Dobson, D.W. Mycotoxin production by Aspergillus, Fusarium and Penicillium species. *Int. J. Food Microbiol.* **1998**, *43*, 141–158. [CrossRef] [PubMed]
4. Lafont, P.; Siriwardana, M.G.; DeBoer, E. Contamination of dairy products by fungal metabolites. *J. Environ. Pathol. Toxicol.* **1990**, *10*, 99–102.
5. Carvajal, M.; Rojo, F.; Méndez, I.; Bolaños, A. Aflatoxin B1 and its interconverting metabolite aflatoxicol in milk: The situation in Mexico. *Food Addit. Contam.* **2003**, *20*, 1077–1086. [CrossRef]
6. Creppy, E.E. Update of survey, regulation and toxic effects of mycotoxins in Europe. *Toxicol. Lett.* **2002**, *127*, 19–28. [CrossRef]
7. Sdogati, S.; Pacini, T.; Bibi, R.; Caporali, A.; Verdini, E.; Orsini, S.; Ortenzi, R.; Pecorelli, I. Co-Occurrence of Aflatoxin B1, Zearalenone and Ochratoxin A in Feed and Feed Materials in Central Italy from 2018 to 2022. *Foods* **2024**, *13*, 313. [CrossRef]
8. Iha, M.H.; Barbosa, C.B.; Okada, I.A.; Trucksess, M.W. Aflatoxin M1 in milk and distribution and stability of aflatoxin M1 during production and storage of yoghurt and cheese. *Food Control* **2013**, *29*, 1–6. [CrossRef]
9. Salhab, A.S.; Edwards, G.S. Production of aflatoxicol from aflatoxin B1 by postmitochondrial liver fractions. *J. Toxicol. Environ. Health* **1977**, *2*, 583–587. [CrossRef]
10. Theumer, M.G.; Henneb, Y.; Khoury, L.; Snini, S.P.; Tadrist, S.; Canlet, C.; Puel, O.; Oswald, I.P.; Audebert, M. Genotoxicity of aflatoxins and their precursors in human cells. *Toxicol. Lett.* **2018**, *287*, 100–107. [CrossRef]
11. Carvajal-Moreno, M.; Vargas-Ortiz, M.; Hernández-Camarillo, E.; Ruiz-Velasco, S.; Rojo-Callejas, F. Presence of unreported carcinogens, aflatoxins and their hydroxylated metabolites, in industrialized Oaxaca cheese from Mexico City. *Food Chem. Toxicol.* **2019**, *124*, 128–138. [CrossRef] [PubMed]
12. Nieto, C.H.D.; Granero, A.M.; Zon, M.A.; Fernández, H. Sterigmatocystin: A Mycotoxin to Be Seriously Considered. *Food Chem. Toxicol.* **2018**, *118*, 460–470. [CrossRef] [PubMed]
13. Pietri, A.; Leni, G.; Mulazzi, A.; Bertuzzi, T. Ochratoxin A and sterigmatocystin in long-ripened Grana cheese: Occurrence, wheel rind contamination and effectiveness of cleaning techniques on grated products. *Toxins* **2022**, *14*, 306. [CrossRef] [PubMed]
14. Italian Ministry of Health. Micotossine non Regolamentate: Metaboliti Dell'aflatossina B1 (Aflatossina M1 e Aflatossicolo) e Sterigmatocistina in Prodotti Lattiero-Caseari. Available online: https://www.salute.gov.it/imgs/C_17_pubblicazioni_3065_allegato.pdf (accessed on 24 February 2025).
15. Iha, M.H.; Barbosa, C.B. Rosa Maria Duarte Favaro Chromatographic Method for the Determination of Aflatoxin M1 in Cheese, Yogurt, and Dairy Beverages. *J. AOAC Int.* **2011**, *94*, 1513–1518. [CrossRef]
16. Hird, S.; Martin, K.; Collette Zabe, N.; Toth, D. *Determination of Aflatoxins in a Wide Range of Food and Agricultural Commodities Using Immunoaffinity Chromatography Column Clean-Up Coupled with UPLC or HPLC with Fluorescence Detection*; Application Notes Waters Corporation; VICAM: Rydalmere, NSW, Australia, 2021.
17. Pietri, A.; Fortunati, P.; Mulazzi, A.; Bertuzzi, T. Enzyme-assisted extraction for the HPLC determination of aflatoxin M1 in cheese. *Food Chem.* **2016**, *192*, 235–241. [CrossRef]
18. Turner, N.W.; Bramhmbhatt, H.; Szabo-Vezse, M.; Poma, A.; Coker, R.; Piletsky, S.A. Analytical methods for determination of mycotoxins: An update (2009–2014). *Anal. Chim. Acta* **2015**, *901*, 12–33. [CrossRef]
19. Veršilovskis, A.; Van Peteghem, C.; de Saeger, S. Determination of sterigmatocystin in cheese by high-performance liquid chromatography-tandem mass spectrometry. *Food Addit. Contam. A* **2009**, *26*, 127–133. [CrossRef]
20. Regal, P.; Díaz-Bao, M.; Barreiro, R.; Cepeda, A.; Fente, C. Application of molecularly imprinted polymers in food analysis: Clean-up and chromatographic improvements. *Cent. Eur. J. Chem.* **2012**, *10*, 766–784. [CrossRef]
21. Ashley, J.; Shahbazi, M.-A.; Kant, K.; Chidambara, V.A.; Wolff, A.; Bang, D.D.; Sun, Y. Molecularly imprinted polymers for sample preparation and biosensing in food analysis: Progress and perspectives. *Biosens. Bioelectron.* **2017**, *91*, 606–615. [CrossRef]
22. Vaz, A.; Cabral Silva, A.C.; Rodrigues, P.; Venâncio, A. Detection Methods for Aflatoxin M1 in Dairy Products. *Microorganisms* **2020**, *8*, 246. [CrossRef]
23. González-jartín, J.M.; Rodríguez-Canás, I.; Alfonso, A.; Sainz, M.; Vieytes, M.R.; Gomes, A.; Ramos, I.; Botana, L.M. Multianalyte Method for the Determination of Regulated, Emerging and Modified Mycotoxins in Milk: QuEChERS Extraction Followed by UHPLC—MS/MS Analysis Es Rodríguez-Ca N. *Food Chem.* **2021**, *356*, 129647. [CrossRef] [PubMed]
24. Wood, J.E.; Gill, B.D.; Indyk, H.E.; Rhemrev, R.; Pazdanska, M.; Mackay, N.; Marley, E. Determination of Aflatoxin M1 in Liquid Milk, Cheese, and Selected Milk Proteins by Automated Online Immunoaffinity Cleanup with Liquid Chromatography—Fluorescence Detection. *J. AOAC Int.* **2021**, *104*, 719–724. [CrossRef] [PubMed]
25. Liu, S.; Fan, J.; Huang, X.; Jin, Q.; Zhu, G. Determination of Sterigmatocystin in Infant Cereals from Hangzhou, China. *J. AOAC Int.* **2016**, *99*, 1273–1278. [CrossRef] [PubMed]

26. Dobson, A.D.W. Mycotoxins in cheese. In *Cheese: Chemistry, Physics and Microbiology*, 4th ed.; McSweeney, P.L.H., Fox, P.F., Cotter, P., Everett, D.W., Eds.; Elsevier: New York, NY, USA, 2017; pp. 595–601.
27. European Commission. Regulation (EC) 2023/915 of 25 April 2023. *Official Journal of the European Union L 119/103 2023*. pp. 103–157. Available online: https://eur-lex.europa.eu/eli/reg/2023/915/oj (accessed on 17 January 2025).
28. European Commission. Regulation No. 2023/2782 of 14 December 2023 Laying Down the Methods of Sampling and Analysis for the Control of the Levels of Mycotoxins in Food and Repealing EC Regulation No. 401/2006. *Official Journal of the European Union*. 2023, pp. 1–44. Available online: https://data.europa.eu/eli/reg_impl/2023/2782/oj (accessed on 15 January 2025).
29. Rodríguez-Cañás, I.; González-Jartín, J.M.; Alvariño, R.; Alfonso, A.; Vieytes, M.R.; Botana, L.M. Detection of mycotoxins in cheese using an optimized analytical method based on a QuEChERS extraction and UHPLC-MS/MS quantification. *Food Chem.* **2023**, *408*, 135182. [CrossRef]
30. Jakšić, S.; Živkov Baloš, M.; Popov, N.; Torović, L.; Krstović, S. Optimisation, validation and comparison of methods for aflatoxin M1 determination in cheese. *Int. J. Dairy Technol.* **2021**, *74*, 681–688. [CrossRef]
31. Braun, D.; Eiser, M.; Puntscher, H.; Marko, D.; Warth, B. Natural contaminants in infant food: The case of regulated and emerging mycotoxins. *Food Control* **2021**, *123*, 107676. [CrossRef]
32. Akinbami, L.J.; Chen, T.-C.; Davy, O.; Ogden, C.L.; Fink, S.; Clark, J.; Riddles, M.K.; Mohadjer, L.K. National Health and Nutrition Examination Survey, 2017–March 2020 Prepandemic File: Sample Design, Estimation, and Analytic Guidelines. 2(190). 2022. Available online: https://stacks.cdc.gov/view/cdc/115434 (accessed on 24 February 2025).
33. CLAL. Per Capita Cheese Consumption—Year 2023. Available online: https://www.clal.it/en/?section=tabs_consumi_procapite#C (accessed on 24 February 2025).
34. European Commission. Regulation (EU) 2022/2379 of the European Parliament and of the Council of 23 November 2022 on Statistics on Agricultural Input and Output, Amending Commission Regulation (EC) No 617/2008 and Repealing Regulations (EC) No 1165/2008, (EC) No 543/2009 and (EC) No 1185/2009 of the European Parliament and of the Council and Council Directive 96/16/EC. Available online: https://eur-lex.europa.eu/eli/reg/2022/2379/oj/eng (accessed on 7 April 2025).
35. European Commission. Regulation (EU) 2017/625 of the European Parliament and of the Council of 15 March 2017 on Official Controls and Other Official Activities Performed to Ensure the Application of Food and Feed Law, Rules on Animal Health and Welfare, Plant Health and Plant Protection Products, Amending Regulations (EC) No 999/2001, (EC) No 396/2005, (EC) No 1069/2009, (EC) No 1107/2009, (EU) No 1151/2012, (EU) No 652/2014, (EU) 2016/429 and (EU) 2016/2031 of the European Parliament and of the Council, Council Regulations (EC) No 1/2005 and (EC) No 1099/2009 and Council Directives 98/58/EC, 1999/74/EC, 2007/43/EC, 2008/119/EC and 2008/120/EC, and Repealing Regulations (EC) No 854/2004 and (EC) No 882/2004 of the European Parliament and of the Council, Council Directives 89/608/EEC, 89/662/EEC, 90/425/EEC, 91/496/EEC, 96/23/EC, 96/93/EC and 97/78/EC and Council Decision 92/438/EEC (Official Controls Regulation). *Official Journal of the European Union*. 2017. L95/1. Available online: https://eur-lex.europa.eu/legal-content/PL/TXT/?uri=CELEX:32017R0625 (accessed on 15 January 2025).
36. European Commission. Directorate-General for Health and Food Safety (DG SANTE). Guidance Document SANTE 11312/2021 v2—Analytical Quality Control and Method Validation Procedures for Pesticide Residues Analysis in Food and Feed. Version 2 Implemented by 01/01/2024. Available online: https://food.ec.europa.eu/document/download/d4786faf-c574-4222-a5c6-45086b3920b8_en?filename=pesticides_mrl_guidelines_wrkdoc_2021-11312.pdf (accessed on 24 February 2025).

molecules

MDPI

Article

Development and Validation of Two Cell-Based Reporter-Gene Assays for Determining the Bioactivity of Recombinant Human Thyroid-Stimulating Hormone Pharmaceutical Products

Lyuyin Wang [1,†], Jing Gao [2,†], Kaixin Xu [2], Jing Li [1,*] and Chenggang Liang [1,*]

[1] National Institutes for Food and Drug Control, No. 31, Huatuo Road, DaXing District, Beijing 102629, China; wanglvyin@nifdc.org.cn

[2] School of Life Science and Technology, China Pharmaceutical University, No. 639, Longmian Road, JiangNing District, Nanjing 210009, China; 17861405031@163.com (J.G.)

* Correspondence: li_jing@nifdc.org.cn (J.L.); liangchenggang@nifdc.org.cn (C.L.)

† These authors contributed equally to this work.

Abstract: To develop a cell-based in vitro thyroid-stimulating hormone (TSH) biological activity assay that can simulate in vivo pharmacodynamic mechanisms, we constructed two HEK293-TSHR cell lines based on two main cell signaling pathways (Gαs-cAMP-PKA and Gαq/11-PLC-Ca^{2+}) that TSH depends on for its in vivo physiological function. These cell lines stably expressed the luciferase reporter driven by the cAMP response element (CRE) and nuclear factor of activated T cells (NFAT) response element, and two reporter-gene assays (RGAs) were correspondingly established and validated. The two transgenic genes could measure signals produced from the simulation of the in vivo effects of TSH from the Gαs-cAMP and Gαq/11-PLC pathways after TSH activation. TSH showed a good dose–response relationship in these two cell lines and conformed to the four-parameter model. We optimized the critical experimental parameters of these two methods and performed comprehensive methodological validation according to the International Council for Harmonization (ICH) Q2 (R1) guidelines, the Chinese Pharmacopoeia, and the United States Pharmacopoeia. The two methods showed good specificity, accuracy, precision, and linearity and can be used to aid in assessments of the biological activity of TSH drugs, product characterization, final product release, stability studies, and comparability studies for biosimilar applications.

Keywords: reporter-gene assay; bioactivity; quality control; thyroid-stimulating hormone; method validation

1. Introduction

Thyroid-stimulating hormone (TSH) is a glycoprotein hormone secreted by the anterior pituitary [1] and is an important endocrine hormone for maintaining the normal physiological function of the thyroid. TSH can stimulate the synthesis and secretion of thyroid hormones, induce iodine uptake in the thyroid, and promote thyroid growth and development [2,3], and it is physiologically important in regulating basal metabolism in the body. The release of TSH is regulated by the classic hypothalamic–pituitary–thyroid axis, which is vital in maintaining the homeostasis of metabolic functions within the body [4]. The hypothalamic thyrotropin-releasing hormone (TRH) causes TSH to be released from the pituitary, and the resulting increased levels of TSH stimulate the thyroid follicular cells to produce more thyroid hormone in the form of triiodothyronine (T3) and thyroxine (T4). This, in turn, inhibits the production of TSH [5]. Recombinant human TSH (rhTSH) is a recombinant protein obtained via genetic recombination and in vitro expression and has mechanisms of action and biological characteristics similar to endogenous human TSH. In clinical practice, rhTSH is mainly used for adjuvant treatment and follow-up diagnosis of differentiated thyroid carcinoma (DTC) patients without distant metastasis after postoperative iodine-131 ablation of residual thyroid tissue [6,7]. Compared with traditional thyroid-hormone withdrawal methods, rhTSH injection can increase serum TSH levels more safely and rapidly and can minimize the suffering of patients caused by discontinuation of thyroid hormones [8,9]. Findings from the last 30 years of clinical application of rhTSH have resulted in it being included in many thyroid cancer-related clinical guideline recommendations around the world [10]. Thyroid cancer is the most common malignant tumor [11]; DTC accounts for more than 95% of all thyroid cancers [12]. Currently, only one rhTSH drug (Thyrogen, Genzyme, Inc., Cambridge, MA, USA) is available globally, one which was approved for marketing in the U.S.A. in 1998 [13]. Many rhTSH drugs are currently in different clinical trial stages [14,15].

Biological activity can reflect the accuracy of the higher order structure of rhTSH molecules, the integrity of biological function, and the inter-batch consistency of products [16], and it is an effective and essential method for assessing the quality of rhTSH products. Hence, developing a rapid, simple, sensitive, accurate, and stable in vitro biological activity assay is essential. The early TSH activity assay measured the increase in iodine-131 in mouse blood after dosing [17]. Because animal experiments involve long cycles, sizeable inter-individual variability, cumbersome operations, and high variability, and, additionally, do not conform to 3R principles [18], a consensus has been reached to develop alternatives to animal experiments. In 2022, Liu et al. developed a luminescence resonance energy transfer (LRET)-based TSH activity assay [19]. Although this assay is simple and has high specificity and sensitivity, it is still a content measurement assay and cannot simulate in vivo pharmacodynamics and thus truly reflect the biological activity of TSH. The use of cell-based in vitro assays to replace animal experiments is a trend in the development of quality control of biologics in China and other countries. Most reported in vitro cellular assays use TSH to stimulate the cAMP pathway and use ELISA to measure the TSH-induced intracellular cAMP increase to evaluate TSH biological activity. Although this is based on TSH's in vivo mechanisms of action (MOA), the operation is cumbersome, variability is high, and sensitivity is low. A cell-based reporter-gene assay uses drug effector signaling pathways. This not only examines live cell function but also is closer to the physiological state, and it intuitively and directly reflects the targeting effects of the drug. This method has high sensitivity, stability, and accuracy, and it has been widely used to measure the in vitro biological activity of therapeutic protein drugs in recent years [20,21].

TSHR belongs to the G-protein coupled 7 transmembrane leucine-rich repeat (LGR) subfamily and is the first critical substrate in the TSH signaling pathway [22]. Moreover, TSHR is one of the key targets for the correction of diseases caused by aberrant TSHR function (such as Grave's hyperthyroidism, non-autoimmune hyperthyroidism, and thyroid cancer) [23]. It is well established that TSH can activate four G protein subtypes ($G\alpha s$, $G\alpha q/11$, $G\beta\gamma$, and $G\alpha 12/13$) that are coupled to the TSH receptor (TSHR) [24]. Among them, the most canonical signaling pathway is the $G\alpha s$-mediated cAMP-PKA pathway [3]. $G\alpha s$ activation causes the intracellular second messenger cAMP concentration to increase, followed by PKA activation and phosphorylation of cAMP response element binding protein (CREB) and other transcription factors [25]. Phosphorylated CREB recognizes the specific response element CRE on gene promoters to regulate the transcription of downstream significant genes that participate in thyroid-hormone synthesis, cell growth, and differentiation, including sodium iodide symporter (NIS), thyroglobulin (TG), and thyroid peroxidase (TPO). $G\alpha q/11$ activation causes phospholipase C (PLC) activation, which enhances the production of the second messenger inositol 1,4,5-trisphosphate (IP3) [26]. Then, this triggers intracellular calcium-ion release and rapid translocation of the transcription factor NFAT into the cell nucleus and binding to a specific response element NFAT-RE on gene promoters to drive the expression of downstream genes. At the same time, protein kinase C (PKC) and its downstream effector gene MAPK are also activated, which activates dual oxidase 2 (DUOX-2), produces hydrogen peroxide, and promotes TG iodination to achieve regulation of critical steps in thyroid hormone synthesis. Hence, this is another essential signaling pathway of TSH activation, the PLC-Ca^{2+} pathway. TSH regulates the transcription and expression of related enzymes and precursor genes through these two signaling pathways [27] to achieve its main biological functions—regulation of iodine uptake, thyroid-hormone synthesis, and targeting of cell growth and differentiation [28,29].

Based on our understanding of these two signal transduction pathways (cAMP-PKA and PLC-Ca^{2+}), which are intimately associated with the biological function of TSH, we selected CRE and NFAT as response elements to construct reporter-gene vectors, which were separately transfected into HEK293 cells together with the TSHR expression vector. Two reporter-gene cell lines were thus obtained, which were the HEK293-TSHR/CRE-Luc cells, which stably express human TSHR and CRE-regulated luciferase, and HEK293-TSHR/NFAT-Luc, which stably expresses human TSHR and NFAT-regulated luciferase. After rhTSH binds to TSHR on the cell membranes of the two aforementioned HEK293 cells, it increases cAMP and Ca^{2+} concentrations, which promotes the expression of CRE and the NFAT-driven luciferase reporter gene. Measuring the intensity of fluorescence signals can thus characterize the biological activity of TSH. We optimized the critical experimental parameters of these methods and performed comprehensive methodological validation according to ICH Q2 (R^2) guidelines, the Chinese Pharmacopoeia, and the United States Pharmacopoeia. The results showed that these two methods are simple, have high sensitivity, good accuracy, and precision, and can be used for evaluation of in vitro pharmacodynamics and quality control of rhTSH products.

2. Results

2.1. *Generation and Selection of the HEK293-TSHR/CRE-Luc and HEK293-TSHR/NFAT-Luc Cell Lines*

After HEK293 cells were infected by packaged viruses, three HEK293-TSHR/CRE-Luc monoclonal cells (C2, C5, and C18) and three HEK293-TSHR/NFAT-Luc monoclonal cells (N9, N4, and N11) were obtained via drug selection and limiting dilution. As shown in Figure 1A, flow cytometry showed that TSHR expression on the surfaces of the three HEK293-TSHR/CRE-Luc cell lines was more than 97%, of which the C18 monoclone had the highest TSHR expression (98.8%). Figure 1C shows that among the three HEK293-TSHR/NFAT-Luc monoclonal cells, the N9 monoclone had the highest TSHR expression (99.4%). As shown in Figure 1B,D, the six monoclonal lines all responded to rhTSH stimulation, among which the C18 and N9 monoclonal lines presented greater responsiveness and a higher S/N ratio. Therefore, we used the C18 HEK293-TSHR/CRE-Luc monoclonal cells and N9 HEK293-TSHR/NFAT-Luc monoclonal cells for subsequent method optimization and validation.

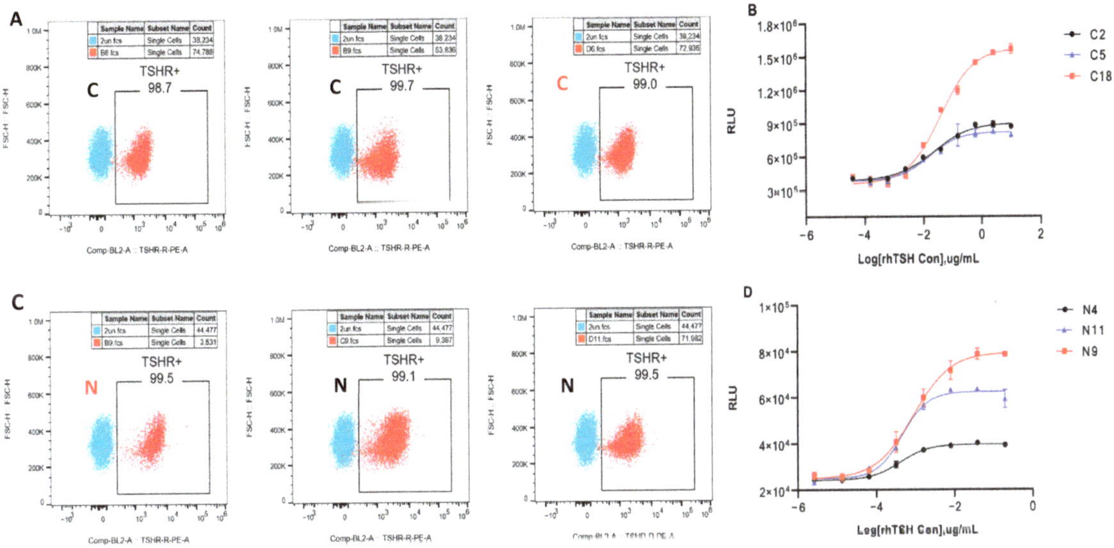

Figure 1. HEK293-TSHR/CRE-Luc and HEK293-TSHR/NFAT-Luc cells responsive to rhTSH. HEK293-TSHR/CRE-Luc and HEK293-TSHR/NFAT-Luc cell lines were analyzed by flow cytometry for the expression of hTSH receptor on the cell surface ((**A**) and (**C**), respectively). Three HEK293-TSHR/CRE-Luc monoclones (C2, C5, and C18) and three HEK293-TSHR/NFAT-Luc monoclones (N4, N11, and N9) were stimulated with serially diluted rhTSH ((**B**) and (**D**), respectively), and the dose–response curves are shown.

2.2. Optimization of the Two Reporter-Gene Assays

2.2.1. CRE Assay

Cells were seeded on a plate at densities of 0.3×10^5, 0.8×10^5, 1.5×10^5, and 2.0×10^5 cells/mL and used to measure the biological activity of the same rhTSH sample. As shown in Figure 2A, when the cell density was 0.8×10^5 cells/mL, the R^2 was 0.991, the SNR was the highest at 11.0, and EC_{50} was the lowest. The rhTSH was used to stimulate HEK293-TSHR/CRE-Luc cells at starting concentrations of 0.5, 2, 10, and 20 μg/mL. As shown in Figure 2B, when the starting concentration was 10 μg/mL, the dose–response curve appeared as a typical S shape, the points on the curve were uniformly distributed, the SNR was the highest, and R^2 was 0.992. Therefore, the optimal starting concentration was determined to be 10 μg/mL. Dilution factor optimization results (Figure 2C) showed that a typical S curve could only be obtained with cells stimulated with six-fold serially diluted rhTSH, and curves obtained using the other three dilution factors did not show a significant upper plateau or lower plateau. Drug stimulation time optimization results (Figure 2D) showed that the S curve could be obtained by rhTSH stimulation of cells for 3, 4, 5, and 6 h. At 4 h, EC_{50} was the lowest, and the SNR was the highest. Therefore, 4 h was selected as the optimal drug stimulation time.

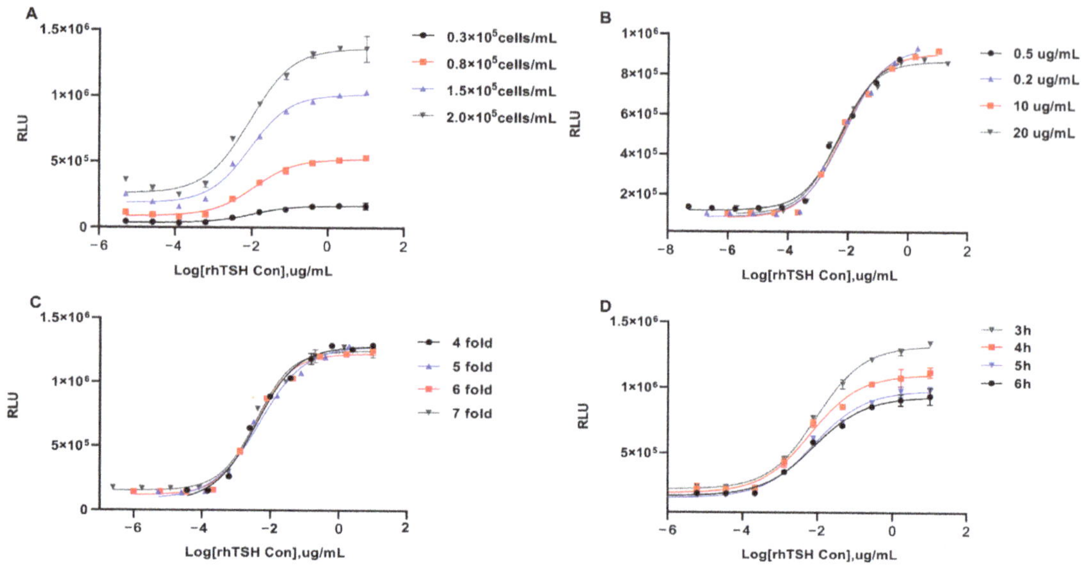

Figure 2. Optimization of the RGAs based on HEK293-TSHR/CRE-Luc cells. HEK293-TSHR/CRE-Luc cells were adjusted to 0.3, 0.8, 1.5, and 2.0 × 10^4 cells per well and incubated with serial dilutions of rhTSH. The dose–response curves were recorded to determine the optimal cell density (**A**). Cells were treated with gradient dilution of rhTSH with different initial concentrations (0.5, 0.2, 10, and 20 μg/mL) (**B**). After fixing the initial concentration at 10 μg/mL, rhTSH was diluted in 4-, 5-, 6-, and 7-fold serial dilutions to 10 concentration levels (**C**). RLU was measured after the test plates were incubated for different times (3, 4, 5, and 6 h) (**D**).

2.2.2. NFAT Assay

Similarly, cells were seeded on a plate at densities of 0.3 × 10^5, 0.8 × 10^5, 1.5 × 10^5, and 2.0 × 10^5 cells/mL, and the NFAT assay was used to measure the biological activity of the same rhTSH sample. As shown in Figure 3A, the typical S curve could be obtained from four cell densities, with an SNR of 3.5. When the cell density was 0.8 × 10^5 cells/mL, the R^2 was the highest at 0.995, and the EC_{50} was the lowest at 0.32. The rhTSH was used to stimulate HEK293-TSHR/NFAT-Luc cells at starting concentrations of 4, 8, 20, and 40 μg/mL. As shown in Figure 3B, when the starting rhTSH concentration was 20 μg/mL, the dose–response curve appeared as a typical S shape, R^2 was 0.995, and the points on the curve were uniformly distributed. Therefore, the optimal starting concentration was determined to be 20 μg/mL. Figure 3C shows the dilution factor optimization results. A typical S curve with complete upper and lower plateaus could only be obtained with HEK293-TSHR/NFAT-Luc cells stimulated with five-fold serially diluted rhTSH. Therefore, five-fold was selected as the optimal serial dilution factor. The stimulation time optimization results (Figure 3D) showed that the SNR of the curve was at its highest at 3.78 when the stimulation time was 9 h. Therefore, the optimal incubation duration for rhTSH stimulation of HEK293-TSHR/NFAT-Luc cells was determined to be 9 h.

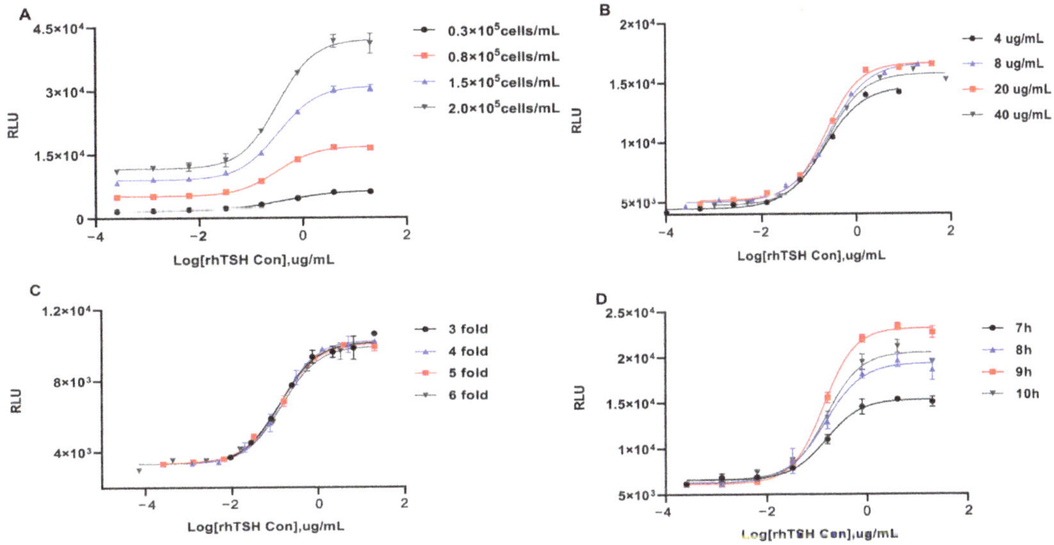

Figure 3. Optimization of the RGAs based on HEK293-TSHR/NFAT-Luc cells. The full dose–response curves are shown with the cells seeded at 8000 per well with various initial concentrations as indicated (**A**). RhTSH was diluted in 3-, 4-, 5-, or 6-fold serial dilutions to generate 10 concentration points, with an initial working concentration of 20 μg/mL. The dose–response curve is shown in (**B**). The dose–response curve shown was measured in the presence of different cell numbers as indicated (**C**) and at incubation times of 7, 8, 9, and 10 h shown in (**D**).

2.3. Methodological Validation

2.3.1. Specificity

After rhTSH was incubated at 60 °C for 0, 2, 8, 18, 36, and 60 h, the two reporter-gene assays were used to measure its biological activity. As shown in Figure 4A, when the CRE assay was used, compared with the untreated reference standard, the relative potencies of the treated reference standard decreased with increasing treatment duration, and they were 86.97, 79.91, 77.86, 65.81, and 56.99% after treatment at 60 °C for 2, 8, 18, 36, and 60 h, respectively. The relative potencies of the heat-degraded reference standard measured using HEK293-TSHR/NFAT-Luc cells were 88.31, 68.39, 65.49, 51.15, and 46.19% after treatment at 60 °C for 2, 8, 18, 36, and 60 h, respectively. The rhTSH contents measured using size exclusion chromatography were 95.36, 88.58, 85.43, 81.24, and 78.16%, respectively. Thus, the rhTSH main component content gradually decreased with increasing heat duration, and the contents of dissociated subunits and aggregates gradually increased, which was consistent with the biological activity variation trend obtained using the two RGAs that showed a positive correlation. In comparing the results of the three methods on the same heated sample, after heating for 8 h, the biological activity measured using the main component content was higher than found with the CRE assay. The biological activity measured using the CRE assay at different time points was stably higher by 12% than that associated with the NFAT assay. To further analyze the cause of the aforementioned experimental phenomenon, we used the heat-inactivated rhTSH content% as the *x*-axis coordinate and the biological activity% obtained from four repeated measurements using the two RGAs as the *y*-axis coordinate for correlation analysis. The results are shown in Figure 4C,D, and the P values are all less than 0.001. The biological activity% measured using the two RGAs was correlated to the heat-inactivated rhTSH content%, and the correlation between the NFAT measurement and heat-inactivated rhTSH% was higher than the CRE assay.

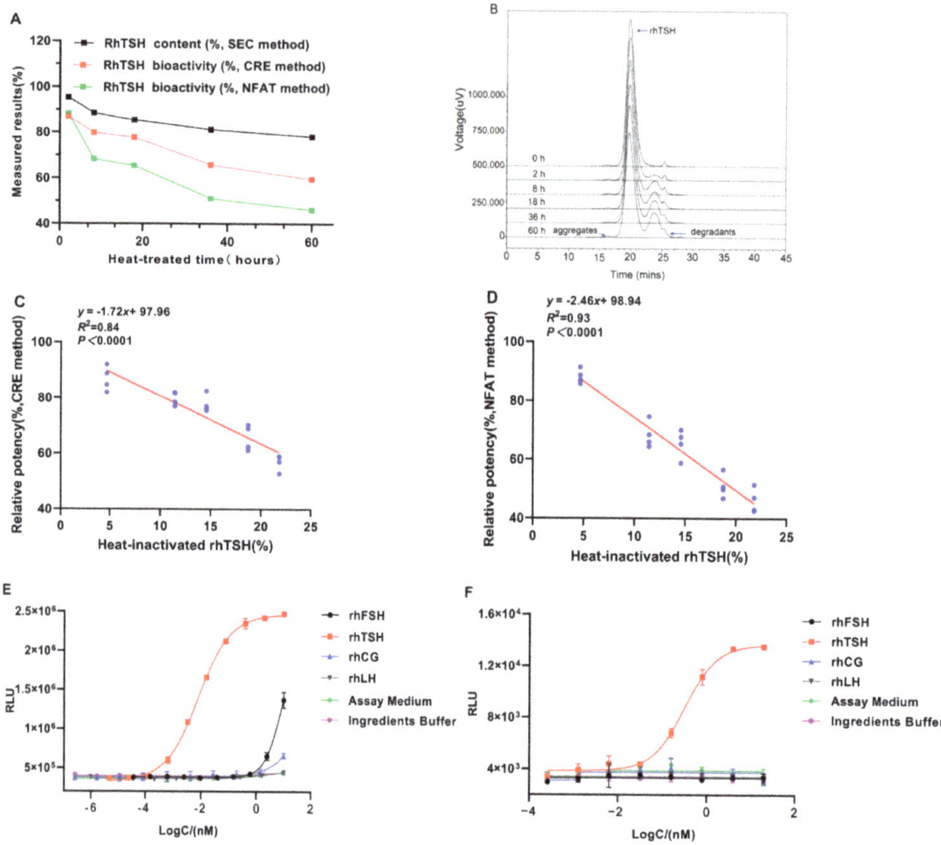

Figure 4. Specificity analysis of two new RGAs. (**A**) Relative potency estimated for CRE RGA and NFAT RGA, and the content measured via size-exclusion high performance liquid chromatography (SE–HPLC) of rhTSH samples kept at 60 °C for different durations of time (h). (**B**) Analysis of aggregates, degradants, and rhTSH contents in heat-inactivated samples via SEC–HPLC. Correlation analysis of the heat-inactivated rhTSH content with biological activity obtained using the CRE (**C**) and NFAT assays (**D**). Dose–response curves of the ingredients buffer, assay medium, rhFSH, rhTSH, rhCG, and rhLH, as determined with CRE (**E**) and NFAT RGAs (**F**).

The rhFSH, rhLH, rhCG, and rhTSH preparations, the buffer solution, and the measurement culture medium were used to stimulate the HEK293-TSHR/CRE-Luc cells. As shown in Figure 4E, there was no response for the rhLH or rhTSH excipient blank (ingredients buffer) or the measurement culture medium, and only rhFSH and rHCG showed a response for this assay. However, the response values did not result in an upper plateau. The HEK293-TSHR/NFAT-Luc cell-based RGA was used, and the rhFSH, rhLH, rhCG, and rhTSH preparations, the buffer solution, and the measurement culture medium showed no responses (Figure 4F). Of the two RGAs, NFAT had better specificity.

2.3.2. Accuracy

The newly constructed CRE and NFAT RGAs were validated. Two testers separately measured the relative potency of test solutions of five different potency levels (50, 80, 100, 125, and 200%), and the results all passed the reliability test. As shown in Tables 1 and 2, the relative bias of relative potency measured within the 50–200% potency range of the two assays was −8.2% to −3.2% (CRE assay) and −2.7 to 6.2% (NFAT assay). This shows that the relative potency measurements were close to the theoretical values and that the two methods have good accuracy.

Table 1. Validation of the accuracy for detecting the relative potency of rhTSH using the CRE assay.

Potency Level	Run	Relative Potency			Relative Bias		
		Mean	Lower Confidence	Upper Confidence	Mean	Lower Confidence	Upper Confidence
50%	8	47.1%	49.8%	44.3%	−5.8%	−11.4%	−0.7%
80%	8	76.7%	81.6%	71.9%	−4.1%	−10.3%	1.6%
100%	8	96.8%	101.1%	92.5%	−3.2%	−7.6%	0.6%
125%	8	117.6%	120.1%	115.0%	−6.0%	−8.1%	−4.0%
200%	8	183.7%	188.7%	178.7%	−8.2%	−10.7%	−5.7%

Table 2. Validation of the accuracy of detecting the relative potency of rhTSH using the NFAT assay.

Potency Level	Run	Relative Potency			Relative Bias		
		Mean	Lower Confidence	Upper Confidence	Mean	Lower Confidence	Upper Confidence
50%	8	51.0%	53.1%	48.9%	2.0%	−2.6%	6.3%
80%	8	85.0%	87.8%	82.1%	6.2%	2.5%	9.7%
100%	8	101.0%	105.9%	96.1%	1.0%	−3.9%	5.7%
125%	8	121.7%	126.6%	116.7%	−2.7%	−6.6%	1.1%
200%	8	208.4%	216.8%	200.0%	4.2%	−0.2%	8.3%

2.3.3. Precision

Two testers measured the relative potency of the five aforementioned test solutions within the same day for 2 days. Each tester had to prepare two sets of test solutions with the same potency level on the same day, and two cell passages were used. Each cell passage was used for two 96-well plate tests, and the geometric mean was calculated. The logarithmic values of the various potency level results were used to calculate the GCV (%), and the Chi-square test was used to calculate the upper limit of the 95% confidence interval of GCV. As shown in Tables 3 and 4, the GCV of relative potencies for each potency level in the eight experiments was 4.2 to 9.8% (CRE assay) and 5.2 to 7.4% (NFAT assay), and the corresponding 95% confidence interval upper limits were less than 20 and 15%, respectively.

Table 3. Intermediate precision validation of the CRE assay.

Potency Level	Run	GSD	CIGSD	GCV	CIGCV
50%	8	1.0890	1.1654	8.9%	16.5%
80%	8	1.0980	1.1828	9.8%	18.3%
100%	8	1.0657	1.1211	6.6%	12.1%
125%	8	1.0332	1.0604	3.3%	6.0%
200%	8	1.0417	1.0762	4.2%	7.6%

Table 4. Intermediate precision validation of the NFAT assay.

Potency Level	Run	GSD	CIGSD	GCV	CIGCV
50%	8	1.0671	1.1238	6.7%	12.4%
80%	8	1.0517	1.0947	5.2%	9.5%
100%	8	1.0737	1.1363	7.4%	13.6%
125%	8	1.0612	1.1127	6.1%	11.3%
200%	8	1.0631	1.1162	6.3%	11.6%

2.3.4. Linearity and Range

Linear regression was carried out on the natural logarithmic values of the theoretical values of the five potency levels and the natural logarithmic values of their corresponding potency measurements. The results are shown in Figure 5. The linear regression formulas of the two assays are Y = 0.904X + 4.038 (R^2 = 0.9913) and Y = 1.020X − 0.468 (R^2 = 0.9897). The slopes were close to 1.0, and R^2 was greater than 0.98. This shows that the two assays have good linearity within the 50–200% potency level.

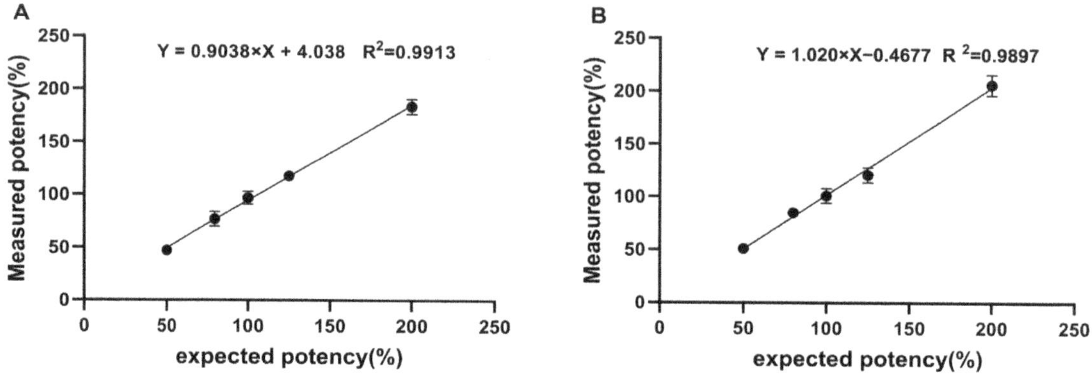

Figure 5. Linearity of the two RGAs. In the CRE (**A**) and NFAT methods (**B**), five samples at different relative bioactivity levels were analyzed, and the measured and expected potency of rhTSH showed a strong linear correlation. Each point indicates the average of eight independent replicates.

2.4. *Application of the Reporter-Gene Assays*

The two newly constructed RGAs were used to measure the biological activity of three batches of rhTSH stock solutions and three batches of rhTSH injections. Each batch was measured three times in parallel. All test results obtained using the two assays passed the reliability test. The relative standard deviation of the three repeated relative potency measurements of three batches of stock solutions (DS1, DS2, and DS3) and three batches of injections (DP1, DP2, and DP3) were all less than 10% (Table 5), showing that these two methods had good repeatability in measuring the in vitro biological activity of rhTSH and can be used for the relative biological activity measurement of rhTSH stock solutions and product samples.

Table 5. Relative bioactivities of the rhTSH stock solutions and injections as determined using two reporter-gene assays (n = 3).

Samples	CRE Assay		NFAT Assay	
	Mean	RSD	Mean	RSD
DS1	111.77%	4.07%	102.09%	3.23%
DS2	98.67%	9.89%	96.21%	2.71%
DS3	95.03%	4.37%	101.56%	4.77%
DP1	98.90%	5.92%	93.58%	3.25%
DP2	94.80%	3.18%	96.14%	4.38%
DP3	99.30%	5.75%	98.45%	5.40%

3. Discussion

In this paper, Gαs-cAMP-PKA and Gαq/11-PLC-Ca^{2+}, two signal transduction pathways that are intimately associated with TSH function, were successfully used to construct double transfected HEK293-TSHR stable lines containing CRE or NFAT response elements

(RE) tagged to a modified form of the luciferase reporter, and two RGAs that are highly correlated with the in vivo MOA of TSH were established.

In the optimization of the two reporter-gene assays, four parameters, namely, cell density, the starting concentration of rhTSH, rhTSH dilution factor, and incubation time, were examined individually to determine the optimal RGA conditions. The experimental parameters of the CRE assay (and NFAT assay) were finally determined, with a starting working concentration of 10 μg/mL (20 μg/mL for NFAT assay) for rhTSH, 10 dilutions in a six-fold (five-fold for NFAT assay) series, and a drug action time of 4 h (9 h for NFAT assay). As shown in Figures 2 and 3, the SNR values of the four-parameter curve obtained using the CRE and NFAT assays were 10 and 4, respectively. Hence, compared with the NFAT assay, the CRE assay produces signals faster with a higher signal intensity. As the drug concentration increases, the detection limit decreases, the detection range becomes broader, and the sensitivity and detection efficiency increase. These results were similar to CHO cell results in past studies [25]. Moreover, this shows that the PKA-Gαs pathway is the main pathway of TSH activation.

After using the analysis of variance and F test for reliability testing of the results of the analysis, we performed sufficient methodological validation of the two assays. To validate the specificity of the two RGAs, we analyzed the activity of the rhTSH reference standard after heating on different days. With increasing degrees of damage, the bioactivity determined by RGA and the main component content determined using size exclusion chromatography showed identical decreasing trends. After rhTSH was incubated at 60 °C for 8 h, the main component content and activity began to decrease considerably. For the same heat-inactivated sample, the biological activity measured using the NFAT assay and the main component content were lower than those measured using the CRE assay (Figure 4A). Hence, the two RGAs can not only distinguish between rhTSH and its heat byproducts, reflecting the correlation between the rhTSH structure and function, but they can more sensitively reflect sample stability than physicochemical methods. In addition, the NFAT assay can more sensitively reflect reduced activity due to heat destruction and has a higher detection capacity for heat-inactivated rhTSH. The correlation analysis results showed that the biological activity obtained using the two RGAs was correlated with the heat-inactivated rhTSH content, and the NFAT assay showed a stronger correlation. This shows that high temperature-induced rhTSH molecular structure changes have some inhibitory effects on the two signaling pathways and have a more significant effect on the Gq/11-mediated PLC-Ca^{2+} signaling pathway. The rhTSH consists of an α-subunit (92 amino acids) and a β-subunit (112 amino acids), bonded non-covalently. At high temperatures, the degradation mechanism of rhTSH is mainly dissociation, and the dissociated subunits are the main heat-destruction products. Studies have reported that free TSH α- and β-subunits can bind to TSHR, and free α-subunits will inhibit TSHR activity [30]. The inhibitory effects of free α subunits on TSHR may be one of the reasons why activation of the two aforementioned signaling pathways was inhibited.

Because TSH, LH, follicle-stimulating hormone (FSH), and HCG all belong to the glycoprotein family, they all contain a common α subunit and a hormone-specific β subunit. Their corresponding receptors, namely, FSHr, LHr, hCGr, and TSHr, are members of the rhodopsin-like G protein-coupled receptor (GPCR) family [31]. Homologous genes code the β subunits and corresponding receptors of the aforementioned glycoprotein hormones. Therefore, rhFSH, rhLH, and rhCG were selected to measure the specificity of the two assays. The results showed no rhFSH, rhLH, or rhCG response in the NFAT assay (Figure 4F). However, high concentrations of rhFSH and rhCG showed responses in the CRE assay, but a typical four-parameter curve could not be obtained (Figure 4E). On the one hand, this shows that the NFAT assay had better specificity. On the other hand, this shows that high concentrations of rhFSH and rhCG could bind to TSHR and activate the Gαs-coupled cAMP-PKA pathway. Our experimental results also showed that rhFSH has stronger activating effects on the cAMP-PKA pathway than does rhCG.

The two RGAs showed good dilution linearity within the relative potency range of 50–200%, with the relative bias not exceeding ±10%, and the relative accuracy was good. As shown in Tables 3 and 4, the two methods did not have high variability, and the corresponding 95% confidence interval upper limits were less than 20% (CRE assay) and 15% (NFAT assay), respectively. The precision of the NFAT assay was slightly better than that of the CRE assay. The RSD of repeated bioactivity measurements of three batches of stock solutions and three batches of injections were all less than 10% (Table 5), indicating that both methods had good repeatability and could be used as routine methods to measure rhTSH biological activity.

4. Materials and Methods

4.1. Cell Lines and Reagents

CRE reporter HEK293 cell line expressing luciferase and NFAT reporter HEK293 cell line expressing luciferase were supplied by Mijia Co., Ltd. (Beijing, China). DMEM culture medium, FBS, phosphate buffered saline (PBS), and puromycin were purchased from Gibco, Inc. (Grand Island, NY, USA). Hygromycin B was purchased from Life Technologies Corporation (Shanghai, China). The international rhTSH standard (03/192) for bioassays was purchased from NIBSC. The Steady-Glo Luciferase Assay System 2520ES Kit was purchased from Promega, Inc. (Madison, WI, USA). The pLV[Exp]-Hygro-EF1A>hTSHR plasmid was purchased from Beijing Mijia Co., Ltd. The rhTSH products and other hormones, including follicle stimulating hormone (FSH), luteinizing hormone (LH), and human chorionic gonadotropin (HCG), were archived biopharmaceuticals that had been preserved at 4 °C or −80 °C in our laboratory.

4.2. Construction of HEK293-TSHR/CRE-Luc and HEK293-TSHR/NFAT-Luc Cell Lines

After pLV[Exp]-Hygro-EF1A>hTSHR plasmid was packaged into lentiviruses, they were used to infect the HEK293/CRE-Luc and HEK293/NFAT-Luc stable cell lines. First, 100 μg/mL hygromycin B was used to select the aforementioned cells until stably transfected cells were obtained. Limiting dilution was used to obtain monoclonal cell lines. The cell lines were named HEK293-TSHR/CRE-Luc and HEK293-TSHR/NFAT-Luc and kept in our laboratory. Finally, flow cytometry was used to analyze TSHR expression on cell surfaces and confirm the positive monoclonal cells.

4.3. Cell Culture

HEK293-TSHR/CRE-Luc and HEK293-TSHR/NFAT-Luc cell lines were cultured in DMEM medium supplemented with 10% FBS, 100 μg/mL Hygromycin B, and 1 μg/mL puromycin. To maintain good condition, both cell lines should be passed every 3 to 4 days, and the densities for passaging should be 200,000 cells/mL for HEK293-TSHR/CRE-Luc cells and 400,000 cells/mL for HEK293-TSHR/NFAT-Luc cells.

4.4. RGA Procedure

The reporter-gene assays based on the HEK293-TSHR/CRE-Luc cell line and the HEK293-TSHR/NFAT-Luc cell line were named the CRE and NFAT assays, respectively. The cells were resuspended in DMEM culture medium containing 10% FBS, 100 μg/mL hygromycin B, and 1 μg/mL puromycin to obtain a cell suspension with a density of 8×10^4 cells/mL. The cells were seeded in a 96-well white cell culture plate at 100 μL/well and cultured for 16–24 h in a 37 °C, 5% CO_2 cell culture incubator. Then, the plates were removed, and the culture medium was discarded. The measurement culture medium

(DMEM culture medium containing 0.1% BSA) was used to dilute the rhTSH sample and reference standard (rhTSH international reference standard) to a starting concentration of 10 μg/mL (20 μg/mL for the NFAT assay), and 6-fold (5-fold for the NFAT assay) serial dilutions were made to obtain 10 (eight for the NFAT assay) dilutions. These dilutions were added at 100 μL/well into the seeded 96-well plates. The culture plates were incubated in a 37 °C, 5% CO_2 cell culture incubator for 4 h (9 h for the NFAT assay) and then removed. Then, 100 μL of Steady-Glo Luciferase Assay reagent.was added to each well, and the plates were allowed to react in the dark for 15 min. The EnSpire microplate reader was used to measure relative luciferase units (RLU).

4.5. Data Analysis

SoftMax Pro software 7.1.2 was used to analyze the experimental data, and GraphPad Prism software 9.5 was used to plot the graph. The logarithm base 10 values of rhTSH concentration were used as the *x*-axis coordinate and RLU was used as the *y*-axis coordinate to plot a four-parameter dose–response curve. The relative bioactivity was expressed as the ratio of the concentration for 50% of the maximal effect (EC_{50}) of the reference to that of the sample in the constraint model. The signal-to-noise ratio (SNR) was expressed as the ratio of the upper asymptote to the lower asymptote. Analysis of variance and an F test were used to test the reliability of the results. The determination criteria were the following: R^2 of the fitted 4-parameter curve of $\geq$0.98, regression item should be highly significant ($p < 0.01$), and deviation from parallel should not be significant ($p \geq 0.01$). The reliability test results can be used to calculate the percentage of relative biological activity.

4.6. Specificity

To validate that the aforementioned reporter-gene assays are specific for rhTSH, a sample specificity test was performed with the rhFSH, rhLH, rhCG, and rhTSH; the blank buffer solution; and the measurement culture medium.

In addition, the specificity test involves interference from possible degradation products as TSH undergoes structural changes under high-temperature treatment. To further validate that this method can reflect the relationship between the TSH structure and function, we aliquoted one batch of rhTSH stock solution into 1000 μL vials. We placed them severally in a 60 °C water bath for 2, 8, 18, 36, and 60 h. The two aforementioned reporter-gene assays were used to measure the relative potency. Size exclusion chromatography was used to measure the content of rhTSH after incubation at 60 °C for different times.

4.7. Accuracy, Precision, and Linearity

The rhTSH international reference standard was used to make a 10 μg (20 μg for the NFAT assay) per mL solution using the measurement culture medium as the method validation reference standard. The measurement culture medium was used to prepare a series of solutions with starting concentrations of 5, 8, 10, 12.5, and 20 μg/mL (10, 16, 20, 25, and 40 μg/mL for the NFAT assay), which were used as test solutions with five different potency levels (50, 80, 100, 125, and 200%). The five relative potency levels have uniform intervals on a logarithmic scale. Two testers used two cell passages to measure the relative potency of the five test solutions within a period of 2 days. Each potency level was measured eight times, and the geometric mean was calculated. To evaluate the accuracy we determined the relative bias via the following formula: RB% = (relative potency measurement/theoretical relative potency $-$ 1) $\times$ 100%. For intermediate precision, the geometric coefficient of variation (GCV, %) of the various potency level results was used, and the Chi-square test was used to calculate the upper limit of the 95% confidence interval of the GCV. The logarithm of the relative potency values of the five test solutions was used as the

x-axis coordinate, and the theoretical logarithmic value of relative potency was used as the y-axis coordinate for linear regression analysis.

5. Conclusions

In recent years, there has been rapid progress in small molecule or antibody agonists and antagonists that target TSHR [1,22,23,25]. These drugs can bind to TSHR and induce selective signal transduction to mediate related physiological effects [32]. The two transgenic cell lines constructed in this study can activate the Gαs and Gαq/11 pathways of TSHR. This provides a tool for screening and MOA studies on the aforementioned small molecules or antagonists and agonists, as well as technical tools for developing novel antibodies or small molecule drugs that target autoimmune diseases caused by abnormal TSHR function.

TSH shows efficacy in osteoporosis, thyroid cancer, polycystic ovary syndrome, and other diseases [33,34]. In the future, modifying the TSH molecular function or developing safe and effective TSH analogs will have high diagnosis and treatment value in clinical practice. The RGAs constructed in this study are simple and efficient, have high precision and accuracy, and have potential applications in the production and R&D of rhTSH drugs, process control, stability studies, finished product releases, and comparability studies for biosimilar applications. Moreover, they could be used as references for biological activity measurements of rhTSH products.

Author Contributions: Conceptualization, C.L.; Data curation, L.W., J.G. and K.X.; Funding acquisition, C.L.; Investigation, J.G.; Methodology, L.W., J.G. and K.X.; Project administration, J.L.; Resources, L.W. and K.X.; Software, J.G.; Supervision, L.W., C.L. and J.L.; Validation, L.W., J.G. and K.X.; Writing—review and editing, L.W. All authors have read and agreed to the published version of the manuscript.

Funding: This research was funded by the State Key Laboratory of Drug Regulatory Science (Grant number 2023SKLDR0108).

Institutional Review Board Statement: Not applicable.

Informed Consent Statement: Not applicable.

Data Availability Statement: Data are contained within the article.

Conflicts of Interest: The authors declare that they have no conflicts of interest.

References

1. Duan, J.; Xu, P.; Luan, X.-D.; Ji, Y.-J.; He, X.-H.; Song, N.; Yuan, Q.-N.; Jin, Y.; Cheng, X.; Jiang, H.-L.; et al. Hormone-and antibody-mediated activation of the thyrotropin receptor. *Nature* **2022**, *609*, 854–859. [CrossRef] [PubMed]
2. Cheng, X.-B.; Zhang, H.; Guan, S.-S.; Zhao, Q.; Shan, Y. Receptor modulator associated with the hypothalamus-pituitary-thyroid axis. *Front. Pharmacol.* **2023**, *4*, 1291856. [CrossRef] [PubMed]
3. Pereira, S.S.; Lobato, C.B.; Monteiro, M.P. Cell signaling within endocrine glands: Thyroid, parathyroids and adrenal glands. In *Tissue-Specific Cell Signaling*; Springer: Cham, Switzerland, 2020; pp. 63–91. [CrossRef]
4. Feldt-Rasmussen, U.; Effraimidis, G.; Klose, M. The hypothalamus-pituitary-thyroid (HPT)-axis and its role in physiology and pathophysiology of other hypothalamus-pituitary functions. *Mol. Cell. Endocrinol.* **2021**, *525*, 111173. [CrossRef] [PubMed]
5. Prayongrat, A.; Lertbutsayanukul, C. Hypothyroidism after radiotherapy for nasopharyngeal carcinoma. *Ann. Nasopharynx Cancer* **2020**, *4*, 3–13. [CrossRef]
6. Castagna, M.G.; Cantara, S.; Pacini, F. Reappraisal of the indication for radioiodine thyroid ablation in differentiated thyroid cancer patients. *J. Endocrinol. Investig.* **2016**, *39*, 1087–1094. [CrossRef]
7. Orlov, S.; Salari, F.; Kashat, L.; Freeman, J.L.; Vescan, A.; Witterick, I.J.; Walfish, P.G. Post-operative stimulated thyroglobulin and neck ultrasound as personalized criteria for risk stratification and radioactive iodine selection in low- and intermediate-risk papillary thyroid cancer. *Endocrine* **2015**, *50*, 130–137. [CrossRef]

8. Moore, M.D.; Postma, E.; Gray, K.D.; Ullmann, T.M.; Hurley, J.R.; Goldsmith, S.; Sobel, V.R.; Schulman, A.; Scognamiglio, T.; Christos, P.J.; et al. Less is more: The impact of multidisciplinary thyroid conference on the treatment of well-differentiated thyroid carcinoma. *World J. Surg.* **2018**, *42*, 343–349. [CrossRef]
9. Kairemo, K.; Kangasmäki, A.; Bom, H.-S. Comparison of I-131 biokinetics after recombinant human TSH stimulation and thyroid hormone withdrawal measured by external detector in patients with differentiated thyroid cancer. *Chonnam Med. J.* **2019**, *55*, 20–24. [CrossRef]
10. Haugen, B.R.; Alexander, E.K.; Bible, K.C.; Doherty, G.M.; Mandel, S.J.; Nikiforov, Y.E.; Pacini, F.; Randolph, G.W.; Sawka, A.M.; Schlumberger, M.; et al. 2015 American thyroid association management guidelines for adult patients with thyroid nodules and differentiated thyroid cancer: The american thyroid association guidelines task force on thyroid nodules and differentiated thyroid cancer. *Thyroid* **2016**, *26*, 1–133. [CrossRef]
11. Ibrahimpasic, T.; Ghossein, R.; Shah, J.P.; Ganly, I. Poorly differentiated carcinoma of the thyroid gland: Current status and future prospects. *Thyroid* **2019**, *29*, 311–321. [CrossRef]
12. Schlumberger, M.; Leboulleux, S. Current practice in patients with differentiated thyroid cancer. *Nat. Rev. Endocrinol.* **2021**, *17*, 176–188. [CrossRef] [PubMed]
13. Schlumberger, M.; Leboulleux, S. Recombinant or endogenous thyroid-stimulating hormone for radioactive iodine therapy in thyroid cancer: State of knowledge and current controversies. *Eur. J. Endocrinol.* **2023**, *188*, R1–R3. [CrossRef]
14. Gu, Y.; Xu, H.; Yang, Y.; Xiu, Y.; Hu, P.; Liu, M.; Wang, X.; Song, J.; Di, Y.; Wang, J.; et al. Evaluation of SNA001, a novel recombinant human thyroid stimulating hormone injection, in patients with differentiated thyroid carcinoma. *Front. Endocrinol.* **2020**, *11*, 615883. [CrossRef] [PubMed]
15. Lin, Y.-S.; Yang, H.; Li, X.-Y.; Wu, L.-Q.; Xu, J.-G.; Yang, A.-M.; Gao, Z. R.; Ding, Y.; Zhang, Y.-Q., Chen, K., et al. Novel recombinant human thyroid-stimulating hormone in aiding postoperative assessment of patients with differentiated thyroid cancer—Phase I/II study. *Eur. J. Nucl. Med. Mol. Imaging* **2022**, *49*, 4171–4181. [CrossRef] [PubMed]
16. Zhao, X.; Qian, W.; Hou, S.; Wu, Y.; Guo, H.; Xu, J.; Zhang, D.; Li, J.; Fu, R.; Xu, M.; et al. Development of a reliable cell-based reporter gene assay to measure the bioactivity of anti-HER2 therapeutic antibodies. *J. Pharm. Biomed. Anal.* **2024**, *245*, 116185. [CrossRef]
17. McKenzie, J.M. The bioassay of thyrotropin in serum. *Endocrinology* **1958**, *63*, 372–382. [CrossRef]
18. Wang, L.-Y.; Liang, C.-G.; Yang, H.-M.; Li, J.; Li, Z.-J.; Zhang, H.; Lv, P.; Li, Y.; Wang, J.-Z. Development of a robust reporter gene-based assay for the bioactivity determination of recombinant human follicle stimulating hormone (rhFSH) pharmaceutical products. *J. Pharm. Biomed. Anal.* **2020**, *177*, 112855. [CrossRef]
19. Liu, C.; Yu, C.; Yang, Y.; Cui, Y.; Zhang, F.; Wang, L.; Wang, J. Development and validation of a reporter gene assay to determine the bioactivity of anti-CTLA-4 monoclonal antibodies. *Int. Immunopharmacol.* **2021**, *101*, 108277. [CrossRef]
20. Tieu, M.V.; Go, A.; Park, Y.J.; Nguyen, H.V.; Hwang, S.Y.; Lee, M.-H. Highly sensitive ELISA using membrane-based microwave-mediated electrochemical immunoassay for thyroid-stimulating hormone detection. *IEEE Sens. J.* **2019**, *19*, 9826–9831. [CrossRef]
21. Tang, D.; Song, X.; Du, Y.; Wang, J.; Lei, Y.; Chen, B. Development of a reporter gene-based assay for the bioactivity determination of rhLH pharmaceutical products. *Anal. Biochem.* **2024**, *686*, 115413. [CrossRef]
22. Fokina, E.F.; Shpakov, A.O. ThyroidStimulating Hormone Receptor: The Role in the Development of Thyroid Pathology and Its Correction. *J. Evol. Biochem. Physiol.* **2022**, *108*, 1094–1111. [CrossRef]
23. Zhang, Y.; Tan, Y.; Zhang, Z.; Cheng, X.; Duan, J.; Li, Y. Targeting Thyroid-Stimulating Hormone Receptor: A Perspective on Small-Molecule Modulators and Their Therapeutic Potential. *J. Med. Chem.* **2024**, *67*, 16018–16034. [CrossRef] [PubMed]
24. Chu, Y.-D.; Yeh, C.-T. The molecular function and clinical role of thyroid stimulating hormone receptor in cancer cells. *Cells* **2020**, *9*, 1730. [CrossRef] [PubMed]
25. Latif, R.; Morshed, S.A.; Ma, R.; Tokat, B.; Mezei, M.; Davies, T.F. A Gq biased small molecule active at the TSH receptor. *Front. Endocrinol.* **2020**, *11*, 372. [CrossRef]
26. Corvilain, B.; Laurent, E.; Lecomte, M.; Vansande, J.; Dumont, J.E. Role of the cyclic adenosine 3′,5′-monophosphate and the phosphatidylinositol-Ca^{2+} cascades in mediating the effects of thyrotropin and iodide on hormone synthesis and secretion in human thyroid slices. *J. Clini. Endocrinol. Metab.* **1994**, *79*, 152–159. [CrossRef]
27. Benvenga, S.; Nordio, M.; Laganà, A.S.; Unfer, V. The role of inositol in thyroid physiology and in subclinical hypothyroidism management. *Front. Endocrinol.* **2021**, *12*, 662582. [CrossRef]
28. Patterson, S.D.; Huang, X.; Jørgensen, H.G.; Michie, A.M. Transcriptional regulation by the NFAT family in acute myeloid leukaemia. *Hemato* **2021**, *2*, 556–571. [CrossRef]
29. Maenhaut, C.; Christophe, D.; Vassart, G.; Dumont, J.; Roger, P.P.; Opitz, R. Ontogeny, anatomy, metabolism and physiology of the thyroid. In *Endotext [Internet]*; MDText.com, Inc.: South Dartmouth, MA, USA, 2015.
30. Morris, J.C., 3rd; Jiang, N.S.; Cherlesworth, M.C.; McCormick, D.J.; Ryan, R.J. The Effects of Synthetic a-Subunit Peptides on Thyrotropin Interaction with its Receptor. *Endocrinology* **1988**, *123*, 456–462. [CrossRef]

31. Faust, B.; Singh, I.; Zhang, K.-H.; Hoppe, N.; Pinto, A.F.M.; Muftuoglu, Y.; Billesbølle, C.B.; Saghatelian, A.; Cheng, Y.-F.; Manglik, A. Autoantibody and hormone activation of the thyrotropin G protein-coupled receptor. *bioRxiv* **2022**. [CrossRef]
32. Xu, Y.; Cai, H.; You, C.; He, X.; Yuan, Q.; Jiang, H.; Cheng, X.; Jiang, Y.; Xu, H.E. Structural insights into ligand binding and activation of the human thyrotropin-releasing hormone receptor. *Cell Res.* **2022**, *32*, 855–857. [CrossRef]
33. Fan, H.; Ren, Q.; Sheng, Z.; Deng, G.; Li, L. The role of the thyroid in polycystic ovary syndrome. *Front. Endocrinol.* **2023**, *14*, 1242050. [CrossRef] [PubMed]
34. Kim, S.-M.; Ryu, V.; Miyashita, S.; Korkmaz, F.; Lizneva, D.; Gera, S.; Latif, R.; Davies, T.F.; Lqbal, J.; Yuen, J.; et al. Thyrotropin, Hyperthyroidism, and Bone Mass. *J. Clin. Endocrinol. Metab.* **2021**, *106*, e4809–e4821. [CrossRef] [PubMed]

molecules

MDPI

Article

Novel Approaches to Monitor Pharmacokinetics and Metabolism of Gemcitabine-Ibandronate Conjugate in Mice and Dogs

Jost Klawitter [1,2,*,†], Mckay Easton [1,†], Alexander Karpeisky [3,†], Kristen B. Farrell [4,†], Douglas H. Thamm [4], Touraj Shokati [1], Uwe Christians [1] and Shawn Patrick Zinnen [3]

1 Department of Anesthesiology, School of Medicine, University of Colorado, Anschutz Medical Campus, Aurora, CO 80045, USA; mckay.easton@gmail.com (M.E.); touraj.shokati@hepquant.com (T.S.); uwe.christians@cuanschutz.edu (U.C.)

2 Department of Psychiatry, School of Medicine, University of Colorado, Anschutz Medical Campus, Aurora, CO 80045, USA

3 MBC Pharma Inc., Aurora, CO 80045, USA; alkarp@mbcpharma.com (A.K.); szinnen@mbcpharma.com (S.P.Z.)

4 Flint Animal Cancer Center, Colorado State University, Fort Collins, CO 80523, USA; kristen.farrell@colostate.edu (K.B.F.); doug.thamm@colostate.edu (D.H.T.)

* Correspondence: jost.klawitter@cuanschutz.edu; Tel.: +1-303-724-5671

† These authors contributed equally to this work.

Abstract: Background. The use of the bone-seeking properties of bisphosphonates (BPs) to target the delivery of therapeutic drugs is a promising approach for the treatment of bone metastases. Currently, the most advanced example of this approach is a gemcitabine-ibandronate conjugate (GEM-IB), where the bone-targeting BP ibandronate (IB) is covalently linked to the antineoplastic agent gemcitabine (GEM) via a spacer phosphate group. In the present study, we describe the development of a new analytical platform to evaluate the metabolism and pharmacokinetics of GEM-IB in mice and dogs and the results of proof-of-concept studies assessing the pharmacokinetics of GEM-IB in dogs and mice. Methods. We validated analytical platforms to analyze GEM-IB and five of its major metabolites IB, gemcitabine-5′-phosphate (GEMMP), gemcitabine (GEM), 2′,2′-difluoro-2′-deoxyuridine-5′-phosphate (dFdUMP), and 2′,2′-difluoro-2′-deoxyuridine (dFdU) and performed proof-of-concept pharmacokinetic studies in mice (5 mg/kg i.p.) and dogs (5 mg/kg i.v.). Results. Intra- and inter-run accuracy and imprecision (3 days) of the assays met the (FDA) acceptance criteria. The proof-of-concept plasma pharmacokinetic studies in mice showed AUCs of 1278, 10,652, 405, 38, 1063, 3389, and 38 h·ng/mL for GEM-IB, IB, GEMMP, dFdU-MP, GEM, and dFdU, respectively. In dog plasma, AUCs of 295, 5725, 83, 11, 1625, and 6569 h·ng/mL were observed for GEM-IB, IB, GEMMP, dFdUMP, GEM, and dFdU. Conclusions. Pharmacokinetic studies in dogs and mice showed that GEM-IB is rapidly converted to IB and GEM; dFdU is formed (from GEM) with a delay. The rapid disappearance of GEM-IB from circulation could be explained by a combination of metabolism and rapid distribution to tissue/bone.

Keywords: bisphosphonates; bone cancer; pharmacokinetics; gemcitabine ibandronate; high-performance liquid chromatography; mass spectrometry; drug metabolism

1. Introduction

Tumor-induced bone disease (TIBD) is one of the major causes of morbidity in cancer patients [1,2]. More than half of all carcinoma patients develop bone metastases [3,4]. TIBD is found in 65–95% of patients with multiple myeloma and advanced breast and

https://doi.org/10.3390/molecules30020354

prostate cancers [2,5]. Because TIBD is associated with considerable morbidity and a median survival time of less than two years [2], the development of more effective therapies is warranted [5,6]. Current TIBD therapies are palliative. Standard anti-cancer chemotherapies at their maximum tolerated doses are unable to reach effective concentrations in the bone and its microenvironment. With the bone microenvironment and metastases to this site being critical to cancer progression [7], there is clearly a need for drugs that can attack and kill bone-associated tumor cells without prohibitive systemic toxicity.

Bisphosphonates (BPs) are bone-specific palliative treatments that reduce tumor-induced skeletal complications [1,8–10]. However, TIBD still progresses in BP-treated cancer patients. The development of drugs with enhanced anti-resorptive and cytotoxic characteristics to improve the treatment of patients with TIBD is greatly needed [1].

The exploitation of the bone-seeking properties of BPs for targeted delivery of cancer therapeutics [1,11–14] is a promising approach to target bone metastases. BPs and cytotoxic agents can be covalently linked, allowing the intact conjugate to leave the circulation and release both drugs in the bone microenvironment [1,13]. These conjugates could combine anti-resorptive and anti-tumor activities while localizing at the site of tumor cell-induced bone destruction. We have shown that **MBC-11** (see Figure 1), a first-in-class conjugate of the bone-targeting BP etidronate covalently linked to the antimetabolite **cytarabine** (**arabinocytidine** or **araC**), was well tolerated in humans and showed significant reductions in the metabolic activity of bone-associated cancer cells [13].

In the present study, we describe the development and validation of analytical platforms to quantify a novel drug in this class, a **gemcitabine–ibandronate conjugate** [15] (**GEM-IB**, see Figure 1A), and its metabolites (Figure 1B) and analyze its metabolism and pharmacokinetics (PK) in dogs and mice. **GEM** is one of the most widely used antineoplastic agents in clinical oncology, used to treat breast, ovarian, bladder, non-small lung, and pancreatic cancer [16,17]. As compared with other cytotoxic chemotherapy drugs, **GEM** is generally considered a tolerable compound for most cancer patients [17]. Treatment-related adverse events are usually clinically manageable and fatal treatment-related events have been reported in only a few (<1% to 4%) patients [17–19]. Side effects include hematological (e.g., neutropenia, anemia, thrombocytopenia, and thromboembolism) and non-hematological (e.g., vomiting, fatigue, and elevated levels of alanine aminotransferase) side effects [17]. The earliest studies of BP use for bone metastases involved trials of clodronate [20,21] and pamidronate [21,22] in the early 1990s. **Ibandronate** (**IB**) is an advanced BP medication used in the prevention and treatment of osteoporosis and metastasis-associated skeletal fractures [23]. **IB** is FDA-approved for the treatment and prevention of osteoporosis in post-menopausal women [24]. **GEM-IB** is a novel conjugate that combines the potent antineoplastic properties of **GEM** with the bone-targeting properties of **IB**. No studies describing the analysis, pharmacokinetics, and stability of this bisphosphonate conjugate have been conducted before.

In a manner analogous to **MBC-11**, **GEM-IB** is hydrolyzed under physiological conditions to form **gemcitabine-5′-phosphate** (**GEMMP**) and **IB**. While **IB** does not undergo further degradation, **GEMMP** is rapidly hydrolyzed to form **GEM** or is deaminated, resulting in **dFdUMP** (see Figure 1B). Hydrolysis of **dFdUMP** or deamination of **GEM** will result in the formation of the essentially inert metabolite **dFdU**. To investigate the absorption distribution and metabolism of **GEM-IB** after administration, it is important to monitor **GEM-IB** and the metabolites **GEMMP**, **IB**, **GEM**, **dFdUMP**, and **dFdU**. In the present study, we describe two novel analytical platforms for the analysis of **GEM-IB** and **IB** (Assay A) and **GEMMP**, **GEM**, **dFdUMP**, and **dFdU** (Assay B) in mouse and dog plasma to support preclinical studies. For the first time, we provide data from two preclinical studies, one in *mice* and one in *dogs*, to prove the feasibility of the approach and present the first set of PK data for this novel drug.

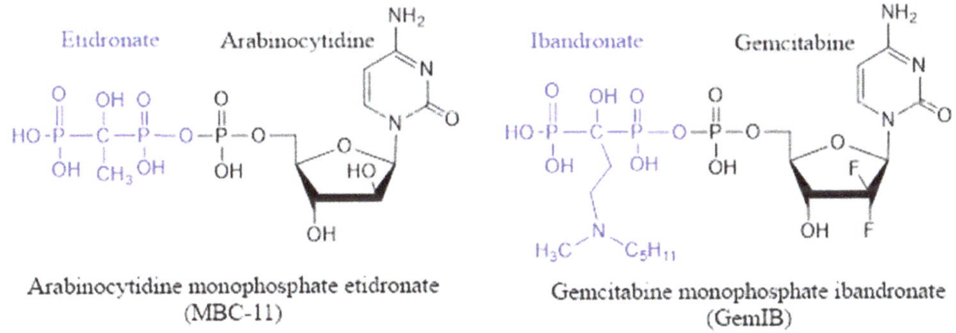

(A)

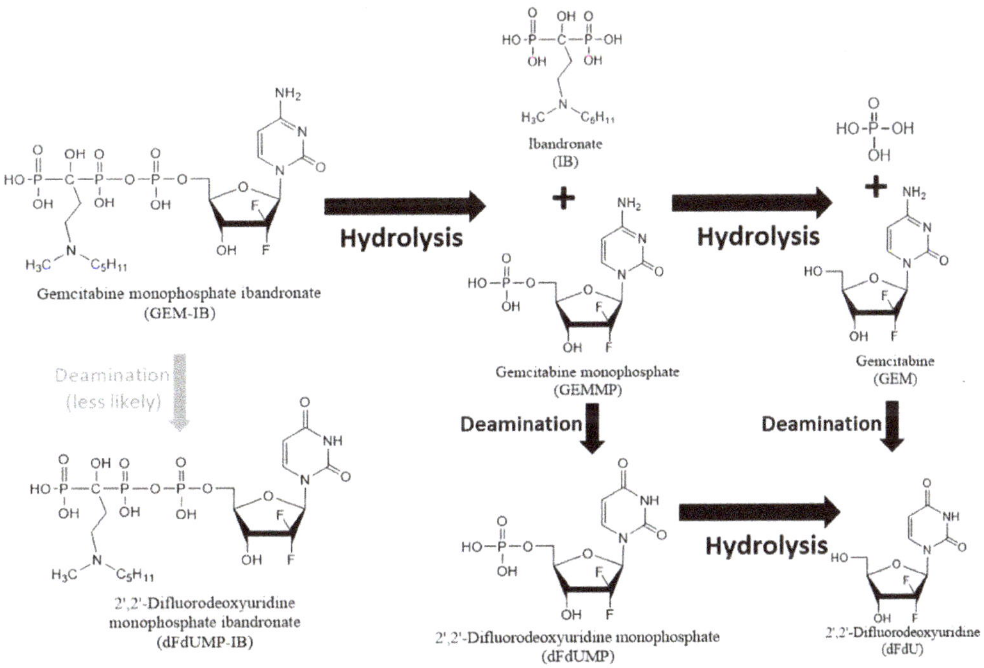

(B)

Figure 1. (**A**) Chemical structures of **MBC-11** [13] and **GEM-IB** [15]. Both drugs are the conjugates of a bisphosphonate bone-targeting component, linked together via phosphate group—**etidronate** and **arabinocytidine** (**MBC-11**) or **ibandronate** and **gemcitabine** (**GEM-IB**). (**B**) Chemical and metabolic breakdown of **GEM-IB**. **GEM-IB** is rapidly hydrolyzed to **gemcitabine-5′-phosphate** (**GEMMP**) and **ibandronate** (**IB**). **GEMMP** can be further hydrolyzed to **gemcitabine** (**GEM**) or enzymatically deaminated to **2′,2′-difluorodeoxyuridine-5′-phosphate** (**dFdUMP**). **2′,2′-Difluorodeoxyuridine** (**dFdU**) is formed either via deamination of **GEM** or hydrolysis of **dFdUMP**.

2. Results

Various conditions were tested to achieve the acceptable chromatographic performance and separation of key components while maintaining optimal sensitivity, reproducibility, and minimal carry-over. Chromatographic performance for pyrimidine nucleosides and nucleotides (**GEM**, **dFdU**, **GEMMP**, and **dFdU-MP**) could be achieved under various common reversed-phase high-performance liquid chromatography (RP-HPLC) conditions. In contrast, achieving acceptable chromatography was challenging for **GEM-IB**. Normal-phase, reversed-phase (RP), and hydrophilic interaction liquid chromatography (HILIC) resulted in poor retention, retention reproducibility, and/or robustness for **IB** and **GEM-IB**. Ion-pair chromatography using various mass spectrometry-compatible ion-pair reagents in combination with RP-HPLC column materials showed an improved peak shape, but lacked reproducibility and sensitivity and showed a significant carry-over effect. This changed with the use of the Hypercarb column material, which consists of 100% porous graphitic carbon (PGC). Using this material and the conditions listed in Section 4 for Assay A, acceptable chromatographic performance was achieved (see Figure 2A). Unfortunately, the less polar metabolites (**GEM** and **dFdU**) did not show sufficient separation using PGC material under these conditions. Thus, a second chromatographic platform (Assay B) using simple RP-HPLC that proved more suitable for said metabolites was used to monitor pyrimidine nucleosides and nucleotides. Complete chromatographic separation of **GEM** and its monophosphate **GEMMP** as well as of **dFdU** from its monophosphate **dFdU-MP** was required due to in-source fragmentation. Figure 2B,C show representative extracted ion chromatograms for the analytes from extracted mouse plasma samples using RP-HPLC Assay B (please see Section 4).

2.1. Linearity, Lower Limit of Quantitation, and Selectivity

The lower limit of quantitation (LLOQ) was defined as the lowest concentration for which the accuracies were within 20% of the nominal concentration for at least half of the samples and the imprecisions were less than 20%. The LLOQ also had a signal-to-noise (S/N) ratio greater than 8. The LLOQ was determined to be 5 ng/mL for **GEM**, **GEMMP**, and **dFdUMP**, while dFdU and **GEM-IB** had an LLOQ of 10 ng/mL and **IB** had an LLOQ of 40 ng/mL in dog and mouse EDTA plasma. The upper limit of quantitation (ULOQ) used in this study was 1000 ng/mL for all compounds, except for **IB**, which had a ULOQ of 4000 ng/mL. Representative extracted ion chromatograms of non-spiked blank mouse and dog plasma and spiked plasma and representative calibration curves for each compound are shown in Supplementary Figure S1a–f. The correlation coefficients for the calibration curves were consistently $r = 0.99$ and better. Figure 3 shows the chromatographic peaks for all compounds at a concentration of 2.5 ng/mL (below the lower limit), as well as the carry-over peaks for the solvent injection immediately following the injection with the highest concentration of sample (1000 ng/mL). The peak intensities of said injections indicate that the carry-over for this assay was less than 1% (See Supplementary Figure S1a–f).

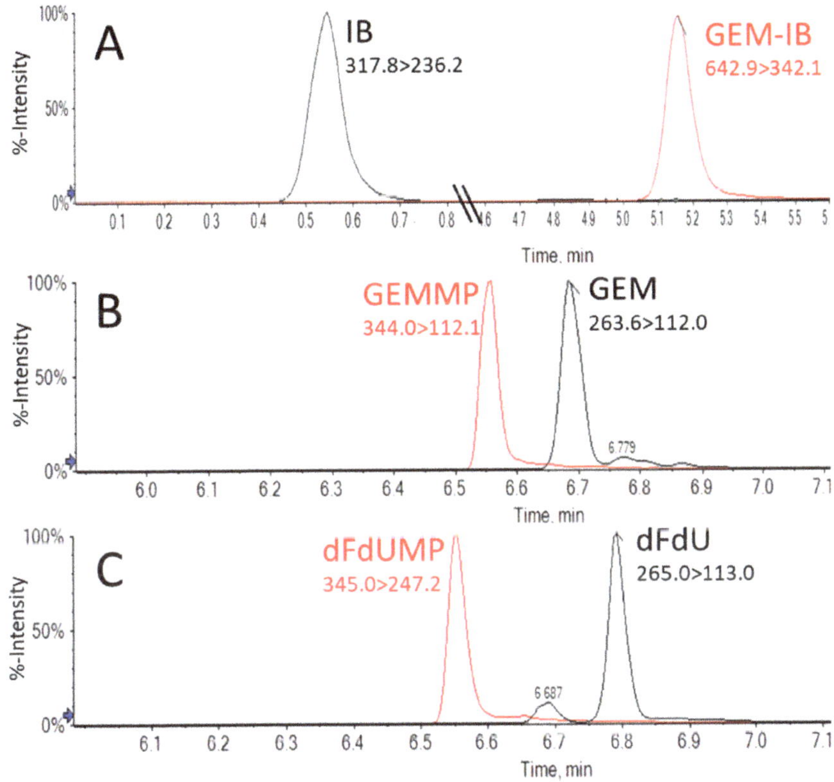

Figure 2. Assay A and Assay B's chromatographic performance. A calibrator sample in mouse plasma is shown in all examples. (**A**) Representative extracted ion chromatograms of the highest calibrator sample for Assay A in mouse plasma for **IB** (4000 ng/mL) and **GEM-IB** (4000 ng/mL). Representative extracted ion chromatograms of the highest calibrator (1000 ng/mL) sample for Assay B in mouse plasma are shown in (**B**,**C**). Separation of **GEMMP** and **GEM** is shown in (**B**) and separation of **dFdU** and **dFdUMP** is shown in (**C**).

2.2. Extraction Recovery and Matrix Effects

Matrix effects and extraction recovery was evaluated using the protocol described by Matuszewski et al. [25]. Six different individual lots of dog plasma and six different individual lots of mouse plasma were used for all recovery and matrix effect experiments. Plasma samples were enriched at three different levels within the working range of the assay and extracted (enriched before) and compared to a set of blank plasma samples that were extracted before the addition of analytes (enriched after). This was performed at three concentration levels corresponding to the high, mid, and low QC levels. The results of both sets of samples were compared and used to calculate the extraction recovery (Recovery in % = Signal area$_{\text{enriched before}}$/Signal area$_{\text{enriched after}}$ × 100). The resulting mean recovery and the coefficient of variation for each compound are shown in Table 1. The recovery of all compounds was calculated to be higher than 80%, with coefficients of variation from 4.4% to 25% (see Table 1). Finally, the six individual lots of dog plasma and the six individual lots of mouse plasma were also used to determine any effects that unique matrices had on the MS/MS signal (ion suppression/ion enhancement). To determine the matrix effect, the blank extracted samples that were enriched after extraction (enriched after) were compared to enriched buffer samples with the same organic/aqueous buffer combination as a sample extract (absolute matrix effect = Signal area$_{\text{enriched after}}$/Signal area$_{\text{buffer sample}}$ × 100, see Table 1). A value of >100% indicates ionization enhancement, and a value of <100% indicates ionization suppression. The relative matrix effect was calculated using the analyte/internal standard area ratio. This value shows if the internal standard can compensate for individual matrix effects (Table 1). The absolute matrix effect relative to the surrogate matrix showed the largest effect for **IB**, with 26.6% ± 2.4% (SD) and 31.2% ± 1.8% (SD) for mouse and dog plasma, respectively. For this analyte, the relative matrix effect showed a much higher value (suppression effect was mitigated by the internal standard), with 76.5% ± 3.2% (SD) and 102% ± 3.2% (SD) for mouse and dog plasma, respectively. The highest standard deviation (inter-individual variation) was observed for **GEM-IB**, with 137.7% ± 74.6% (SD) and 77.5% ± 34.3% (SD) for mouse and dog plasma, respectively. This variation was also compensated by the internal standard used, as shown by the relative matrix effect for **GEM-IB**, with 77.8% ± 11.7% (SD) and 77.1% ± 9.9% (SD) for mouse and dog plasma, respectively.

Table 1. Recovery, matrix effects, and matrix interference in mouse and dog EDTA plasma.

	Recovery		Absolute Matrix Effect		Relative Matrix Effect		Matrix Interference	
	Mouse	Dog	Mouse	Dog	Mouse	Dog	Mouse	Dog
	%-Recovery	%-Recovery	%-ME	%-ME	%-ME	%-ME	%-Accuracy	%-Accuracy
GEM	103.3 ± 10.6	102.8 ± 17.4	73 ± 17.6	66.9 ± 12.3	76.4 ± 7	56.1 ± 5.4	97.7 ± 11.7	102.8 ± 17.4
dFdU	101.1 ± 14.2	96.6 ± 19.5	79.4 ± 10.3	81 ± 8.1	80.6 ± 12.6	55.6 ± 7.3	99.3 ± 14.6	96.6 ± 19.5
GEMMP	97 ± 13.6	107 ± 17.6	72.5 ± 12	65.2 ± 12.7	76.7 ± 11.4	55 ± 6.8	101.3 ± 13.9	107 ± 17.6
dFdUMP	87.2 ± 15.1	97.2 ± 16.5	62.8 ± 11.8	55.7 ± 10.6	63.4 ± 7.3	38.1 ± 5.3	91.3 ± 12.4	97.2 ± 16.5
IB	103.1 ± 11.1	99 ± 4.4	26.6 ± 2.4	31.2 ± 1.8	76.5 ± 3.2	102 ± 3.2	102 ± 2.8	102.7 ± 2
GEM-IB	80.7 ± 25	94.4 ± 6.3	137.7 ± 74.6	77.5 ± 34.3	77.8 ± 11.7	77.1 ± 9.9	103.7 ± 9.1	101.1 ± 13.4

Data are presented as means ± standard deviation.

2.3. Matrix Interference

The six individual lots of both dog and mouse plasma were used to assess matrix interferences. For each lot, three levels of quality control samples that covered the working range of the assay were analyzed and the result was compared with the nominal concentration, for a total of n = 18 data points per compound. The calculated matrix interference of each compound is shown in Table 1. All compounds fulfilled the predefined acceptance criteria of a mean accuracy of 80–120% and mean imprecision of less than 20% for this experiment.

2.4. Accuracy and Imprecision

The accuracy and imprecision of this assay were tested by extracting calibrators and quality control samples on three separate days (runs). For each run, two sets of calibrators and six sets of quality control samples were extracted and analyzed. Five concentration levels were chosen for the analysis of **GEMMP**, **dFdU-MP**, and **GEM** (5, 10, 15, 200, and 800 ng/mL), while **dFdU** was tested at four concentration levels (10, 15, 200, and 800 ng/mL). dFdU was not tested at 5 ng/mL since it falls below the LLOQ. Both intra-batch and inter-batch accuracies and precisions were determined (see Supplementary Table S1). At least two-thirds of the calibration samples fell within 15% of the nominal concentration (or within 20% for the lowest concentration level). For quality control samples, at least half of the samples at each level and at least two-thirds of the samples overall fell within 15% of the expected concentration (or within 20% for the lowest concentration level). The coefficient of variation also fell within 15% (or 20% for the lowest concentration level) for all but two inter-batch quality control levels of **GEMMP**, which fell between 15% and 16%.

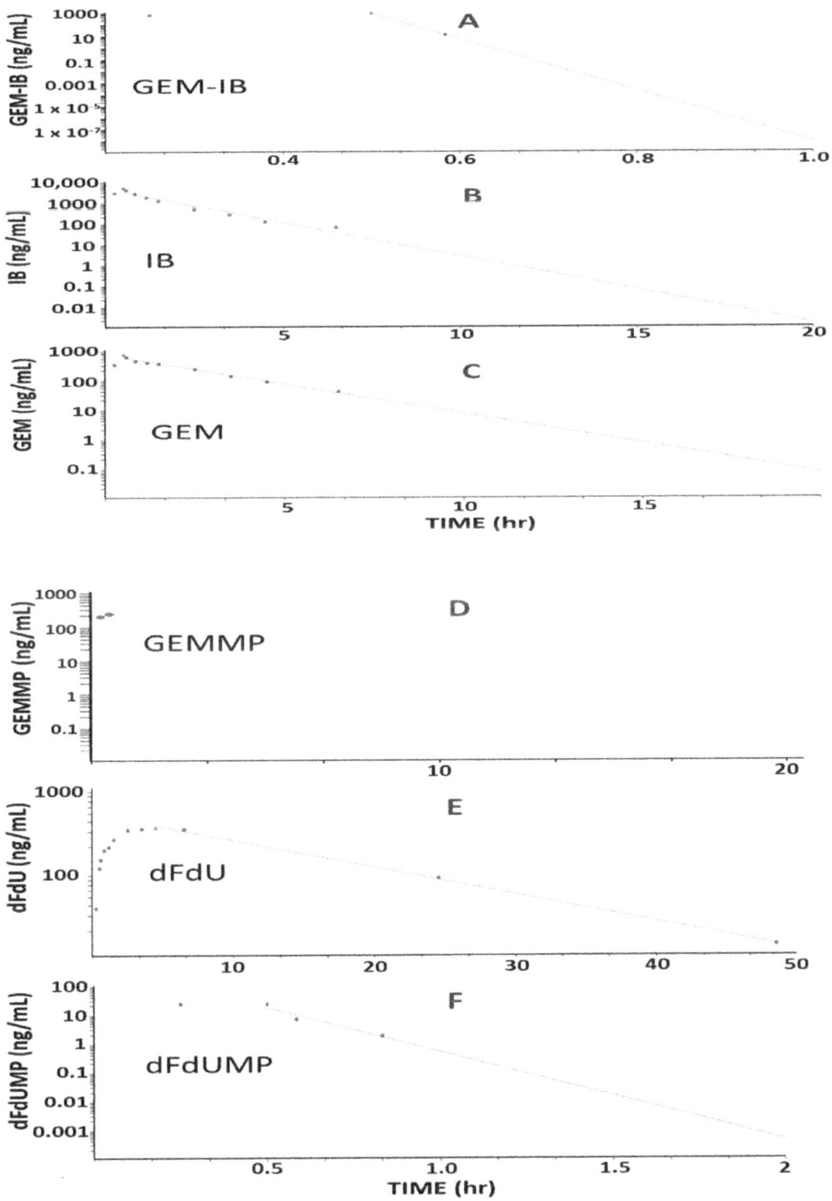

Figure 3. (Dog PK) Plot of PK profiles for **GEM-IB** and metabolites in dog plasma. For dogs (n = 3), **GEM-IB** (5 mg/kg) was infused i.v. for 30 min. Subfigures (**A**)–(**F**) are plots of the mean PK profiles for **GEM-IB**, **IB**, **GEM**, **GEMMP**, **dFdU**, and **dFdUMP**, respectively. Area under the concentration (AUC) versus time curves (green lines) are shown with extrapolation. Blue dots represent the mean concentrations observed. PK was calculated with Phoenix WinNonlin version 8.3, Certara, Princeton, NJ, USA.

2.5. Stability

For stability testing, QC samples were prepared by spiking plasma with QC stock solutions. These stability samples were stored for the predetermined time periods before extraction (accelerated stability and freeze–thaw stability) or after extraction (autosampler stability) and were run with freshly prepared calibrator standards. For accelerated stability, samples were stored at 4 °C and at room temperature for 4 h and 24 h prior to sample extraction. Except for **GEM-IB**, all analytes were stable (±20% of nominal) for 24 h at room temperature for at least 24 h (see Supplementary Tables S2 and S3). Assay A compounds (**GEM-IB** and **IB**) were also tested at 1 h and 2 h exposed to 4 °C and at room temperature to evaluate for which duration **GEM-IB** is stable. While **IB** was stable under all conditions, **GEM-IB** was stable for up to 1 h on ice in mouse plasma with 89.2% ± 7.0% (mean ± SD) of the nominal concentration. After 2 h, **GEM-IB** had degraded to 64.2% ± 11.8% (mean ± SD). In contrast, **GEM-IB** in dog plasma was stable for up to 24 h on ice.

Freeze–thaw cycle stability was assessed by storing the stability samples overnight in the freezer (<−70 °C) and thawing them the next day on the bench. Once thawed, samples were placed back into the freezer. This was carried out for three freeze–thaw cycles for all compounds. Except for **GEM-IB** in mouse plasma, all study compounds were within ±20% of the nominal enriched concentration (see Supplementary Tables S5 and S6). Due to the instability of **GEM-IB**, one and two freeze–thaw cycles were tested for the Assay A compounds (**GEM-IB** and **IB**). While **IB** was stable under these conditions, **GEM-IB** showed instability after one freeze–thaw cycle (53.7% ± 6.9%).

The stability of extracted dog and mouse plasma samples in the autosampler at 4 °C was established for 24 and 48 h. Except for **GEM-IB**, in mouse and dog plasma extracts, all analytes were within ±20% of the nominal enriched concentration for up to 48 h in the autosampler. **GEM-IB** was stable for up to 24 h under these conditions.

To assess the stability of study samples, PK samples were re-extracted after storage at <−70 °C for 1.5 years (see Supplementary Table S4). Under these conditions, **GEM-IB** and IB showed an accuracy of 105.5% ± 40.4% (mean ± standard deviation) and 111.4% ± 38.0% for **GEM-IB** and 107.8% ± 2.5% and 113.7% ± 23.9% for **IB** in dog plasma and mouse plasma, respectively. Standard deviations were higher than the anticipated ±25% for **GEM-IB** but passed for IB. **GEM** was shown to be stable under these conditions, with accuracies of 90.4% ± 23.9% (mean ± standard deviation) and 94.2% ± 11.8% for dog and mouse plasma, respectively. Sample stability assessment for **GEMMP**, **dFdU**, and **dFdU-MP** failed the criteria in dog and mouse plasma at <−70 °C for 1.5 years (see Supplementary Table S4).

2.6. *Proof-of-Concept PK Studies*

Studies in mice (n = 24, eight time points, n = 3 per time point) and in dogs (n = 3, twelve time points) were performed to determine key PK parameters for **GEM-IB** and its metabolites in mice and dogs (Tables 2 and 3 and Figures 3–5). After i.p. injection in mice, the maximal concentrations (C_{max}) were 4185, 22,777, 1309, 130, 2122, and 832 ng/mL for **GEM-IB, IB, GEMMP, dFdUMP, GEM**, and **dFdU**, respectively. All analytes showed a time of maximal concentration (t_{max}) of 5 min (the first measured time point), with the exception of **dFdU**, which was formed with a delay (t_{max} = 1 h). The areas under the time concentration curve over the observation period ($AUC_{0\text{-}T}$) in mouse plasma for **GEM-IB, IB, GEMMP, GEM, dFdU**, and **dFdU-MP** were 1278 h·ng/mL, 10,652 h·ng/mL, 405 h·ng/mL, 38 h·ng/mL, 1063 h·ng/mL, and 3389 h·ng/mL, respectively. In dog plasma, $AUC_{0\text{-}T}$s of 295 h·ng/mL, 5725 h·ng/mL, 83 h·ng/mL, 11 h·ng/mL, 1625 h·ng/mL, and 6569 h·ng/mL were observed for **GEM-IB, IB, GEMMP, dFdU-MP, GEM**, and **dFdU**, respectively. The half-life of **GEM-IB** was determined to be 8 min after i.p. injection in mouse plasma and less than 1 min after the end of the infusion in dog plasma. Table 2 lists the key PK parameters determined in mice and Table 3 lists the key PK parameters in dogs. **GEMMP** could not be completely evaluated in dog plasma, since all values after the infusion period were below the LLOQ.

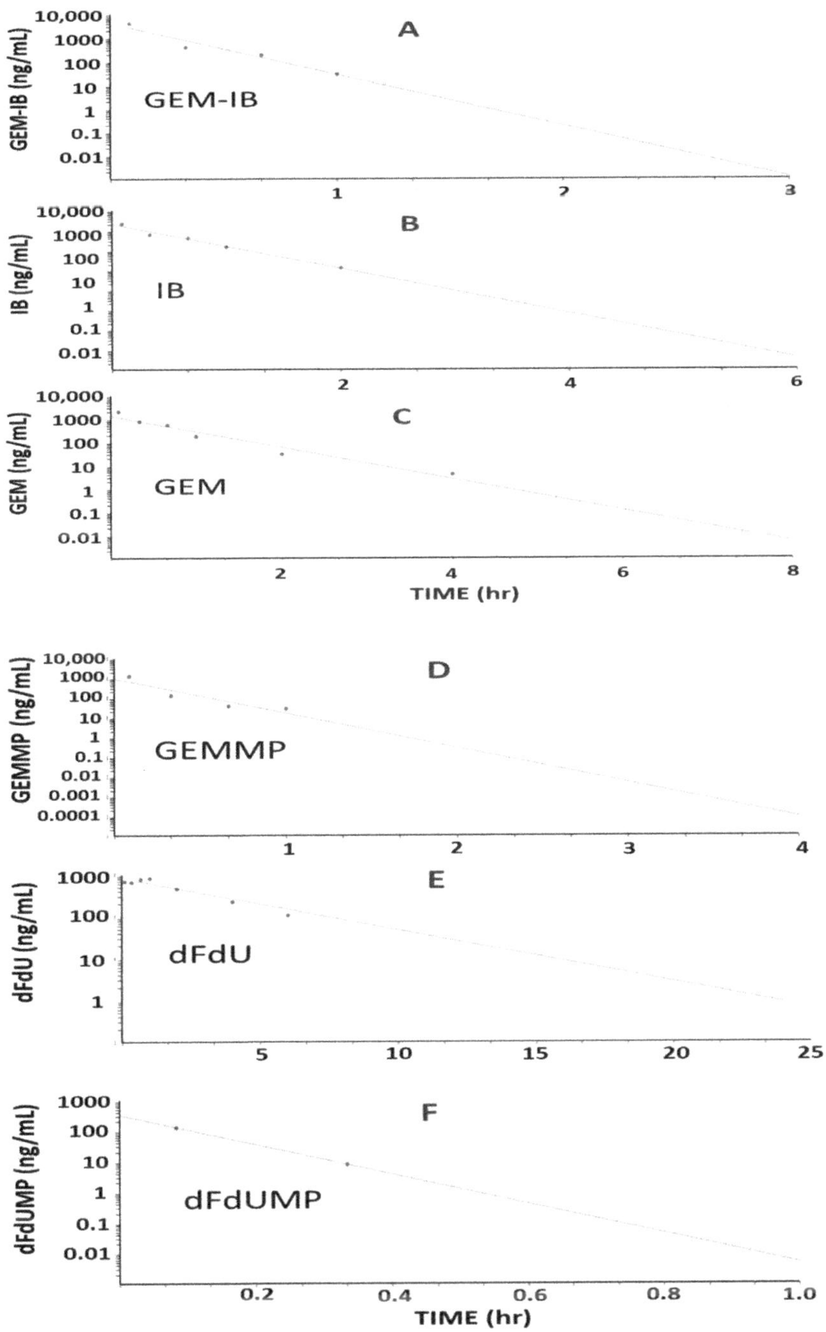

Figure 4. (Mouse PK) Plot of PK profiles for **GEM-IB** and metabolites in mouse plasma. *Mice* (n = 24) were injected i.p. with **GEM-IB** at 5 mg/kg. Subfigures (**A**)–(**F**) are plots of the mean PK profiles for **GEM-IB**, **IB**, **GEM**, **GEMMP**, **dFdU**, and **dFdUMP**, respectively. Area under the concentration (AUC) versus time curves (green lines) are shown with extrapolation. Blue dots represent the mean concentrations observed. PK was calculated using Phoenix WinNonlin (version 8.3, Certara, Princeton, NJ, USA).

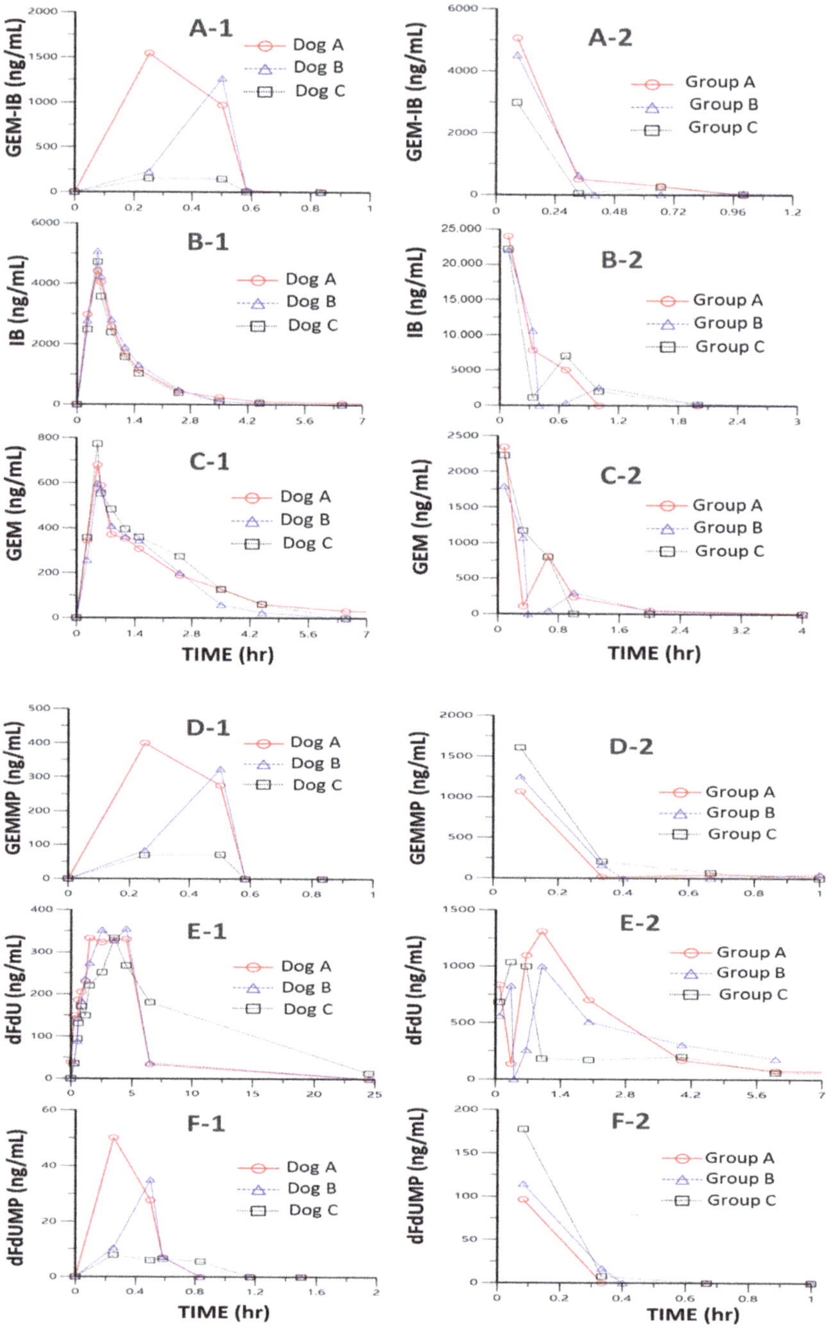

Figure 5. (Individual dog and mouse drug and metabolite levels.) Subfigures (**A-1**,**B-1**,**C-1**,**D-1**,**E-1**,**F-1**) show plots of the observed individual dog plasma drug and metabolite levels for **GEM-IB**, **IB**, **GEM**, **GEMMP**, **dFdU**, and **dFdUMP**, respectively. The mice were grouped into 3 groups (Group A, Group B, and Group C) and subfigures (**A-2**,**B-2**,**C-2**,**D-2**,**E-2**,**F-2**) show plots of the observed plasma drug and metabolite levels for **GEM-IB**, **IB**, **GEM**, **GEMMP**, **dFdU**, and **dFdUMP**, respectively.

Table 2. Key PK parameters after **GEM-IB** administration (i.p.) in *mice* (n = 3 *mice* per time point, 8 timepoints, 24 mice total). The non-compartmental PK analysis was carried out based on the mean concentrations for each time point. In mice, each time point was collected from a different animal.

Parameter	Units	GEM-IB	IB	GEM	GEMMP	dFdU	dFdUMP
C_{max}	ng/mL	4185	22,777	2122	1309	832	130
T_{max}	h	0.083	0.083	0.083	0.083	1.0	0.083
AUC	h·ng/mL	1278	10,652	1063	405	3389	38
R-Squared		0.946	0.987	0.954	0.848	0.991	1.000
No of points		4	5	6	4	5	2
Rate Constant	1/h	5.0	2.6	1.5	4.0	0.4	11.1
$T_{1/2}$	h	0.138	0.268	0.448	0.172	1.798	0.063
Route		IV Bolus	IV Infusion	IV Infusion	IV Infusion	IV Infusion	IV Infusion
CL	mL/h/kg	3939	509	4714	12,647	1861	135,477
Vd_{ss}	mL/kg	469	198	2063	1573	4869	8994

Abbreviations: C_{max}, maximum (peak) plasma concentration; T_{max}, time to peak drug concentration; AUC, area under the curve, representing the total drug exposure across time; $T_{1/2}$, half-life, which is the time it takes for half the drug to be eliminated; CL, systemic clearance; Vd_{ss}, volume of distribution.

Table 3. Key PK parameters after **GEM-IB** (30 min i.v.) administration in dogs (means, n = 3).

Parameter	Units	GEM-IB	IB	GEM	GEMMP	dFdU	dFdUMP
C_{max}	ng/mL	795	4737	684	224	330	23
T_{max}	h	0.5	0.5	0.5	0.5	4.5	0.5
AUC	h·ng/mL	295	5725	1625	83	6569	11
R-Squared		1.000	0.944	0.985	ND	0.991	0.873
No points		2	9	9	0	6	3
Rate Constant	1/h	50.6	0.7	0.5	ND	0.1	7.0
$T_{1/2}$	h	0.014	0.932	1.511	ND	9.845	0.099
Route		IV Infusion	IV Infusion	IV Infusion	IV Infusion	IV Infusion	IV Infusion
Infusion Time	h	0.5	0.5	0.5	0.5	0.5	0.5
CL	mL/h/kg	17,021	956	3705	ND	738	448,855
Vd_{ss}	mL/kg	1934	1037	7852	ND	9533	66,286

Abbreviations: C_{max}, maximum (peak) plasma concentration; T_{max}, time to peak drug concentration; AUC, area under the curve, representing the total drug exposure across time; $T_{1/2}$, half-life, which is the time it takes for half the drug to be eliminated; CL, systemic clearance; Vd_{ss}, volume of distribution. ND—not detected.

3. Discussion

GEM-IB is a novel **GEM-BP** conjugate that combines anti-resorptive and anti-tumor activities while localizing at the site of tumor cell-induced bone destruction. The exploitation of the bone-seeking properties of BPs (IB) for targeted delivery of chemotherapy (**GEM**) is a promising approach to target bone metastases. It was the aim of this study to develop and validate analytical platforms for the quantitative determination of **GEM-IB** plasma levels and consecutive analysis of the PK of **GEM-IB** and its metabolites in mouse and dog PK studies. The active metabolites **IB** and **GEM**, as well as another intermediate (**GEMMP**) and degradation products (**dFdU** and **dFdUMP**), were included.

3.1. Chromatographic Performance

Initially, various chromatographic conditions were tested for the separation of **GEM-IB, IB, GEM, GEMMP, dFdU**, and **dFdUMP**. While the separation of **GEM, GEMMP, dFdU**, and **dFdUMP** could be achieved using various different column materials, chromatographic analysis of **GEM-IB** and **IB** was challenging. For the chromatography of **GEM, GEMMP, dFdU**, and **dFdUMP** (Assay B), a Kinetex XB-C18 column (150 mm × 4.60 mm, 2.6 μm particle size, Phenomenex, Torrance, CA, USA) was chosen and 0.3% formic acid in HPLC-grade water (Buffer A) and acetonitrile (Buffer B) were used as mobile phases, resulting in good chromatographic separation of the study compounds, high sensitivity in positive ion mode, and a robust and reproducible chromatographic platform. As mentioned before, the development of a chromatographic assay for analytes **GEM-IB** and **IB** was more challenging. For Assay B, separation of the monophosphates from the nucleoside compounds (**GEM** from **GEMMP** and **dFdU** from **dFdUMP**) was critical because **GEMMP** can be converted to **GEM** and **dFdUMP** to **dFdU** during the ionization process in the electrospray ionization source. Ion-pairing has been used in numerous other studies involving the quantitation of nucleotides and oligonucleotides [26–28]. The use of ion-pairing allows chromatographic separation of polar compounds to be accomplished on a non-polar column, such as the C18 column used in this study. The amine moiety of **diethylamine** (**DEA**) binds to the polar groups of the analyte, in this case phosphate groups, while the hydrocarbon chains of **DEA** bind to the column media [29]. Initially, when ion-pair reagent was used for **IB** and **GEM-IB**, reasonable chromatographic performance could be achieved with minor tailing, but carry-over was excessive. Thus, several modifications were made to optimize the mobile phase buffer and different columns were tested. The use of a 10 mM **hexafluoroisopropanol** (**HFIP**) and 0.05% (*v*/*v*) **DEA** at pH 9.4 in combination with the novel Hypercarb (porous graphite) column stationary phase, which consists of 100% PGC, was the key to achieving good chromatographic separation with minimal carry-over. The use of PGC material was superior to the C8- and C18-based reversed-phase materials usually used in ion-pair chromatography. This could be due to the less lipophilic but more ionic nature of this relatively novel material. This platform resulted in the best separation and peak shape and minimized carry-over among the tested alternatives.

3.2. *Assay Acceptance Criteria*

As shown in Supplementary Table S1 and Table 1, the assay met the predefined acceptance criteria for intra- and inter-batch accuracy and imprecision. The specificity of the assay was evaluated by analysis of blank extracted plasma samples from six different individuals. The endogenous signal for the analytes was less than 20% of the LLOQ and there was no detectable signal for any of these compounds in the blank sample (see Supplementary Figure S1a–e). The extraction recovery was above 80% for all compounds (Table 1). Matrix interference also fell within ±15% of the nominal concentration for each compound (Table 1), indicating that the assays report accurate results independent of the individual matrix. Matrix interference testing demonstrated that the analytes could be analyzed accurately in individual matrices and that the internal standards compensated for matrix effects efficiently.

Matrix effects (ion suppression or enhancement) were studied because it is vital that all matrices behave similarly during extraction and analysis so that calibration samples that are created from bulk mouse and dog plasma will quantitatively represent the true concentrations of compounds in study samples. Possible matrix effects include ion suppression or enhancement, wherein components of the extracted plasma affect the ability of the compounds of interest to ionize within the source [25,27,30]. There was a pronounced absolute matrix effect for **IB** in mouse and dog plasma, with 26.6% ± 2.4% and 31.2% ± 1.8%, respectively. This effect can be expected for phosphorylated compounds, as described by our group for endogenous nucleotides [27] and by others [30]. However, standard deviations for the observed effect were low (2.4% and 1.8% for mouse and dog plasma, respectively), indicating that the effect is comparable/ similar amongst different matrices. In addition, the relative matrix effect, which normalized the area of the analytes by the area of the internal standards, showed improved values of 76.5% ± 3.2% and 102.0% ± 3.2% for **IB** in mouse and dog plasma, respectively. This indicated that the internal standards were able to compensate for the matrix effect. The matrix interference test ultimately determines if the measured matrix effects have an impact on the accurate determination of the analytes in individual matrices. All compounds fulfilled the predefined acceptance criteria for this test. Therefore, the observed matrix effects are comparable in all mouse and dog plasma samples and have no impact on the accurate determination of **GEM-IB**, **IB**, **GEM**, **GEMMP**, **dFdU**, and **dFdUMP** in mouse and dog plasma matrices, as demonstrated by the acceptable intra- and inter-batch accuracies and imprecisions.

Intra- and inter-batch accuracies and imprecision were determined over the duration of three days for mouse and dog plasma. The predefined acceptance criteria included 85% to 115% for accuracy (80–120% at the LLOQ) and a coefficient of variation (CV) of less than 15% for imprecision. These were applied to the mean accuracy and imprecision (n = 6 per level and day). In addition, two-thirds of all validation quality control samples overall and 50% of samples at a given level had to fulfill the accuracy acceptance criteria. All analytes met these predefined acceptance criteria for the three validation days.

3.3. Stability Assessments

An accelerated stability evaluation for up to 24 h on ice or at ambient temperature revealed that all analytes except for **GEM-IB** were stable under the tested conditions. **GEM-IB** was designed to be hydrolyzed under physiological conditions (see Figure 1B), with the **IB** moiety guiding and anchoring **GEM-IB** to the bone and hydrolysis enabling the release of the potent antineoplastic **GEM**. This drug design feature required extra attention to be paid to **GEM-IB** stability throughout the development of analytical methods. The number of freeze–thaw cycles that samples can be exposed to before the analytes start degrading and whether there is an effect of freeze–thaw cycles was investigated. Except for **GEM-IB**, all compounds were stable after exposure to three freeze–thaw cycles. Repeated experiments showed that **GEM-IB** was unstable after one freeze–thaw cycle. This could be due to the experimental design, which included thawing samples on the benchtop and then placing these back into the freezer. The instability of **GEM-IB** on the benchtop at ambient temperature was due to hydrolysis. In addition, **GEM-IB** showed instability after one freeze–thaw cycle (54%). However, extracts of **GEM-IB** remain stable for extended durations. Therefore, it is recommended to divide **GEM-IB** mouse plasma prior to freezing into 100 μL aliquots and add the protein precipitation solution to the frozen aliquot prior to thawing at 4 °C on ice. This will eliminate hydrolysis when thawing in plasma. Extracted samples were stable in the autosampler at 4 °C for 24 h (**GEM-IB**) and even up to 48 h (all other analytes). Reanalysis of PK study samples after 1.5 years (**GEM-IB** and **IB**) and 2.5 years (**GEM**, **GEMMP**, **dFdU**, and **dFdUMP**) revealed that only **GEM** and **IB** could be considered stable for this duration. Although mean accuracies for **GEM-IB** were within the acceptable range of 80–120% accuracy, standard deviations were high (40.4%), indicating variability across individual samples during long-term storage. Thus, extended storage (1.5 years) results in inaccuracy for **GEM-IB** in individual samples and is not recommended. The data for **GEM** are in alignment with the literature, since others have shown that **GEM** is stable in human plasma for a duration of up to 30 days at −70 °C [31] and also that **GEM** is stable in mouse plasma for a duration of up to 42 days [32]. It has also been shown that **IB** is stable in human plasma for up to 4 months at −20 °C [33]. However, **dFdU** has been shown to be stable for up to 42 days in enriched mouse plasma [32], which was not the case for the re-extracted samples that were stored at <−70 °C in the present study. Comparable stability data for **GEM-IB**, **GEMMP**, and **dFdU** are currently not available in the literature.

3.4. PK Studies

GEM-IB [15] was designed to improve upon the therapeutic benefits observed in clinical trials [12,13] of **MBC-11**. While direct measurement in the bone matrix was not made of either **MBC-11** nor released cytarabine, bone localized effects proved 2–20-fold greater compared to free cytarabine [13]. This observation suggests that the conjugation of etidronate to cytarabine not only drives bone localization, but also inhibits the inactivating deamination of cytarabine. **MBC-11** plasma PK also indicated that hydrolysis occurred on the minute time scale while in circulation. We hypothesize that vascular endonucleotidases catalyze the hydrolysis. The larger steric bulk of **GEM-IB**'s **IB** moiety compared to **MBC-11's** etidronate moiety is thought to reduce the rapid hydrolysis and further deamination of the **GEM** moiety, all with the potential to increase the dose fraction delivered to the bone. The PK studies presented here determine key plasma PK parameters of **GEM-IB** and metabolites after the administration of 5 mg/kg of the study drug (**GEM-IB**) via i.p. and 30 min i.v. injection in mice and dogs, respectively. The minute-long half-lives for **GEM-IB** (8 min after i.p. injection in mouse plasma and less than 1 min after the end of the infusion in dog plasma) indicated that **GEM-IB** was rapidly distributed or hydrolyzed, similar to what was observed with **MBC-11**. The major metabolite was **IB**, the hydrolysis product of **GEM-IB**. **IB** had estimated half-lives of 16 min in mice and 55 min in dogs (Table 2). Maximal plasma concentrations (C_{max}) differed between dogs and mice and were generally higher in mice. However, this can also be explained by the route of administration (i.p. vs. i.v.). Surprisingly, relatively high $AUC_{0\text{-}T}$s of **IB** were observed after administration in mice (mean: 10,652 h·ng/mL), while the sum of all **GEM**-related metabolites added up to 4895 h·ng/mL. This was not the case for dogs with $AUC_{0\text{-}T}$s for **IB** of 5725 h·ng/mL and 8289 h·ng/mL as the sum of all **GEM**-related metabolites. This might be indicative of differences in the metabolism of **GEM-IB** between these species. Mice may be able to produce **GEM** metabolites that were not captured by this assay. Moreover, the observed differences in the pharmacokinetic parameters are likely at least in part explained by the different routes of administration (i.p. vs. i.v.), resulting in different distribution and hydrolysis rates. The plasma PK behaviors of **GEM**, **dFdU**, and **IB** after the hydrolysis of **GEM-IB** are consistent with the well-known PK values established by others when **GEM** or **IB** are administered as free drugs [34,35]. The development and use of the presented plasma PK methods provides the foundation for future work, asking the critical questions of the dose fraction that is bone-localized and the bone-localized levels of **GEM** vs the inactivated deaminated **dFdU** that drive efficacy.

4. Materials and Methods

To analyze **GEM-IB** and its five metabolites, two analytical assays were developed. Assay A captures the polyphosphate-containing compounds **GEM-IB** and **IB**. Assay B was designed to analyze the nucleotide analog metabolites including **GEMMP**, **dFdUMP**, **GEM**, and **dFdU**.

4.1. Materials

Gemcitabine hydrochloride and **2′,2′-difluoro-2′-deoxyuridine** were purchased from Thermo Fisher Scientific (Waltham, MA, USA). **GEM-IB** was synthesized according to a published procedure [15] and **2′,2′-Difluoro-2′-deoxyuridine-monophosphate** was prepared by standard deamination of **GEMMP**. **GEMMP** and the internal standards, **gemcitabine-^{13}C,^{15}N$_2$ (GEM-^{13}C)** and **2′,2′-difluoro-2′-deoxyuridine-^{13}C,^{15}N$_2$ (dFdU-^{13}C)**, were purchased from Toronto Research Chemicals (North York, ON, Canada). Ibandronate sodium salt was purchased from Sigma Aldrich (St. Louis, MO, USA). **Adenosine-^{13}C$_{10}$,^{15}N$_5$ 5′-triphosphate (^{13}C$_{10}$-^{15}N$_5$-ATP)** was purchased from Sigma Aldrich (St. Louis, MO, USA). K_2EDTA canine Beagle plasma and mouse plasma used for the assay validation were purchased from Innovative Research (Novi, MI, USA). HPLC-grade **water, methanol, acetonitrile, formic acid,** and **dibutylamine (DBA)** were purchased from Thermo Fisher Scientific (Waltham, MA, USA).

4.2. Stock Solutions

Individual stocks of all compounds (**GEM-IB**, **GEMMP**, **IB**, **GEM**, **dFdU**, and **dF-dUMP**) were prepared by weighing out solid material and dissolving it in HPLC-grade **water** to achieve a final concentration of 1 mg/mL.

4.3. Calibrators and Quality Controls

Calibration and quality control samples were prepared by spiking 180 µL of plasma with 20 µL of the corresponding calibrator or quality control stock, as listed above.

For **Assay A** (analysis of **GEM-IB** and **IB**), combined stock solutions with 10 µg/mL of **GEM-IB** and 40 µg/mL **IB** were prepared. These were used to prepare calibrator stock solutions used for the enrichment of mouse or dog plasma. Calibrator standards were prepared at concentrations of 2.5, 5, 10, 25, 50, 100, 250, 500, and 1000 ng/mL for **GEM-IB** and 10, 20, 40, 100, 200, 400, 1000, 2000, and 4000 ng/mL for **IB**. Quality control samples (QC_{LLOQ}, QC_{low-1}, QCl_{ow-2}, QC_{mid}, and QC_{high}) were prepared at 5, 10, 30, 200, and 800 ng/mL for **GEM-IB** and 20, 40, 120, 800, and 3200 ng/mL for **IB**.

For **Assay B** (analysis of **GEMMP**, **GEM**, **dFdUMP**, and **dFdU**), combined stock solutions with all four compounds were prepared at a concentration of 10 µg/mL from the individual 1 mg/mL stocks. These were used to prepare the calibrator stock solutions used for the enrichment of mouse/dog plasma. Calibrator standards were prepared at concentrations of 2.5, 5, 10, 25, 50, 100, 250, 500, and 1000 ng/mL. Quality control samples (QC_{LLOQ}, QC_{low-1}, QCl_{ow-2}, QC_{mid}, and QC_{high}) were prepared at concentrations of 5, 10, 15, 200, and 800 ng/mL. A protein precipitation solution was prepared by adding 250 ng/mL of each internal standard (**GEM-^{13}C** and **dFdU-^{13}C**) to 20 mM DBA in methanol. All stocks were stored at −20 °C.

4.4. Sample Extraction

Assay A (**GEM-IB** and **IB**): The extraction of calibration standards, quality control samples, and PK study samples was identical. Plasma, quality control stocks, and calibrator stocks were removed from the storage freezer and kept at room temperature until fully thawed. Aliquots of 100 µL of standards/quality control or blank sample were transferred into a 1.5 mL low-binding polypropylene vial with a conical bottom and snap-on lid. An amount of 10 µL of the of the internal standard solution (15 and 10 µg/mL of **IB-D$_3$** and **ATP-IS**, respectively) was added and mixed. For protein precipitation, 50 µL of methanol was added and then samples were vortexed for 10 min. This was followed by centrifugation at 25,000× *g* for ten minutes at 4 °C (Thermo Scientific MR 23i Centrifuge, Thermo Scientific, Waltham, MA, USA). To remove remaining protein, the supernatant was transferred to 10,000 molecular weight cut off (MWCO) centrifugal filters (800 µL volume, PES 10,000 with 2.0 mL receiver tubes, Analytical Sales and Services, Flanders, NJ, USA) and centrifuged again at 25,000× *g* for ten minutes at 4 °C. The filtrate was transferred into HPLC vials and placed into the HPLC autosampler maintained at 4 °C.

Assay B (**GEMMP**, **GEM dFdUMP** and **dFdU**): The extraction of calibration standards and quality control samples was identical to that of study samples. An amount of 200 µL of plasma sample was added to a 1.5 mL snap-top Eppendorf tube with 100 µL of protein precipitation solution. The samples were then vortexed for five minutes followed by centrifugation at 25,000× *g* for ten minutes at 4 °C. Hereafter, supernatants were transferred to 10K Amicon centrifugal filters (Merck-Millipore, Burlington, MA, USA) and centrifuged at the same settings for 60 min. The filtrates were then transferred into HPLC vials for analysis.

4.5. HPLC-MS/MS Analysis

Assay A (**GEM-IB** and **IB**): Quantification of the analytes was carried out using high-performance liquid chromatography–tandem mass spectrometry (HPLC-MS/MS). Chromatography was performed using an Agilent 1260 Infinity II bio-inert HPLC system (Agilent Technologies, Santa Clara, CA, USA) on a Hypercarb Porous Graphitic Carbon HPLC column (3 µm, 2.1 × 50 mm, Thermo Scientific, Waltham, MA, USA). The HPLC system consisted of an Agilent 1260 Infinity II bio-inert quaternary pump, an Agilent 1260 Infinity II bio-inert column oven, and an Agilent 1260 Infinity II bio-inert autosampler (Agilent Technologies, Santa Clara, CA, USA). An amount of 10 µL of the sample extracts was injected. The column was maintained at 22 °C. HPLC-MS-grade water supplemented with 10 mM hexafluoroisopropanol (HFIP) and 0.05% (*v*/*v*) diethyl amine (DEA) at pH 9.4 was used for the aqueous mobile phase (Buffer A) and HPLC-grade acetonitrile was used for the organic mobile phase (Buffer B). The flow rate was set to 400 µL/min throughout the assay. The initial settings were 98% Buffer A and 2% organic Buffer B for the first two minutes. During the following 0.7 min, the organic Buffer B was increased to 30%. At minute 3.5, the gradient reached 98% Buffer B, which was kept for 0.5 min, and at minute 4.2 the column was re-equilibrated to starting conditions for 2.8 min until the next injection. The HPLC system was connected to a Sciex a 5500+ triple quadrupole mass spectrometer (SCIEX, Concord, ON, Canada) via a turbo electrospray ionization source (SCIEX) operated in the negative electrospray ionization (ESI) mode. Data were acquired in negative multiple reaction monitoring (MRM) mode and all compounds were detected as $[M–H]^-$. The ion transitions used during the multiple reaction monitoring mode were m/z = 318→236, 321→239, 643→342, and 521→79 for **IB**, **IB-D$_3$**, **GEM-IB**, and **adenosine-$^{13}C_{10}$,$^{15}N_5$ 5′-triphosphate** (**$^{13}C_{10}$-$^{15}N_5$-ATP**), respectively. The dwell time for analysis was set to 50 ms. The curtain gas was kept at 20 psi and the ion source gas at 55 psi. The ion spray voltage was −4500 volts, and the temperature of the ion source was set to 400 °C. Compounds were quantified using the analyte/internal standard area ratios based on calibration curves that were constructed with 1/x weighting and a linear regression fit. Quantification was carried out using the Sciex Analyst 1.7.1 Software (SCIEX, Foster City, CA, USA).

Assay B (**GEMMP**, **GEM**, **dFdUMP** and **dFdU**): Quantification of the analytes was carried out using HPLC-MS/MS. Chromatography was performed using an Agilent 1100 series HPLC system (Agilent Technologies, Santa Clara, CA, USA) equipped with a Kinetex XB-C18 column (150 mm × 4.60 mm, 2.6 µm particle size, Phenomenex, Torrance, CA, USA). The HPLC system consisted of an Agilent 1100 series binary pump, an Agilent 1100 degasser, an Agilent 1260 column oven, and a LEAP PAL autosampler (CTC Analytics/Archer Science, Lake Elmo, MN, USA) that was set to inject 8 µL of sample. The column was maintained at 30 °C. HPLC-grade water with 0.3% (*v*/*v*) formic acid was used for the aqueous mobile phase (Buffer A) and HPLC-grade acetonitrile (Buffer B) was used for the organic mobile phase. The flow rate was set to 600 µL/min throughout the assay. The initial settings were 99% Buffer A and 1% organic Buffer B for the first minute. During the following 1.2 min, the organic Buffer B was increased to 10% and within the following 1.3 min to 40%. At minute 4.5, the gradient reached 85% Buffer B and within the following 0.2 min the gradient was ramped up to 99% Buffer B. The 99% Buffer B was kept for 1.8 min and at minute 6.6, the column was re-equilibrated at the starting conditions for 0.9 min until the next injection. The HPLC system was connected to a Sciex a 6500 QTRAP mass spectrometer (SCIEX, Concord, ON, Canada) via a turbo electrospray ionization source (SCIEX) operated in positive electrospray ionization (ESI) mode. The MS/MS was run in the positive multiple reaction monitoring mode and all compounds were detected as $[M+H]^+$. The ion transitions used during the multiple reaction monitoring mode were m/z = 344→112, 345→247, 264→112, 265→113, 267→115 and 268→116 for

GEMMP, dFdUMP, GEM, dFdU, $^{13}C_3$-GEM, and **$^{13}C_3$-dFdU**, respectively. The dwell time was set to 20 milliseconds. The curtain gas was kept at 20 psi and the ion source gas at 40 psi. The ion spray voltage was 5500 volts, and the temperature of the ion source was set to 550 °C. The compounds were quantified using the analyte/internal standard area ratios based on calibration curves that were constructed with 1/x weighting and a linear regression fit. Quantification was carried out using the Sciex Multiquant OS Software version 1.7 or higher (SCIEX, Foster City, CA, USA).

4.6. Assay Validation

Method development and validation was conducted following applicable FDA guidelines for bioanalytical assays [36], as considered fit for purpose. **Specificity.** Endogenous interferences were excluded by analysis of blank plasma from 6 different individual lots of plasma. The lower limit of quantification (LLOQ) was determined as the lowest concentration consistently achieving an accuracy better than ±20% of the nominal concentration, with imprecision ≤20%. **Predefined acceptance criteria.** The performance of the assay was considered acceptable if intra- and inter-day imprecision (coefficient of variance, %CV) at each concentration was ≤15%, except at the LLOQ (≤20%). Intra- and inter-day accuracy had to be ±15% of the nominal value, except at the LLOQ (±20%). Calibration curves had to have a correlation coefficient (r) of 0.99 or better. **Analytical recovery and imprecision.** Intra- and inter-day analytical accuracy and imprecision were tested at least at four concentration levels depending on the LLOQ for each compound. Intra-day imprecision and accuracy were determined with $n = 6$ per QC level. Inter-day imprecision and accuracy were measured over 3 days, with six replicates for each QC concentration ($n = 18$). **Extraction efficiency and matrix effect.** Extraction efficiency for **GEM-IB** and metabolites was established by comparing the signals of the analytes after the extraction of QCs ($n = 6$/concentrations), with the signals of the extracted blank matrix spiked with the respective concentrations of analytes after the extraction procedure [25]. The matrix effect was determined by comparing the analyte LC-MS/MS signal after spiking of the analyte into the extracted blank matrix (matrix from 6 different individual lots, analyte concentrations as for the QC samples) with the HPLC-MS/MS signal of the same amount of analyte in neat solution [25]. **Carry-over.** Carry-over was assessed by injecting blank buffer samples after analysis of the highest calibrators. Carry-over was considered insignificant if there was no signal at the relevant retention times exceeding 20% of the analyte signals at the LLOQ. **Stability.** QC samples were freshly prepared. One set was extracted and was analyzed immediately to determine baseline signals. Analyte stability in plasma was evaluated at two QC concentration levels ($n = 3$/concentration level) under different conditions: 1, 2, and 24 h at room temperature and at 4 °C. In addition, stability after extraction was determined after 24 h and 48 h at 4 °C in the autosampler and for up to 3 freeze (−70 °C)–thaw (ambient) cycles. Stability was assessed by comparing analyte concentrations with baseline (t_0) concentrations. Stability was assumed when the results were within ±20% of the baseline concentrations. Long-term stability estimates were performed by re-extraction of samples used during the PK study ($n = 6$). Samples were initially analyzed for the PK assessment and were stored for 1.5 years (**GEM-IB** and **IB**) and 2.5 years (**GEM, GEMMP, dFdU,** and **dFdU-MP**) at <−70 °C prior to reanalysis. Stability was assessed by comparing analyte concentrations with the initial assessments. Stability was acceptable when the results were within ±20% of the baseline concentrations and the percent standard deviation (%-SD) was below 25%.

4.7. Proof-of-Concept Studies

Dog PK study. After completion of the validation, the assays were used to analyze plasma samples that were collected for a study investigating the PK of **GEM-IB** in dogs. For this proof-of-concept study, healthy beagle dogs (n = 3) received 5 mg/kg **GEM-IB** i.v. over the duration of 30 min. Plasma samples for PK analysis were collected before and during the infusion period (30 min) at 0, 15, and 30 min and after infusion ended at minutes 5, 20, 40, 60, 120, 180, 240, and 360. Blood samples were collected in K_2EDTA containers with tetrahydrouridine 0.25 mg/mL to prevent deamination and plasma was separated. Samples were frozen and stored below −70 °C until HPLC-MS/MS analysis. **GEM-IB** and **IB** were assessed using the procedure described for Assay A and **GEMMP**, **dFdUMP**, **GEM**, and **dFdU** were analyzed using the protocol described under Assay B.

Mouse PK study. Another study was performed as a proof-of-concept study to investigate the pharmacokinetics of **GEM-IB** in *mice*. Twenty-four mice were randomly assigned to three sequences. Each sequence consisted of eight mice and the plasma of each mouse was collected at one of the eight consecutive time points. Mice were injected with 5 mg/kg **GEM-IB** in saline i.p. Eight time points (with n = 3 mice per time point) were investigated and 0.25 mg/mL K_2EDTA-tetrahydrouridine to prevent the deamination of blood samples was drawn via cardiac puncture at 5, 20, 40, 60, 120, 240, and 360 min. Plasma was separated, and samples were frozen and stored below −70 °C until HPLC-MS/MS analysis. **GEM-IB** and **IB** were assessed using the procedure described for Assay A and **GEMMP**, **dFdUMP**, **GEM**, and **dFdU** were analyzed using the protocol described under Assay B. The experimental protocols and animal care were in accordance with the Guide for the Care and Use of Laboratory Animals (Institute of Laboratory Animal Resources, 1996) and were approved by the Institutional Animal Care and Use Committee of Colorado State University (Ft. Collins, CO, USA). The animals were housed in an AAALAC-accredited and USDA-registered facility and the studies were carried out under the supervision of a board-certified veterinarian.

4.8. Statistics

All values are expressed as mean ± standard deviation (SD). Quantitative data were compared with analysis of variance (ANOVA), followed by Tukey's post hoc test for multiple comparisons, using SPSS statistics version 28.0.1.0 (IBM, Armonk, NY, USA). A value of $p < 0.05$ was considered statistically significant. Intra- and inter-batch accuracies and precisions were calculated using Watson LIMS software version 7.6 (Thermo Fisher Scientific, Philadelphia, PA, USA). Non-compartmental PK parameters were calculated using Phoenix WinNonlin version 8.3 (Certara, Princeton, NJ, USA) or by the integrated PK module of Thermo Fisher Scientific Watson LIMS software (version 7.6, SP1).

5. Conclusions

Overall, the present assays for the quantification of **GEM-IB**, **IB**, **GEMMP**, **dFdUMP**, **GEM**, and **dFdU** in mouse and canine plasma met the pre-defined acceptance criteria for sensitivity, selectivity, specificity, precision, and accuracy. Extraction recovery and matrix interference met the acceptance criteria, and matrix effects could be compensated for by the internal standards. The assay was successfully applied to the analysis of study samples from two proof-of-concept PK studies in mice and dogs. Key PK parameters for **GEM-IB** and its metabolites were determined. The rapid distribution and metabolism of **GEM-IB** will require further investigation in follow-up studies. In said studies, tissue and bone material needs to be collected and analyzed to further investigate the bone accumulation of **GEM-IB** and its metabolites.

Supplementary Materials: The following supporting information can be downloaded at https://www.mdpi.com/article/10.3390/molecules30020354/s1: Figure S1: Representative calibration curves and representative extracted ion chromatograms. Figures S1.a and S1.b show calibration curves and representative extracted ion chromatograms for **IB** and **GEM-IB** in mouse plasma (S1.a) and dog plasma (S1.b). A and B show calibration curves for **IB** and **GEM-IB**, C and D show extracted ion chromatograms of blank extracted mouse samples for **IB** and **GEM-IB** (S1.a C and D) and dog plasma for **IB** and **GEM-IB** (S1.b C and D). The row below (E and F) shows calibrator samples at the lower limit of quantitation 40 ng/mL and 10 ng/mL for **IB** and **GEM-IB** respectively. H and G show representative extracted ion chromatograms of **IB** and **GEM-IB** in study samples after infusion with **GEM-IB**. Similarly, Figure S1.c A-B and Figure S1.d A-B show representative calibration curves for **GEM** (A) and **GEMMP** (B). The following 3 rows show blank extracted matrix (C and D), the lower limit of quantitation samples (E and F) and study samples for **GEM** and **GEMMP** (G and H) in mouse (S1.c C to H) and dog plasma (S1.d C to H). Finally, the **dFdU** and **dFdUMP** in mouse and dog plasma is displayed in Figure S1.e and S1.f, respectively. Figure S1.e A-B and Figure S1.f A-B show representative calibration curves for **dFdU** (A) and **dFdUMP** (B) and the following rows show blank extracted matrix (C and D), the lower limit of quantitation samples (E and F) and study samples for dFdU and **dFdUMP** (G and H) in mouse (S1.e C to H) and dog plasma (S1.f C to H); Table S1: Intra- and inter-run accuracy and imprecision; Table S2: Accelerated stability in mouse plasma; Table S3: Accelerated stability in dog plasma; Table S4: Frozen PK sample stability in dog plasma and mouse plasma; Table S5: Freeze-thaw and autosampler stability for mouse plasma; Table S6: Freeze-thaw and autosampler stability for dog plasma.

Author Contributions: A.K., S.P.Z., K.B.F. and D.H.T. planned and executed all animal dosing and pharmacokinetics experiments. J.K., M.E., T.S. and U.C. were responsible for the planning and execution of all analytical experiments, data analysis, and statistics. All authors have read and agreed to the published version of the manuscript.

Funding: This research was supported by the National Cancer Institute of the National Institutes of Health under grant numbers R43CA203166 and R44CA203166 and the State of Colorado Advanced Industries Accelerator Program under grant number CTGG1 2022-3001 awarded to MBC Pharma, Inc. The content is solely the responsibility of the authors and does not necessarily represent the official views of the National Institutes of Health or the State of Colorado.

Institutional Review Board Statement: The experimental protocols and animal care were in accordance with the Guide for the Care and Use of Laboratory Animals (Institute of Laboratory Animal Resources, 1996) and were approved (protocol number 15-6283A, approval date 21 December 2015) by the Institutional Animal Care and Use Committee of Colorado State University (Ft. Collins, CO, USA). The animals were housed in an AAALAC-accredited and USDA-registered facility and the studies were carried out under the supervision of a board-certified veterinarian.

Informed Consent Statement: Not applicable.

Data Availability Statement: All data from this study are available at the archives and secure servers of iC42 Clinical Research and Development, Department of Anesthesiology, University of Colorado, Aurora, Colorado, USA. Data can be made available upon reasonable request.

Acknowledgments: Thanks to (1) the Colorado State University Flint Animal Cancer Center for the use of facilities and Colorado State University Laboratory Animal Resources for animal care, and (2) The University of Colorado is a Certara Center of Excellence. The Center of Excellence program supports leading institutions with Certara's state-of-the-art, model-informed drug development software.

Conflicts of Interest: The authors declare the following competing financial interest(s): Alexander Karpeisky and Shawn Patrick Zinnen are employees of MBC Pharma, Inc. and are Company stock holders. Kristen B. Farrell was an employee of MBC Pharma, Inc. Douglas H. Thamm is MBC Pharma Scientific Advisory Board member and Company stock option holder.

References

1. Reinholz, M.M.; Zinnen, S.P.; Dueck, A.C.; Dingli, D.; Reinholz, G.G.; Jonart, L.A.; Kitzmann, K.A.; Bruzek, A.K.; Negron, V.; Abdalla, A.K.; et al. A promising approach for treatment of tumor-induced bone diseases: Utilizing bisphosphonate derivatives of nucleoside antimetabolites. *Bone* **2010**, *47*, 12–22. [CrossRef] [PubMed]
2. Coleman, R.E. Clinical features of metastatic bone disease and risk of skeletal morbidity. *Clin. Cancer Res.* **2006**, *12*, 6243s–6249s. [CrossRef] [PubMed]
3. Ardakani, A.H.G.; Faimali, M.; Nystrom, L.; Mesko, N.; Mughal, M.; Ware, H.; Gikas, P. Metastatic bone disease: Early referral for multidisciplinary care. *Cleve Clin. J. Med.* **2022**, *89*, 393–399. [CrossRef] [PubMed]
4. Dunbar, K.J.; Efe, G.; Cunningham, K.; Esquea, E.; Navaridas, R.; Rustgi, A.K. Regulation of metastatic organotropism. *Trends Cancer* 2024, *in press*. [CrossRef] [PubMed]
5. Ucci, A.; Giacchi, L.; Rucci, N. Primary Bone Tumors and Breast Cancer-Induced Bone Metastases: In Vivo Animal Models and New Alternative Approaches. *Biomedicines* **2024**, *12*, 2451. [CrossRef]
6. Lipton, A.; Berenson, J.R.; Body, J.J.; Boyce, B.F.; Bruland, O.S.; Carducci, M.A.; Cleeland, C.S.; Clohisy, D.R.; Coleman, R.E.; Cook, R.J.; et al. Advances in treating metastatic bone cancer: Summary statement for the First Cambridge Conference. *Clin. Cancer Res.* **2006**, *12*, 6209s–6212s. [CrossRef]
7. Kreps, L.M.; Addison, C.L. Targeting Intercellular Communication in the Bone Microenvironment to Prevent Disseminated Tumor Cell Escape from Dormancy and Bone Metastatic Tumor Growth. *Int. J. Mol. Sci.* **2021**, *22*, 2911. [CrossRef]
8. Aapro, M.; Abrahamsson, P.A.; Body, J.J.; Coleman, R.E.; Colomer, R.; Costa, L.; Crino, L.; Dirix, L.; Gnant, M.; Gralow, J.; et al. Guidance on the use of bisphosphonates in solid tumours: Recommendations of an international expert panel. *Ann. Oncol.* **2008**, *19*, 420–432. [CrossRef]
9. Body, J.J. Bisphosphonates for malignancy-related bone disease: Current status, future developments. *Support. Care Cancer* **2006**, *14*, 408–418. [CrossRef]
10. Widyadharma, I.P.E.; Tertia, C.; Vania, A.; Tiffani, P.; Wiratnaya, I.G.E. The effect of denosumab vs. zoledronic acid in preventing skeletal-related events, including pain-related bone metastasis: A systematic review. *Postep. Psychiatr. Neurol.* **2024**, *33*, 172–177. [CrossRef]
11. Farrell, K.B.; Karpeisky, A.; Thamm, D.H.; Zinnen, S. Bisphosphonate conjugation for bone specific drug targeting. *Bone Rep.* **2018**, *9*, 47–60. [CrossRef] [PubMed]
12. Zinnen, S.; Thamm, D.; Vail, D.; Fan, T.; Karpeisky, A. Evaluation of a novel bone targeted aracytidine therapy in dogs with spontaneous osteosarcoma. *Bone* **2011**, *1*, S51. [CrossRef]
13. Zinnen, S.P.; Karpeisky, A.; Von Hoff, D.D.; Plekhova, L.; Alexandrov, A. First-in-Human Phase I Study of MBC-11, a Novel Bone-Targeted Cytarabine-Etidronate Conjugate in Patients with Cancer-Induced Bone Disease. *Oncologist* **2019**, *24*, 303-e102. [CrossRef] [PubMed]
14. Yoon, I.C.; Xue, L.; Chen, Q.; Liu, J.; Xu, J.; Siddiqui, Z.; Kim, D.; Chen, B.; Shi, Q.; Laura Han, E.; et al. Piperazine-Derived Bisphosphonate-Based Ionizable Lipid Nanoparticles Enhance mRNA Delivery to the Bone Microenvironment. *Angew. Chem. Int. Ed. Engl.* **2024**, e202415389. [CrossRef]
15. Farrell, K.B.; Zinnen, S.; Thamm, D.H.; Karpeisky, A. Gemcitabine-Ibandronate Conjugate Enables the Bone-Targeted Combination Therapy in Bone Cancer: Synthesis and Efficacy in Combination with Docetaxel. *Bioconjug. Chem.* **2021**, *32*, 2530–2539. [CrossRef]
16. Toschi, L.; Finocchiaro, G.; Bartolini, S.; Gioia, V.; Cappuzzo, F. Role of gemcitabine in cancer therapy. *Future Oncol.* **2005**, *1*, 7–17. [CrossRef]
17. Beutel, A.K.; Halbrook, C.J. Barriers and opportunities for gemcitabine in pancreatic cancer therapy. *Am. J. Physiol. Cell Physiol.* **2023**, *324*, C540–C552. [CrossRef]
18. Von Hoff, D.D.; Ervin, T.; Arena, F.P.; Chiorean, E.G.; Infante, J.; Moore, M.; Seay, T.; Tjulandin, S.A.; Ma, W.W.; Saleh, M.N.; et al. Increased survival in pancreatic cancer with nab-paclitaxel plus gemcitabine. *N. Engl. J. Med.* **2013**, *369*, 1691–1703. [CrossRef]
19. Conroy, T.; Desseigne, F.; Ychou, M.; Bouche, O.; Guimbaud, R.; Becouarn, Y.; Adenis, A.; Raoul, J.L.; Gourgou-Bourgade, S.; de la Fouchardiere, C.; et al. FOLFIRINOX versus gemcitabine for metastatic pancreatic cancer. *N. Engl. J. Med.* **2011**, *364*, 1817–1825. [CrossRef]
20. Paterson, A.H.; Powles, T.J.; Kanis, J.A.; McCloskey, E.; Hanson, J.; Ashley, S. Double-blind controlled trial of oral clodronate in patients with bone metastases from breast cancer. *J. Clin. Oncol.* **1993**, *11*, 59–65. [CrossRef]
21. Raghu Subramanian, C.; Talluri, S.; Mullangi, S.; Lekkala, M.R.; Moftakhar, B. Review of Bone Modifying Agents in Metastatic Breast Cancer. *Cureus* **2021**, *13*, e13332. [CrossRef] [PubMed]
22. Conte, P.F.; Latreille, J.; Mauriac, L.; Calabresi, F.; Santos, R.; Campos, D.; Bonneterre, J.; Francini, G.; Ford, J.M. Delay in progression of bone metastases in breast cancer patients treated with intravenous pamidronate: Results from a multinational randomized controlled trial. The Aredia Multinational Cooperative Group. *J. Clin. Oncol.* **1996**, *14*, 2552–2559. [CrossRef] [PubMed]

23. D'Oronzo, S.; Wood, S.; Brown, J.E. The use of bisphosphonates to treat skeletal complications in solid tumours. *Bone* **2021**, *147*, 115907. [CrossRef] [PubMed]
24. Bauss, F.; Schimmer, R.C. Ibandronate: The first once-monthly oral bisphosphonate for treatment of postmenopausal osteoporosis. *Ther. Clin. Risk Manag.* **2006**, *2*, 3–18.
25. Matuszewski, B.K.; Constanzer, M.L.; Chavez-Eng, C.M. Strategies for the assessment of matrix effect in quantitative bioanalytical methods based on HPLC-MS/MS. *Anal. Chem.* **2003**, *75*, 3019–3030. [CrossRef]
26. Klawitter, J.; Bek, S.; Zakaria, M.; Zeng, C.; Hornberger, A.; Gilbert, R.; Shokati, T.; Klawitter, J.; Christians, U.; Boernsen, K.O. Fatty acid desaturation index in human plasma: Comparison of different analytical methodologies for the evaluation of diet effects. *Anal. Bioanal. Chem.* **2014**, *406*, 6399–6408. [CrossRef]
27. Klawitter, J.; Schmitz, V.; Klawitter, J.; Leibfritz, D.; Christians, U. Development and validation of an assay for the quantification of 11 nucleotides using LC/LC-electrospray ionization-MS. *Anal. Biochem.* **2007**, *365*, 230–239. [CrossRef]
28. Bartlett, M.G. Current state of hydrophilic interaction liquid chromatography of oligonucleotides. *J. Chromatogr. A* **2024**, *1736*, 465378. [CrossRef]
29. Roussis, S.G.; Pearce, M.; Rentel, C. Small alkyl amines as ion-pair reagents for the separation of positional isomers of impurities in phosphate diester oligonucleotides. *J. Chromatogr. A* **2019**, *1594*, 105–111. [CrossRef]
30. Williams, M.L.; Olomukoro, A.A.; Emmons, R.V.; Godage, N.H.; Gionfriddo, E. Matrix effects demystified: Strategies for resolving challenges in analytical separations of complex samples. *J. Sep. Sci.* **2023**, *46*, e2300571. [CrossRef]
31. Li, G.; Zhao, M.; Zhao, L. Ultra-performance liquid chromatography-tandem mass spectrometry for simultaneous determination of 12 anti-tumor drugs in human plasma and its application in therapeutic drug monitoring. *J. Pharm. Biomed. Anal.* **2021**, *206*, 114380. [CrossRef] [PubMed]
32. Zhong, B.; Gibson, E.G.; Davis, A.; Roussel, M.F.; Stewart, C.F. LC-MS/MS method for quantitation of gemcitabine and its metabolite 2′,2′-difluoro-2′-deoxyuridine in mouse plasma and brain tissue: Application to a preclinical pharmacokinetic study. *J. Pharm. Biomed. Anal.* **2021**, *198*, 114025. [CrossRef] [PubMed]
33. Tarcomnicu, I.; Gheorghe, M.C.; Silvestro, L.; Savu, S.R.; Boaru, I.; Tudoroniu, A. High-throughput HPLC-MS/MS method to determine ibandronate in human plasma for pharmacokinetic applications. *J. Chromatogr. B Anal. Technol. Biomed. Life Sci.* **2009**, *877*, 3159–3168. [CrossRef] [PubMed]
34. Barrett, J.; Worth, E.; Bauss, F.; Epstein, S. Ibandronate: A clinical pharmacological and pharmacokinetic update. *J. Clin. Pharmacol.* **2004**, *44*, 951–965. [CrossRef] [PubMed]
35. Beumer, J.H.; Eiseman, J.L.; Parise, R.A.; Joseph, E.; Covey, J.M.; Egorin, M.J. Modulation of gemcitabine (2′,2′-difluoro-2′-deoxycytidine) pharmacokinetics, metabolism, and bioavailability in mice by 3,4,5,6-tetrahydrouridine. *Clin. Cancer Res.* **2008**, *14*, 3529–3535. [CrossRef]
36. U.S. Department of Health and Human Services. *M10 Bioanalytical Method Validation and Study Sample Analysis. Guidance for Industry, November 2022*; Food and Drug Administration, FDA: Silver Spring, MD, USA, 2022.

molecules

MDPI

Article

Thermooxidation of Four Sartans: Kinetic Analysis Based on Thermo-Gravimetric Data

Adriana Ledeţi [1], Bianca Baul [2], Amalia Ridichie [1,2], Denisa Ivan [1], Titus Vlase [3], Carmen Tomoroga [1,*], Anca Dragomirescu [1], Gabriela Vlase [3], Răzvan Adrian Bertici [4], Dana Emilia Man [4] and Ionuţ Ledeţi [1,2]

1 Advanced Instrumental Screening Center, Faculty of Pharmacy, Victor Babeş University of Medicine and Pharmacy, 2 Eftimie Murgu Square, 300041 Timisoara, Romania; afulias@umft.ro (A.L.); amalia.ridichie@umft.ro (A.R.); circioban.denisa@umft.ro (D.I.); dragomirescu.anca@umft.ro (A.D.); ionut.ledeti@umft.ro (I.L.)

2 Faculty of Industrial Chemistry and Environmental Engineering, University Politehnica Timisoara, 2 Victoriei Square, 300006 Timisoara, Romania; bianca.baul@student.upt.ro

3 Research Centre for Thermal Analysis in Environmental Problems, West University of Timisoara, Pestalozzi Street 16, 300115 Timisoara, Romania; titus.vlase@e-uvt.ro (T.V.); gabriela.vlase@e-uvt.ro (G.V.)

4 Faculty of Medicine, Victor Babeş University of Medicine and Pharmacy, 2 Eftimie Murgu Square, 300041 Timisoara, Romania; razvan.bertici@umft.ro (R.A.B.); man.dana@umft.ro (D.E.M.)

* Correspondence: axente.carmen@umft.ro

Abstract: Angiotensin II receptor antagonists are tetrazole derivatives used in the treatment of high blood pressure, and are also indicated for the treatment of heart failure (NYHA class II-IV). They are used alone or in combination with other classes of antihypertensives or diuretics for the effective management of high blood pressure. In this study, we aim to evaluate the thermal stability and degradation kinetics for the principal compounds used in therapy from this class, namely telmisartan, valsartan, olmesartan medoxomil, and losartan potassium. To obtain the thermoanalytical data for the kinetic investigations, the TG and DTG curves were registered at five different heating rates (β = 2, 4, 6, 8, and 10 °C min^{-1}). The kinetic methods used were a preliminary ASTM E698 method and two isoconversional methods: Flynn–Wall–Ozawa and Friedman. For each molecule, the results showed complex decomposition processes consisting of complex reaction sequences.

Keywords: angiotensin II receptor antagonists; sartan; kinetic analysis; thermal stability

1. Introduction

Angiotensin II receptor antagonists (sartans) are tetrazole derivatives used in the treatment of high blood pressure, and are also indicated for the treatment of heart failure (NYHA class II-IV) in adults with left ventricular systolic dysfunction (ejection fraction = 40%) to reduce hospitalizations and death from cardiovascular causes [1–3]. They are used alone or in combination with other classes of antihypertensives or diuretics, e.g., hydrochlorothiazide, for the effective management of high blood pressure [1–4].

As their mechanism of action, they bind selectively to the angiotensin type 1 receptor (AT1), preventing the binding of angiotensin II protein (a protein which exerts a hypertensive effect, by vasoconstriction, stimulation and synthesis of aldosterone and antidiuretic hormone, cardiac stimulation, and the increase of renal sodium reabsorption). Thus, sartans present a protective effect on the heart by improving the cardiac function, reducing the afterload and increasing the cardiac output, and preventing ventricular hypertrophy and remodeling. This class of compounds also affects the renin–angiotensin–aldosterone system, which plays an important role in hemostasis and regulation of renal, vascular, and cardiac functions [5,6].

After oral administration, the active substances are quickly absorbed, but their bioavailability is low, due to the effect of the first hepatic passage. To increase their bioavailability (from 4.5% to 28.6% in the case of olmesartan), esterified forms of the drugs have been developed; for example, olmesartan is available as a prodrug in the form of olmesartan medoxomil, which is rapidly transformed in vivo in pharmacologically active olmesartan; also, losartan is conditioned as a potassium salt, called losartan potassium. Binding to plasma proteins is strong (>99%) and constant at plasma concentrations well above the range achieved with recommended doses. Regarding the half-life, there are differences between the molecules from this class: for olmesartan it is 13 h, for telmisartan it is 24 h, for losartan between 1.5 and 2.5 h (it presents a more potent metabolite, namely the E3174 metabolite, for which the half-life is between 6 and 9 h), and for valsartan between 6 and 9 h. Elimination is mainly in the unchanged form, predominantly biliary and digestive, with a small part of the administered dose being excreted in the urine [5,7–11]. The physicochemical properties of the selected sartans are presented in Table 1.

Table 1. Physicochemical properties of olmesartan medoxomil, telmisartan, losartan potassium, and valsartan.

	Olmesartan Medoxomil	Telmisartan	Valsartan	Losartan Potassium
Chemical formula	$C_{29}H_{30}N_6O_6$	$C_{33}H_{30}N_4O_2$	$C_{24}H_{29}N_5O_3$	$C_{22}H_{22}ClKN_6O$
Molar mass/ g mol^{-1}	558.6	514.6	435.5	461.0
Aggregation state	Solid	Solid	Solid	Solid
Melting point/°C	128–148 153–165 177–180 [12]	261–263	116–117	268–271 [13]
Solubility * at 25 °C /mg L^{-1}	Water: p.i. (<2.32·10^{-7}) [14]	Water: p.i. (2.8·10^{-6}); Strong acids: s.s. (except for HCl: p.i.) Strong bases: s. [10]	Water: s.s.; EtOH, DMSO, DMF: s. [15]	Water: s. MeOH, EtOH, 1-PrOH, 2-PrOH DMSO, DMF: s. [13,16]

* p.i. = practically insoluble; s.s. = slightly soluble; s = soluble.

The aim of this study was to evaluate the thermal stability and degradation kinetics for a series of sartans, namely telmisartan (TELM), valsartan (VLS), olmesartan medoxomil (OLM), and losartan potassium (LOS), to gather information regarding the thermal stability of the principal compounds used in therapy from this class. The kinetic methods used were a preliminary ASTM E698 method and, respectively, two isoconversional methods: Flynn–Wall–Ozawa (FWO) and Friedman (FR). Also, to obtain the kinetic triplet and for a better understanding of the mechanism underlying the degradation process, the modified non-parametric (NPK) kinetic method was applied. The chemical structures of these sartans are shown in Figure 1. Considering that TELM does not present a tetrazole ring, it presents some differences in terms of pharmacokinetic properties, namely having the highest affinity for the AT1 receptor among the available angiotensin II receptor blockers and the lowest affinity for the angiotensin type 2 receptor (AT2). TELM may also have PPAR γ (peroxisome proliferator-activated receptor gamma) agonist properties that could confer beneficial metabolic effects, as PPAR γ is a nuclear receptor that regulates specific gene transcription and whose target genes are involved in the regulation of glucose and lipid metabolism, as well as in anti-inflammatory responses [1,4,17].

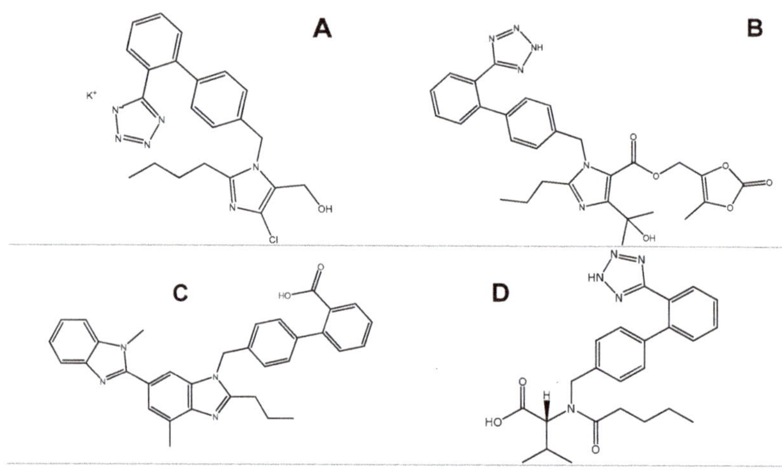

Figure 1. Chemical structures of the selected sartans: (**A**) LOS, (**B**) OLM, (**C**) TELM, (**D**) VLS.

2. Results and Discussion

2.1. Results of the Thermal Investigations

In Figure 2, the obtained thermoanalytical curves (TG/DTG and DSC) are represented, while in Table 2 the interpretation of the curves can be observed.

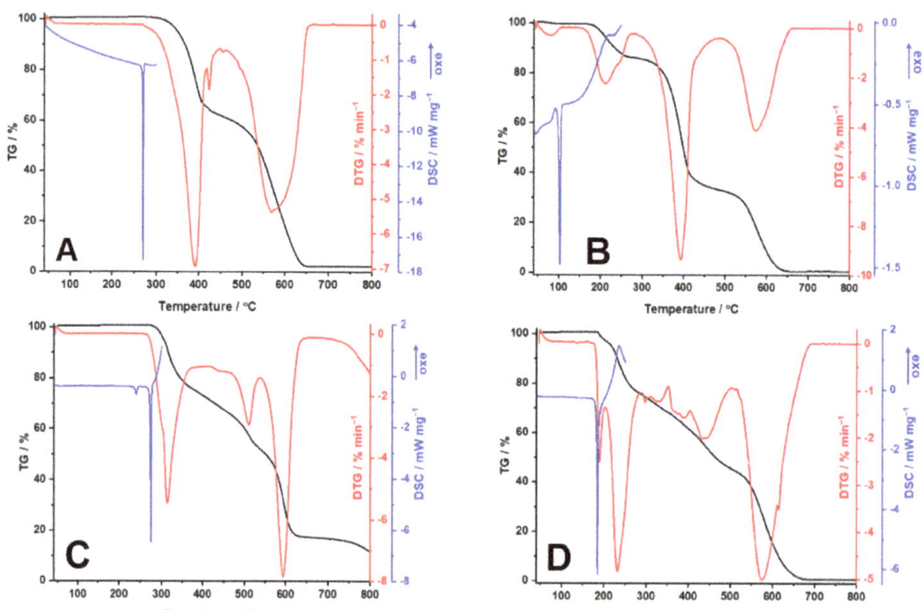

Figure 2. Thermoanalytical curves of (**A**) TELM, (**B**) VLS, (**C**) LOS, (**D**) OLM.

TELM (Figure 2A) shows high thermal stability (up to 274 °C) when the melting process highlighted by the DSC curve between 264 and 277 °C with a peak at 271 °C (an endothermic event) accompanies the beginning of the mass loss, so there is a phase transition accompanied by decomposition. The decomposition process takes place in two steps identifiable on the TG/DTG curves. The first process takes place in the range of

274–419 °C with a mass loss of ~40%. The second decomposition process commences at 450 °C and ends at 668 °C, presenting a Δm = 59.7%.

Table 2. The interpretation of the thermoanalytical curves.

Sample	Process	T_{onset} (°C)	T_{offset} (°C)	$T_{onset\ DTG}$ (°C)	$T_{max\ DTG}$ (°C)	$T_{onset\ DSC}$ (°C)	$T_{peak\ DSC}$ (°C)	ΔH_{fus} (J g^{-1})	Δm (%)
TELM	I	274	419	274	392; 424	256	271	103.5	38.8
	II	450	668	466	569	-	-	-	59.7
OLM	I	171	202	171	189	172	185	122.8	3.4
	II	202	287	202	232	-	-	-	21.7
	III	287	353	287	297; 327	-	-	-	7.4
	IV	353	504	353	367; 387; 441	-	-	-	22.9
	V	504	690	515	575	-	-	-	44.7
LOS	I	258	380	258	314	232	240	9.3	25.4
						263	275	99.4	
	II	418	538	420	510	-	-	-	19.3
	III	538	652	540	594	-	-	-	34.6
VLS	I	45	108	45	80	92	103	26.4	0.5
	II	148	277	148	212	-	-	-	13.3
	III	284	486	287	393	-	-	-	52.9
	IV	486	662	493	575	-	-	-	32.4

For VLS on the TG/DTG curves, four steps of the decomposition process are observed (Figure 2B). The first process, which starts at an ambient temperature, represents the loss of the absorbed water (Δm = 0.5%). The endothermic peak observed at 103 °C indicates the melting temperature of VLS, in good agreement with the data presented in two patents [18,19]. The first degradation process of VLS starts at 148 °C and continues until 277 °C, presenting a maximum on the DTG curve at 212 °C and a slight loss of its mass. The second process starts at 284 °C, and is the major decomposition step considering that VLS loses more than half of its mass. The third step of decomposition of anhydrous VLS is between 486 and 662 °C, highlighted on the DTG curve by one peak at 575 °C.

Regarding LOS, thermal stability up to 258 °C can be observed, as well as a degradation process consisting of three steps (Figure 2C). The first one is characterized by Δ = 25.4%, accompanied on the DSC curve by two endothermic peaks at 240 and 275 °C. The first endothermic event can be associated with an enantiotropic polymorphic transition, as presented in the literature [20], while the second one represents the melting point of losartan potassium, in good agreement with the data presented in the literature. The second degradation step begins at 418 °C, presenting a loss of 19.3% of its mass and one maxim on the DTG curve at 510 °C. The third degradation step is between 538 and 652 °C, correlated with a loss of ~35% of the sample mass.

In the case of OLM decomposition (Figure 2D), five distinct processes could be identified in the selected temperature range. The first one, which takes place in the range 171–202 °C, corresponds to a dehydration process (theoretical water content 0.5 mol/mol OLM, determined water content 0.52 mol/mol OLM), being a result in agreement with the hemihydrate mentioned in patent EP1801111 [21]. It is known that the dehydration of pharmaceutical active hydrates generally occurs at temperatures close to the normal boiling temperature of water, considering that the hydrogen bonds that are formed in the molecular network are usually broken at this temperature [22]. For this hemihydrate, it is observed that the loss of water from the network occurs at considerably higher temperatures, indicating the strong binding of it in the crystalline network. In the same temperature range, an endothermic event can be observed on the DSC curve, which characterizes the melting point of OLM (peak at 185 °C). The following processes observed represent the decomposition of OLM. It can be concluded that OLM is thermally stable up to 202 °C, the temperature at which it begins to degrade, losing up to ~22% of its mass. Afterwards, the degradation continues with a slight loss of mass (7.4%) between 287 and 353 °C. Later, two major decompositions are observed in the temperature ranges of 353–504 °C and 504–690 °C, with losses of mass of 22.9% and ~45%, respectively.

2.2. Results of the Kinetic Investigations

For the kinetic investigations for each sartan, the first degradation process noticed on the DTG curve was selected, with an exception for OLM, for which two degradation processes were investigated, the dehydration and the decomposition of OLM. In Table 3 are presented the selected temperature ranges for each sartan.

Table 3. The temperature ranges for the first decomposition processes for each sartan for the kinetic investigations.

β (°C min^{-1})	The Temperature Range for the Selected Process (°C)				
	LOS	VLS	TELM	OLM Dehydration	OLM Decomposition
2	248–348	136–251	266–359	168–191	191–249
4	255–363	138–260	268–387	169–196	196–262
6	256–371	141–268	272–400	170–198	198–275
8	258–375	144–272	274–406	171–202	202–279
10	258–380	148–277	274–419	171–202	202–287

ASTM E698 is a non-isoconversional kinetic method based on the Arrhenius equation, being described as a linear regression correlating the conversion degree to the heating rate [23]. It is used only as a preliminary method, since it is more suitable for single-step processes [24–27]. The following is the equation used to determine the activation energy:

$$\beta\frac{d\alpha}{dT} = k_o \exp\left(\frac{E_a}{RT}\right)(1-\alpha) \tag{1}$$

FWO is an isoconversional integral kinetic model which permits the calculation of the activation energy without knowing the reaction order. In order to obtain the value of it the Doyle approximation is used, the equation being as follows [28–33]:

$$\log\beta = \log\frac{AE_a}{g(\alpha)R} - 2.315 - 0.457\frac{E_a}{RT} \tag{2}$$

By plotting the FWO method (the natural logarithm from the heating rate vs. 1/T), parallel lines are obtained for each conversion degree (from 0.01 to 0.99), and the activation energy is determined from the slope of each line [28].

The FR kinetic model is an isoconversional differential method considered to be the most accurate one, since no approximation is used to calculate the value of the activation energy. It is determined after the plotting of (ln $\frac{d\alpha}{dt}$ vs. 1/T) and with the help of the following equation [25,26,34–36]:

$$\ln\left(\frac{d\alpha}{dt}\right)_{\alpha,t} = \ln[A_\alpha f(\alpha)] - \frac{E_a}{RT_{\alpha,t}} \tag{3}$$

The NPK method is a kinetic model that does not use any approximation and in which the reaction rate results from two independent functions: f(α), a function of the degree of conversion, and k(T), a temperature dependence. Furthermore, this method indicates the type of processes that take place during the degradation (chemical or physical), with the help of the reaction orders (m and n), as indicated by the model of Šesták and Berggren:

$$f(\alpha) = \alpha^m (1 - \alpha)^n \tag{4}$$

Another advantage of this kinetic model is that it provides the value of the variance, which gives information regarding the contribution of each process to the global degradation process [37–39].

In Table 4 are represented the values of the activation energy obtained for each sartan by employing the FWO and FR isoconversional kinetic methods.

Table 4. The obtained values for the activation energy using the FWO and FR isoconversional methods [40].

	E_a (kJ mol^{-1}) vs. α for									
	TELM		VLS		OLM				LOS	
					Dehydration		Decomposition			
α	FWO	FR	FWO	FR	FWO	FR	FWO	FR	FWO	FR
0.05	107.8	92.4	134.1	139.8	518.8	393.1	168.8	137.9	212.0	124.8
0.10	103.3	81.5	136.8	137.1	488.1	403.6	160.8	128.6	186.6	122.1
0.15	99.5	76.5	137.1	136.7	470.4	359.5	155.6	125.7	174.3	132.0
0.20	96.8	76.3	137.5	138.0	448.9	275.6	151.7	123.5	167.4	137.5
0.25	95.1	78.3	137.7	136.2	426.7	220.1	148.5	120.3	163.9	151.6
0.30	93.9	78.6	137.3	133.5	406.5	188.3	145.6	121.9	161.8	131.2
0.35	93.1	79.7	136.4	130.4	388.3	168.0	143.2	117.3	155.9	104.8
0.40	92.6	81.6	135.3	128.6	371.9	157.9	140.6	115.3	152.1	122.1
0.45	92.2	82.4	134.2	127.2	357.2	155.5	138.0	108.5	150.1	130.2
0.50	92.0	83.8	133.2	127.8	343.8	154.4	135.1	104.9	148.6	131.8
0.55	91.9	86.0	132.5	129.0	331.4	158.5	132.3	104.2	147.2	132.2
0.60	91.9	87.6	132.2	130.7	319.9	163.7	129.7	104.8	146.1	138.7
0.65	92.0	90.0	132.6	135.2	309.1	170.2	127.2	103.2	145.8	143.3
0.70	92.2	91.5	134.1	141.1	298.8	176.7	124.9	103.6	145.8	148.3
0.75	92.6	93.5	136.8	148.8	288.8	183.2	123.0	107.4	146.5	148.4
0.80	93.0	96.0	141.8	158.7	278.9	189.6	120.7	100.3	147.6	154.1
0.85	93.6	98.4	148.2	166.1	268.8	195.8	118.4	92.4	149.6	159.8
0.90	94.4	101.1	155.5	173.2	257.9	199.9	115.4	95.9	153.0	166.9
0.95	95.6	106.6	163.0	177.4	244.2	197.8	109.9	84.2	159.4	175.3
$\bar{E}_a$ (kJ mol^{-1})	94.9 ± 1.0	87.5 ± 2.0	138.8 ± 1.9	141.9 ± 3.6	358.9 ± 18.9	216.4 ± 18.5	136.3 ± 3.7	110.5 ± 3.1	158.6 ± 3.9	139.7 ± 3.9

The thermal decomposition process that was investigated in the case of TELM (Figure 3A–F) is the one that takes place between 266 and 359 °C at β = 2 °C min^{-1}, a process that moves to a higher temperature with the increase in the heating rate, reaching the range of 274–419 °C at β = 10 °C min^{-1}. The ASTM E698 method indicates an activation energy of 78.3 kJ mol^{-1}, which highlights the relatively low stability of TELM in relation to the other sartans analyzed, as can be seen from Table 4.

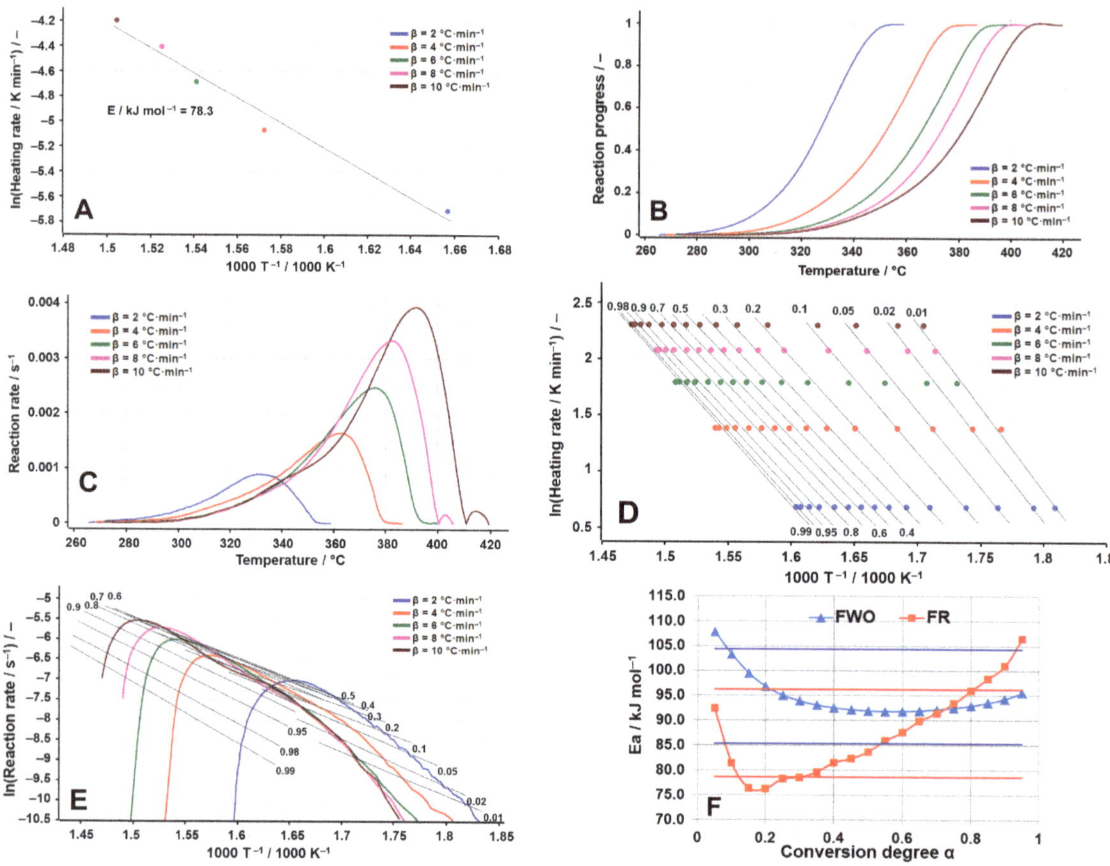

Figure 3. Results of the kinetic analysis performed on TELM: (**A**) kinetic method ASTM E698; (**B**) reaction progress vs. temperature; (**C**) reaction rate vs. temperature; (**D**) kinetic method FWO; (**E**) kinetic method Fr; (**F**) E_a vs. α according to the FWO and FR methods.

Thus, for TELM, the results of the isoconversional methods disagree with those of the preliminary method (ASTM E698), indicating considerably higher activation energies. From the analysis of the variation in E_a vs. α, it is observed that most of the individual values of E_a are found within the deviation limit of $\pm 10\%$ around the mean; thus, the FWO method presents at the beginning of the process higher activation energies, which then tend towards a smaller value, while the FR method indicates the minimum of the activation energy at $\alpha = 0.2$, after which a sharp increasing tendency towards the end of the investigated process can be noted, with the values obtained for $\alpha > 0.80$ being outside the selected range. This tendency suggests a change in the decomposition mechanism with the Increase in the heating rate, observations also supported by the aspect of the dependence of the reaction rate on temperature (Figure 3C), which indicates at temperatures above 400 °C (for the processes recorded at $\beta = 8\ °C\ min^{-1}$ and $\beta = 10\ °C\ min^{-1}$) the appearance of a secondary process of a much lower amplitude than the primary process. The complexity of the decomposition process is also suggested by the value of the difference between E_{max} and E_{min}, which is greater than 20% of the average value for the activation energy only for the Friedman method, since this method is a differential one and analyzes the process point by point, unlike the FWO method, which is an integral isoconversional method that provides an overview of the entire process analyzed (see Table 5).

Table 5. The estimation of the complexity of the analyzed processes based on the values of the activation energies.

	Kinetic Method	E_{max}	E_{min}	$\overline{E}_a$	$E_{max}-E_{min}$	$20\%\cdot\overline{E}_a$
TELM	FWO	107.8	91.9	94.9	15.9	19.0
	FR	106.6	76.3	87.5	30.3	17.5
VLS	FWO	163	132.2	138.8	30.8	27.8
	FR	177.4	127.2	141.9	50.2	28.4
OLM Dehydration	FWO	518.8	244.2	358.9	274.6	71.8
	FR	403.6	154.4	216.4	249.2	43.3
OLM Degradation	FWO	168.8	109.9	136.3	58.9	27.3
	FR	137.9	84.2	110.5	53.7	22.1
LOS	FWO	212	145.8	158.6	66.2	31.7
	FR	175.3	104.8	139.7	70.5	27.9

For TELM, it can be observed that two simultaneous processes take place during the degradation, the value of the activation energy being the result of the following equation:

$$\overline{E} = \lambda_1 \cdot E_1 + \lambda_2 \cdot E_2 \quad (5)$$

The experimental points obtained at five heating rates are represented in a tridimensional system and interpolated as a continuous reaction rate surface (see Figure 4). The first process contributes the most to the degradation process (λ = 87.5%), consisting of both physical (m = 1) and chemical (n = 1/2) transformations. Considering that the value of the variance of the second process is significantly lower than that of the first process, its contribution to the global process is minor. During this step, both physical (m = 3/2) and chemical (n = 1) transformations are noted (see Table 6).

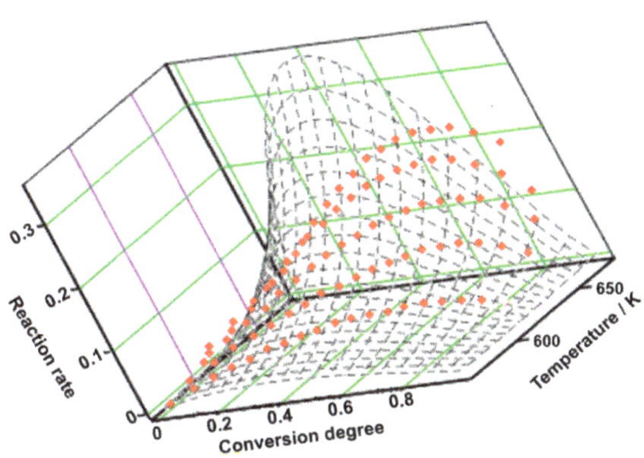

Figure 4. The 3D transformation surface for TELM from the NPK method.

In the case of VLS (Figure 5A–F), the ASTM E698 method presents a greater dispersion of the experimental points, which is an indication of the change in the decomposition mechanism with the increase in the heating rate. This is also supported by the other data obtained: the sigmoid that represents the variation in the reaction progress vs. temperature (Figure 5B) is irregular, similar to the dependence of the reaction rate on temperature, which presents a secondary process characterized by a local maximum between 220 and 270 °C. The FWO and FR methods confirm the existence of a complex decomposition process, consisting of several overlapping processes, the concave aspect of E_a vs. α indicating a process with a reversible stage (it decreases with α until α = 0.45 (FR) or until α = 0.60

(FWO)), followed by a thermooxidation process involving parallel or consecutive reactions (it increases strongly with α until the end of the investigated process) [41]. This aspect is also represented by the values of the difference between E_{max} and E_{min}, which are greater than 20% of the average value for the activation energy (Table 5).

Table 6. Results of the NPK method for analysis of TELM and comparison with isoconversional methods.

Sample	Process	λ/%	E/kJ mol^{-1}	A/min^{-1}	n	m	R^2	f(α)	$\overline{E}_a$ (kJ mol^{-1})		
									NPK	FWO	FR
TELM	1	87.5	85.7 ± 1.2	$3.4 \cdot 10^7 \pm 1.1 \cdot 10^4$	1/2	1	0.950	$(1-x)^{1/2} \cdot x^1$	82.6 ± 1.2	87.5 ± 2.0	94.9 ± 1.0
	2	8.6	89.5 ± 0.2	$1.3 \cdot 10^7 \pm 6.6$	1	3/2	0.955	$(1-x)^1 \cdot x^{3/2}$			

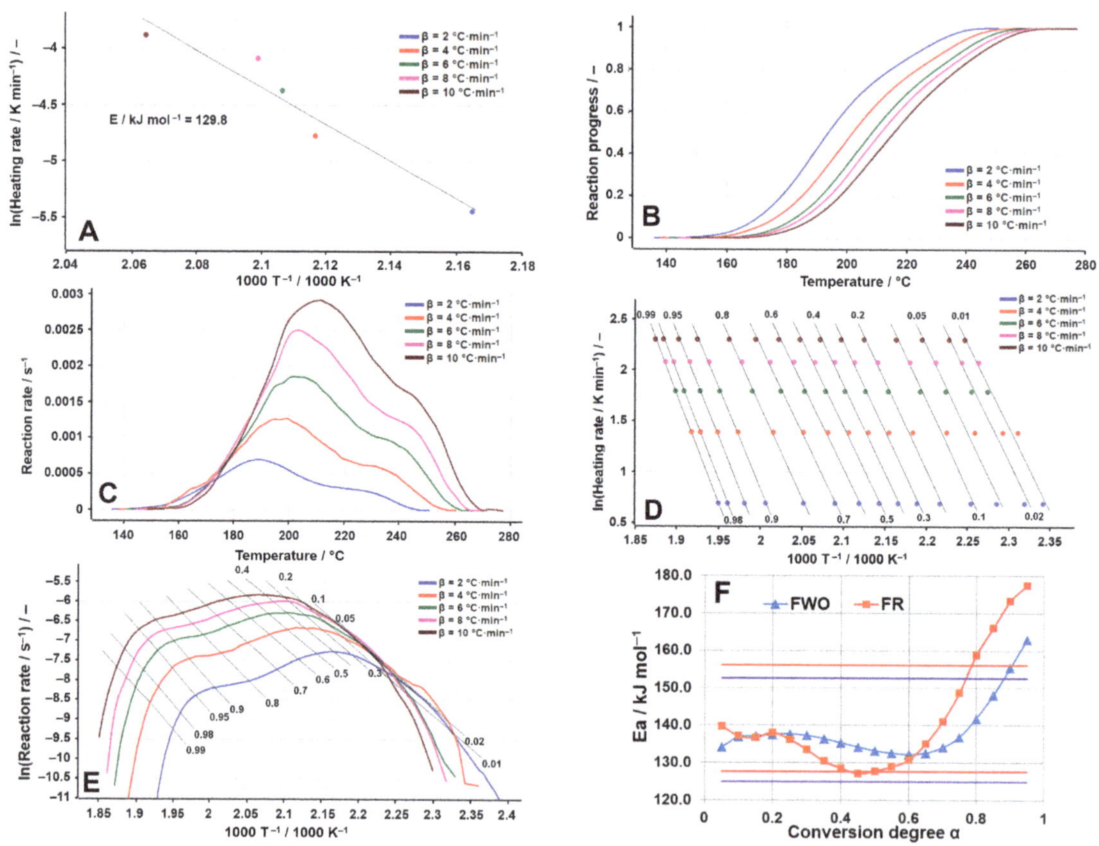

Figure 5. Results of the kinetic analysis performed on VLS: (**A**) kinetic method ASTM E698; (**B**) reaction progress vs. temperature; (**C**) reaction rate vs. temperature; (**D**) kinetic method FWO; (**E**) kinetic method Fr; (**F**) E_a vs. α according to the FWO and FR methods.

Regarding the graphical representation (Figure 6) and the results of the NPK analysis of VLS (Table 7), the degradation process consists of two parallel processes, the obtained value of the activation energy being the result of using the two values of variance (λ).

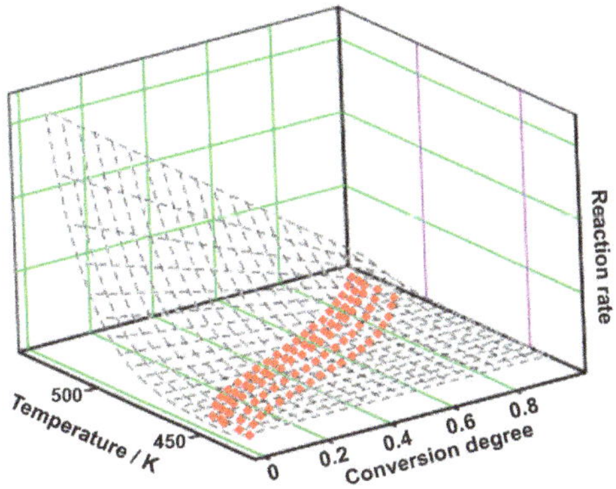

Figure 6. The 3D transformation surface for VLS from the NPK method.

Table 7. Results of the NPK method for analysis of VLS and comparison with isoconversional methods.

Sample	Process	λ/%	E/kJ mol^{-1}	A/min^{-1}	n	m	R^2	f(α)	$\overline{E}_a$ (kJ mol^{-1})		
									NPK	FWO	FR
VLS	1	91.4	136.3 ± 2.5	$3.0{\cdot}10^{15} \pm 6.8{\cdot}10^{10}$	1	0	0.993	$(1-x)^1$	136.4 ± 2.9	138.8 ± 1.9	141.9 ± 3.6
	2	8.3	142.4 ± 0.3	$1.9{\cdot}10^{16} \pm 8.7{\cdot}10^{6}$	0	1	0.986	x^1			

The greatest contribution to the degradation process is presented by the first process, which displays a variance of λ = 91.4% and which is a chemical reaction with a reaction order of n = 1. For the second process (λ = 8.3%), even if its contribution to the global process is remarkably reduced, the value of the activation energy is similar, and it consists only of physical transformations (m = 1).

With reference to LOS, the analyzed process is in the range of 248–348 °C at a value of the heating rate of β = 2 °C min^{-1}, with a value of the activation energy of 128.8 kJ mol^{-1} observed by using the ASTM E698 method, similar to the one obtained for VLS, being an indication of comparable stability between the two. The results of the kinetic investigation are presented in Figure 7A–F.

The isoconversional methods also indicate the complex degradation mechanism of the LOS: if for the FWO method a rapid decrease in E_a values is observed up to α = 0.25, after which the values remain within ±10% around the mean, for the FR method the variation is non-monotonic, confirming the complexity of the decomposition process, a hypothesis also sustained by the values presented in Table 5. Those results are expected, namely a complex decomposition mechanism, given its salt structure.

In Figure 8 and Table 8, the results of the NPK method are presented. As can be observed, the main degradation process consists of two simultaneous processes, the first one being the main contributor to the process, with a variance of 89.4%. The value of the global activation energy is determined by considering the contribution of each stage to the main process, following the equation $\overline{E} = \lambda_1{\cdot}E_1 + \lambda_2{\cdot}E_2$. The beginning of the degradation consists only of chemical degradations, given the reaction order of n = 2/5, which take place along with the chemical degradations (n = 3/2) and physical transformations (m = 1) of the second process.

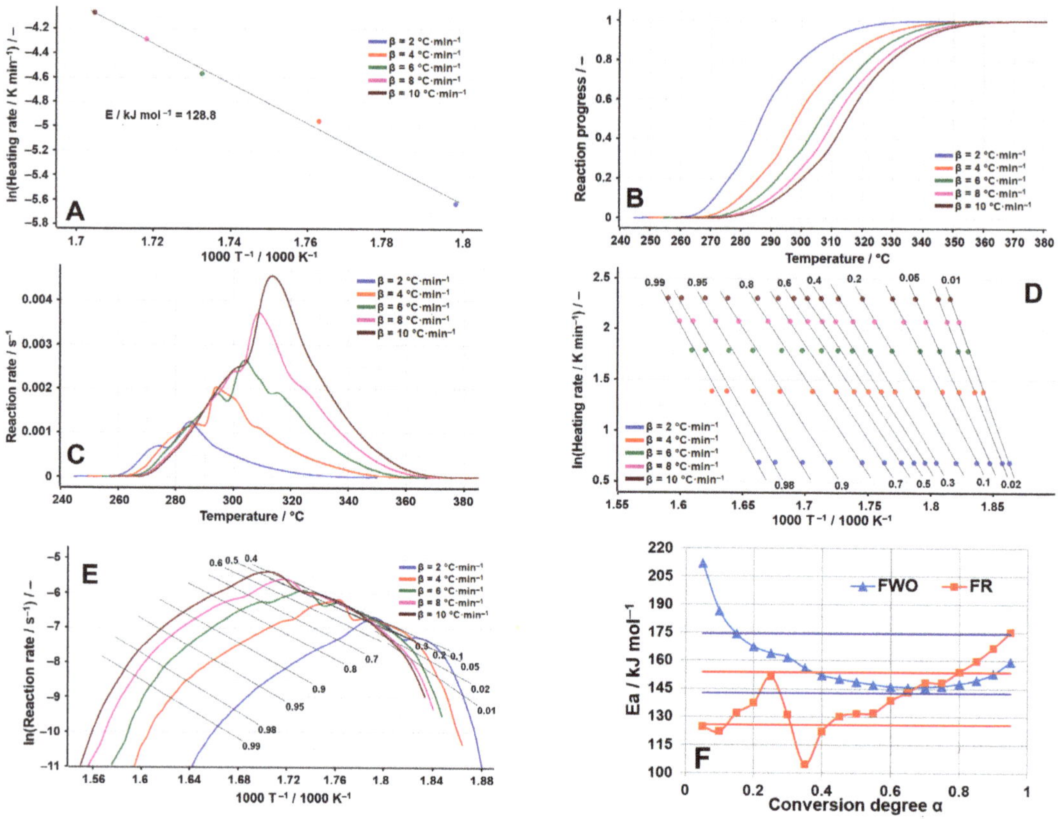

Figure 7. Results of the kinetic analysis performed on LOS: (**A**) kinetic method ASTM E698; (**B**) reaction progress vs. temperature; (**C**) reaction rate vs. temperature; (**D**) kinetic method FWO; (**E**) kinetic method Fr; (**F**) E_a vs. α according to the FWO and FR methods.

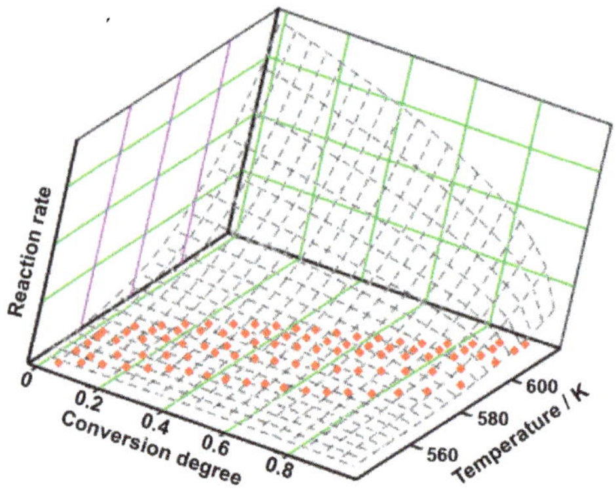

Figure 8. The 3D transformation surface for LOS from the NPK method.

Table 8. Results of the NPK method for analysis of LOS and comparison with isoconversional methods.

Sample	Process	λ/%	E/kJ mol^{-1}	A/min^{-1}	n	m	R^2	f(α)	$\overline{E}_a$ (kJ mol^{-1})		
									NPK	FWO	FR
LOS	1	89.4	158.6 ± 2.2	$6.1 \cdot 10^{14} \pm 8.8 \cdot 10^{7}$	2/5	0	0.983	$(1-x)^{2/5}$	151.3 ± 2.4	139.7 ± 3.9	158.6 ± 4.1
	2	7.0	138.5 ± 0.3	$2.6 \cdot 10^{11} \pm 1.4 \cdot 10^{11}$	3/2	1	0.995	$(1-x)^{3/2} \cdot x^1$			

Regarding OLM, two distinct processes were analyzed from a kinetic point of view to obtain a better understanding of the degradation mechanism. The first analyzed process is represented by dehydration, since OLM is used as a hemihydrate salt, and the results are presented in Figure 9A–F.

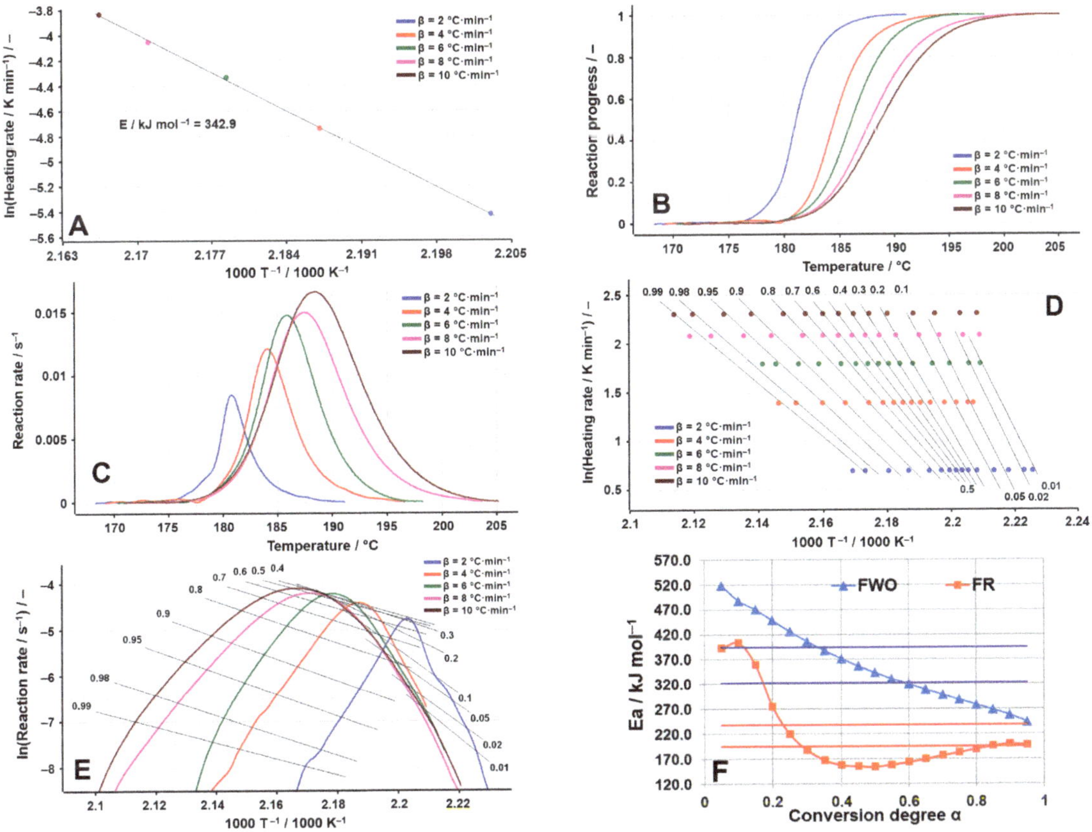

Figure 9. Results of the kinetic analysis performed on OLM dehydration: (**A**) kinetic method ASTM E698; (**B**) reaction progress vs. temperature; (**C**) reaction rate vs. temperature; (**D**) kinetic method FWO; (**E**) kinetic method Fr; (**F**) E_a vs. α according to the FWO and FR methods.

The ASTM E698 method reveals an unusually high activation energy of 342.9 kJ mol^{-1}, which is not typical for oxidative decomposition processes in pharmaceutical compounds. Two isoconversional methods, FR and FWO, indicate even higher values, as presented in Table 4 and Figure 9F. Specifically, the FWO method shows a progressive decrease in E_a values as the reaction proceeds, ranging from an extreme 518.8 kJ mol^{-1} at α = 5% to 244.2 kJ mol^{-1} at α = 95%. For this hemihydrate, the substantial and consistent variation in E_a can be attributed to the effects of mass and heat transfer. Examining the olmesartan medoxomil–water system at a macroscopic scale suggests that a single water molecule is associated with two active pharmaceutical molecules. In molar terms, 18 g of water is "bound" or "released" by 2 × 558.6 g of active ingredient, totaling 1117.2 g. This integration of water molecules within the OLM molecular network complicates the initiation of mass and heat transfer, thus necessitating higher energies until a steady-state diffusion of water molecules from the solid can be reached. During dehydration, local fluctuations in water vapor pressure and temperature may occur, potentially distorting kinetic analysis results and leading to misleading activation energy dependencies as conversion increases [42,43].

By applying the NPK method, the tridimensional graphic of the experimental points together with the interpolation of the reaction rate as a continuous surface was obtained (Figure 10). The dehydration of OLM consists of two concomitant steps, the first one with λ = 91% and the second one with λ = 8.4% (see Table 9). In the matter of these two processes, they are the result of both physical transformations (m = 1/2) and chemical degradations (n = 1).

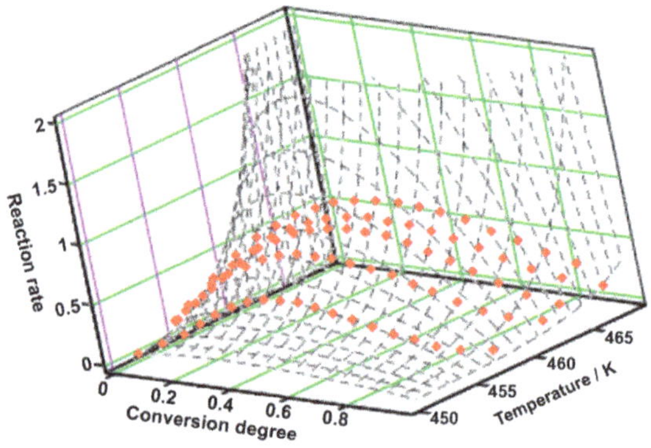

Figure 10. The 3D transformation surface for dehydration of OLM from the NPK method.

Table 9. Results of the NPK method for analysis of dehydration of OLM and comparison with isoconversional methods.

Sample	Process	λ/%	E/kJ mol^{-1}	A/min^{-1}	n	m	R^2	f(α)	$\overline{E}_a$ (kJ mol^{-1})		
									NPK	FWO	FR
Dehydration of OLM	1	91.0	461.8 ± 2.7	$1.0 \cdot 10^{53} \pm 2.8 \cdot 10^{12}$	1	1/2	0.987	$(1-x)^1 \cdot x^{1/2}$	441.4 ± 2.9	358.9 ± 18.9	216.4 ± 18.5
	2	8.4	252.4 ± 0.2	$2.0 \cdot 10^{28} \pm 1.5 \cdot 10^{8}$	1	1/2	0.943	$(1-x)^1 \cdot x^{1/2}$			

By analyzing the decomposition process of OLM (Figure 11A–F), a pattern similar to that observed during dehydration is noted when examining the variation in activation energies through the FWO method. The decomposition of anhydrous OLM involves multiple competing processes, a hypothesis further supported by the graphical representation of the reaction rate curves vs. T^{-1} (Figure 11C). The variation seen in Figure 11F can be attributed to the diffusion of water vapor from the solid matrix of the active pharmaceutical ingredient. This diffusion leads to continuous changes in the reactivity of the reactant particles, Influenced by alterations In the crystalline structure, the emergence of defects at interstitial sites, their migration, and other related phenomena.

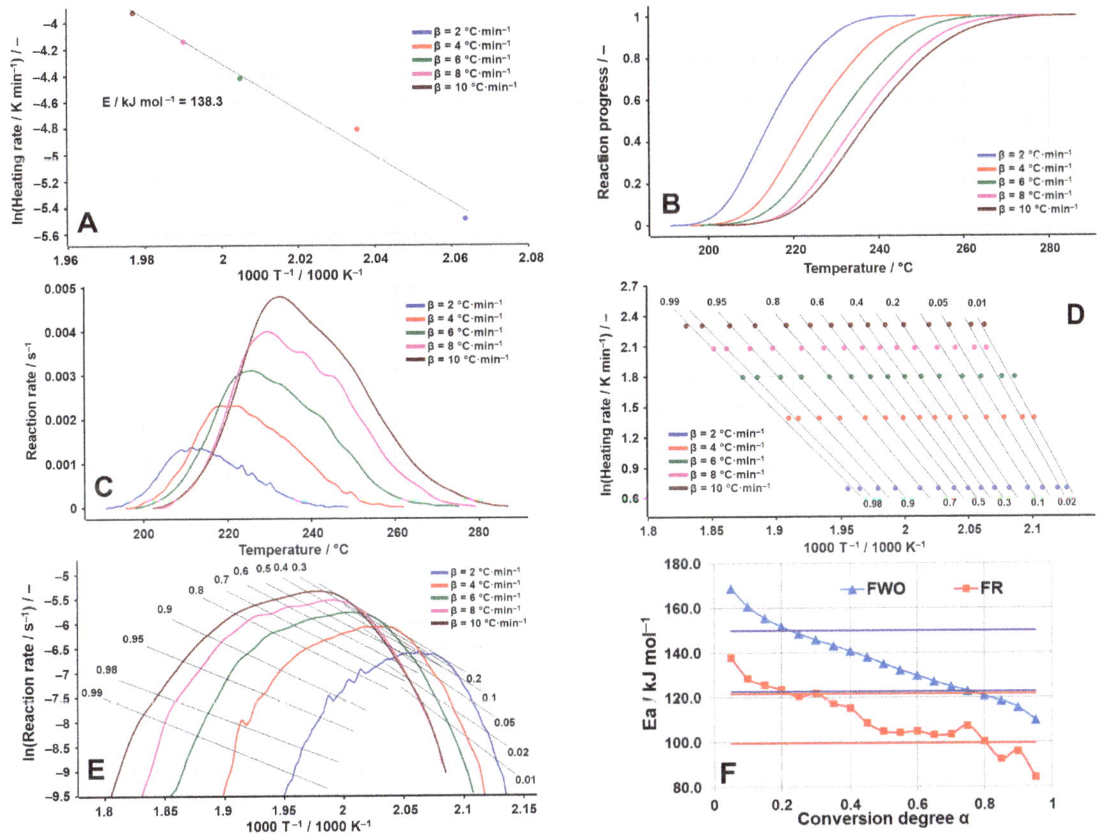

Figure 11. Results of the kinetic analysis performed on OLM degradation: (**A**) kinetic method ASTM E698; (**B**) reaction progress vs. temperature; (**C**) reaction rate vs. temperature; (**D**) kinetic method FWO; (**E**) kinetic method Fr; (**F**) E_a vs. α according to the FWO and FR methods.

By applying the NPK method, it was obtained the tridimensional graphic of the experimental points together with the interpolation of the reaction rate as a continuous surface (Figure 12). Concerning the degradation of OLM, it can be observed that the major contributor to the process is the first step, presenting a variance of 94.2% (see Table 10). The transformations during this step are only chemical (n = 4/5), while for the second step both are observed (physical, m = 2/3, and chemical, n = 1).

Table 10. Results of the NPK method for analysis of degradation of OLM and comparison with isoconversional methods.

Sample	Process	λ/%	E/kJ mol^{-1}	A/min^{-1}	n	m	R^2	f(α)	$\bar{E}_a$ (kJ mol^{-1})		
									NPK	FWO	FR
Degradation of OLM	1	94.2	136.5 ± 1.6	$4.0{\cdot}10^{14} \pm 2.1{\cdot}10^{6}$	4/5	0	0.997	$(1-x)^{4/5}$	133.9 ± 13.6	110.5 ± 3.1	136.3 ± 3.7
	2	3.8	141.4 ± 0.1	$5.7{\cdot}10^{13} \pm 1.2{\cdot}10^{4}$	1	2/3	0.991	$(1-x)^{1} \cdot x^{2/3}$			

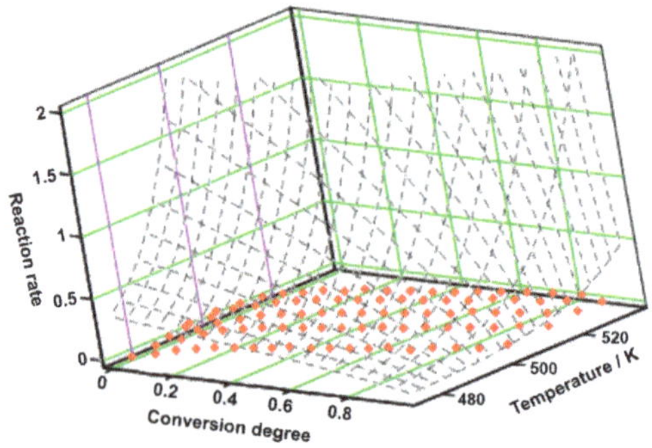

Figure 12. The 3D transformation surface for degradation of OLM from the NPK method.

3. Materials and Methods

3.1. Samples

The active pharmaceutical ingredients used were commercially produced and used without any further purification in this study, presenting the following provenance and purity:

- TELM was obtained from Sigma-Aldrich (St. Louis, MO, USA), product code PHR1855, secondary pharmaceutical standard;
- VLS was obtained from Sigma-Aldrich (St. Louis, MO, USA), product code PHR1315, secondary pharmaceutical standard, certified reference material;
- OLM hemihydrate was obtained from Sigma-Aldrich (St. Louis, MO, USA), product code SML1391, presenting a purity of ≥98% (HPLC);
- LOS was obtained from Sigma-Aldrich (St. Louis, MO, USA), product code PHR1602, secondary pharmaceutical standard, certified reference material.

3.2. Thermal Investigations

The thermal investigations were performed on the Setline TGA (SETARAM, Caluire, France) instrument using an open alumina crucible in a dynamic air atmosphere (100 mL min^{-1}), to obtain the TG and DTG data. Five different heating rates were selected, namely β = 2, 4, 6, 8, and 10 °C min^{-1} from ambient temperature up to 800 °C, for a sample weighing approximately 7 mg. Regarding the DSC curves, a NETZSCH DSC 204F1 Phoenix (NETZSCH, Selb, Germany) instrument and sealed aluminum crucibles were used, in an inert nitrogen medium with a flow rate of 20 mL min^{-1} and a heating rate of 10 °C min^{-1}.

3.3. Kinetic Analysis

The kinetic study (using the ASTM E698, FR, and FWO methods) was performed on the main decomposition step using AKTS Thermokinetics software, Version 4.46 (AKTS AG TechnoArk, Siders, Switzerland). The mathematical background and importance of using isoconversional kinetic methods have been extensively reported in the literature [44–48]. Also, to obtain the values for the kinetic triplet, the modified non-parametric method (NPK) was used, through a protocol that was previously reported by our research group [37–39].

The abbreviations used are those recommended and accepted by the ICTAC committee [49,50]: α—degree of conversion, t—time, β—linear heating rate (°C min^{-1}), A—pre-exponential factor, according to the Arrhenius kinetic model (min^{-1}), f(α)—differential conversion function, g(α)—integral conversion function, E_a—activation energy (kJ mol^{-1}), R—universal gas constant (J mol^{-1} K^{-1}), T—absolute temperature (K), m_o—mass of the sample at the beginning of decomposition, m_f—mass of the sample at the end of decomposition, and m_T—mass of the sample at a certain temperature T.

4. Conclusions

From the results of this study, it can be concluded that all the selected sartans undergo a complex decomposition process, consisting of complex reaction sequences (processes involving parallel and successive reactions, processes with reversible steps, and processes with a transition to the diffusion regime); the curves E_a vs. α show minimal and maximal values and ranges where E_a is independent of the conversion. It is also worth mentioning that the change in the decomposition mechanism with the increase in the heating rate occurs for all investigated sartans, the variation in the individual-point E_a values being in wide intervals outside the limit interval of 10% variation around the average, and therefore, the average value tabulated for the activation energy only has the mathematical meaning of an average value and does not represent the average energy of a unitary, individualized process.

Author Contributions: Conceptualization, B.B., A.R., D.E.M. and I.L.; methodology, A.L., C.T., D.I. and G.V.; validation, C.T., R.A.B., A.D. and T.V.; investigation, B.B., A.R., A.D. and D.I.; resources, D.E.M., R.A.B., A.L., T.V., G.V. and I.L.; writing—original draft preparation, I.L., D.I., R.A.B., D.E.M. and B.B.; writing—review and editing, A.L., C.T., A.R. and G.V.; supervision, A.L., T.V., A.D. and I.L.; project administration, B.B. All authors have read and agreed to the published version of the manuscript.

Funding: We would like to acknowledge Victor Babeș University of Medicine and Pharmacy Timișoara for their support in covering the costs of publication for this research paper.

Institutional Review Board Statement: Not applicable.

Informed Consent Statement: Not applicable.

Data Availability Statement: Raw data are available upon request from the corresponding author of this work.

Conflicts of Interest: The authors declare no conflicts of interest.

References

1. Muszalska, I.; Sobczak, A.; Dołhań, A.; Jelińska, A. Analysis of Sartans: A Review. *J. Pharm. Sci.* **2014**, *103*, 2–28. [CrossRef] [PubMed]
2. Dams, I.; Ostaszewska, A.; Puchalska, M.; Chmiel, J.; Cmoch, P.; Bujak, I.; Białońska, A.; Szczepek, W.J. Synthesis and physicochemical characterization of the process-related impurities of olmesartan medoxomil. Do 5-(Biphenyl-2-yl)-1-triphenylmethyltetrazole intermediates in sartan syntheses exist? *Molecules* **2015**, *20*, 21346–21363. [CrossRef]
3. Gościniak, A.; Stasiłowicz-Krzemień, A.; Michniak-Kohn, B.; Fiedor, P.; Cielecka-Piontek, J. One Molecule, Many Faces: Repositioning Cardiovascular Agents for Advanced Wound Healing. *Molecules* **2024**, *29*, 1938. [CrossRef] [PubMed]
4. Van Liefde, I.; Vauquelin, G. Sartan-AT1 receptor interactions: In vitro evidence for insurmountable antagonism and inverse agonism. *Mol. Cell Endocrinol.* **2009**, *302*, 237–243. [CrossRef] [PubMed]
5. Olmesartan Medoxomil. Available online: https://pubchem.ncbi.nlm.nih.gov/compound/130881 (accessed on 13 July 2024).
6. Naik, P.; Murumkar, P.; Giridhar, R.; Yadav, M.R. Angiotensin II receptor type 1 (AT1) selective nonpeptidic antagonists—A perspective. *Bioorganic Med. Chem.* **2010**, *18*, 8418–8456. [CrossRef] [PubMed]
7. Dobrescu, D.; Negres, S.; Dobrescu, L.; Mckinnon, R. *Memomed*, 30th ed.; Editura Universitara: Bucharest, Romania, 2024; ISBN 2069-2447.
8. Prasad, P.P.; Yeh, C.M.; Gurrieri, P.; Glazer, R.; McLeod, J. Pharmacokinetics of multiple doses of valsartan in patients with heart failure. *J. Cardiovasc. Pharmacol.* **2002**, *40*, 801–807. [CrossRef]
9. Sica, D.A.; Gehr, T.W.B.; Ghosh, S. Clinical pharmacokinetics of losartan. *Clin. Pharmacokinet.* **2005**, *44*, 797–814. [CrossRef]
10. Telmisartan. Available online: https://pubchem.ncbi.nlm.nih.gov/compound/Telmisartan (accessed on 13 July 2024).
11. Cozma, G.V.; Apostu, A.; Macasoi, I.; Dehelean, C.A.; Cretu, O.M.; Dinu, S.; Gaiță, D.; Manea, A. In Vitro and In Ovo Evaluation of the Potential Hepatoprotective Effect of Metformin. *Medicina* **2022**, *58*, 705. [CrossRef]
12. Ashwini Kumar, K.; Hiten Sharadchandra, M.; Nath, A.; Prasad, M. Polymorphic Form of Olmesartan Medoxomil. US20120184751A1, 19 July 2012. Available online: https://patents.google.com/patent/US20120184751A1/en (accessed on 13 July 2024).
13. Losartan Potassium. Available online: https://pubchem.ncbi.nlm.nih.gov/compound/11751549 (accessed on 13 July 2024).
14. Shakeel, F.; Haq, N.; Alanazi, F.K.; Alsarra, I.A. Measurement and correlation of solubility of olmesartan medoxomil in six green solvents at 295.15-330.15 K. *Ind. Eng. Chem. Res.* **2014**, *53*, 2846–2849. [CrossRef]
15. Valsartan. Available online: https://pubchem.ncbi.nlm.nih.gov/compound/Valsartan (accessed on 13 July 2024).

16. Guo, K.; Yin, Q.; Yang, Y.; Zhang, M.; Wang, J. Solubility of losartan potassium in different pure solvents from (293.15 to 343.15) K. *J. Chem. Eng. Data* **2008**, *53*, 1467–1469. [CrossRef]
17. Husain, A.; Azim Md Sabir, M.S.; Mitra, M.; Bhasin, P.S. A review on Candesartan: Pharmacological and pharmaceutical profile. *J. Appl. Pharm. Sci.* **2011**, *1*, 12–17.
18. Erwin, M.; Hans Rudolf, O.; Peter, B.; Wolfgang, M. Valsartan Salts. US9499499B2, 22 November 2016. Available online: https://patents.google.com/patent/US9499499B2/en (accessed on 14 July 2024).
19. Liu, F.; Wu, G.; Jiang, W.; Lin, C.C.; Cai, X.; Lin, P.; Lu, Y.; Liu, L. Crystal Forms of Valsartan Disodium Salt. US010745363B2, 18 August 2020. Available online: https://patents.google.com/patent/US10745363B2/en (accessed on 14 July 2024).
20. Al-Majed, A.R.A.; Assiri, E.; Khalil, N.Y.; Abdel-Aziz, H.A. *Losartan: Comprehensive Profile*, 1st ed.; Elsevier: Amsterdam, The Netherlands, 2015. [CrossRef]
21. Antoncic, L. Novel Polymorph Forms of Olmesartan Medoxomil. EP1801111B1, 16 July 2014. Available online: https://patents.google.com/patent/EP1801111B1/ru (accessed on 14 July 2024).
22. Jurczak, E.; Mazurek, A.H.; Szeleszczuk, Ł.; Pisklak, D.M.; Zielińska-Pisklak, M. Pharmaceutical hydrates analysis—Overview of methods and recent advances. *Pharmaceutics* **2020**, *12*, 1–25. [CrossRef] [PubMed]
23. Osman, A.I.; Fawzy, S.; Farrell, C.; Al-Muhtaseb, A.H.; Harrison, J.; Al-Mawali, S.; Rooney, D.W. Comprehensive thermokinetic modelling and predictions of cellulose decomposition in isothermal, non-isothermal, and stepwise heating modes. *J Anal Appl Pyrolysis* **2022**, *161*, 1–13. [CrossRef]
24. Fawzy, S.; Osman, A.I.; Farrell, C.; Al-Muhtaseb, A.H.; Harrison, J.; Al-Fatesh, A.S.; Fakeeha, A.H.; Doran, J.; Yang, H.; Rooney, D.W. Characterization and kinetic modeling for pyrolytic conversion of cotton stalks. *Energy Sci. Eng.* **2021**, *9*, 1908–1918. [CrossRef]
25. Ouyang, S.M.; Chen, Y.J.; Chen, C.F.; Chen, W.C. Oil in porous substrates: A thermogravimetric and simultaneous thermal analysis of spontaneous combustion risk. *J. Therm. Anal. Calorim.* **2023**, *148*, 4669–4679. [CrossRef]
26. Chen, W.C.; Lin, Y.H.; Ouyang, S.M.; Lee, Y.R. Effect of relative humidity on the emission height and reaction force of single-tube fireworks. *J. Therm. Anal. Calorim.* **2023**, *148*, 4997–5007. [CrossRef]
27. Ridichie, A.; Bengescu, C.; Ledeţi, A.; Rusu, G.; Bertici, R.; Vlase, T.; Vlase, G.; Peter, F.; Ledeţi, I.; Rădulescu, M. Thermal stability, preformulation, and kinetic degradation studies for gestrinone. *J. Therm. Anal. Calorim.* **2024**, 1–15. [CrossRef]
28. Mandapati, R.N.; Ghodke, P.K. Kinetics of pyrolysis of cotton stalk using model-fitting and model-free methods. *Fuel* **2021**, *303*, 121285. [CrossRef]
29. Liu, H.; Hong, R.; Xiang, C.; Wang, H.; Li, Y.; Xu, G.; Chang, P.; Zhu, K. Thermal decomposition kinetics analysis of the oil sludge using model-based method and model-free method. *Process Saf. Environ. Prot.* **2020**, *141*, 167–177. [CrossRef]
30. Ozawa, T. A New Method of Analyzing Thermogravimetric Data. *Bull. Chem. Soc. Jpn.* **1965**, *38*, 1881–1886. [CrossRef]
31. Flynn, J.H.; Wall, L.A. A quick, direct method for the determination of activation energy from thermogravimetric data. *J. Polym. Sci. Part B Polym. Lett.* **1966**, *4*, 323–328. [CrossRef]
32. Alvarado Flores, J.J.; Pintor Ibarra, L.F.; Mendez Zetina, F.D.; Rutiaga Quiñones, J.G.; Alcaraz Vera, J.V.; Ávalos Rodríguez, M.L. Pyrolysis and Physicochemical, Thermokinetic and Thermodynamic Analyses of Ceiba aesculifolia (Kunth) Britt and Baker Waste to Evaluate Its Bioenergy Potential. *Molecules* **2024**, *29*, 4388. [CrossRef] [PubMed]
33. Muravyev, N.V.; Luciano, G.; Ornaghi, H.L.; Svoboda, R.; Vyazovkin, S. Artificial neural networks for pyrolysis, thermal analysis, and thermokinetic studies: The status quo. *Molecules* **2021**, *26*, 3727. [CrossRef] [PubMed]
34. Ridichie, A.; Ledeţi, A.; Peter, F.; Ledeţi, I.; Muntean, C.; Rădulescu, M. Kinetic Investigation of the Oxidative Thermal Decomposition of Levonorgestrel. *Processes* **2023**, *11*, 3210. [CrossRef]
35. Muravyev, N.V.; Vyazovkin, S. The Status of Pyrolysis Kinetics Studies by Thermal Analysis: Quality Is Not as Good as It Should and Can Readily Be. *Thermo* **2022**, *2*, 435–452. [CrossRef]
36. Sidek, F.N.; Saleh, S.; Abdul Samad, N.A.F. Kinetic parameter estimation for pyrolysis of empty fruit bunch using model-fitting and model-free methods. *Mater. Today Proc.* **2022**, *57*, 1241–1247. [CrossRef]
37. Ledeţi, I.; Fuliaş, A.; Vlase, G.; Vlase, T.; Bercean, V.; Doca, N. Thermal behaviour and kinetic study of some triazoles as potential anti-inflammatory agents. *J. Therm. Anal. Calorim.* **2013**, *114*, 1295–1305. [CrossRef]
38. Ledeti, I.; Vlase, G.; Vlase, T.; Doca, N.; Bercean, V.; Fulias, A. Thermal decomposition, kinetic study and evolved gas analysis of 1,3,5-triazine-2,4,6-triamine. *J. Therm. Anal. Calorim.* **2014**, *118*, 1057–1063. [CrossRef]
39. Ledeţi, I.; Vlase, G.; Vlase, T.; Fuliaş, A. Kinetic analysis of solid-state degradation of pure pravastatin versus pharmaceutical formulation. *J. Therm. Anal. Calorim.* **2015**, *121*, 1103–1110. [CrossRef]
40. Baul, B.B. Contributions Concerning the Preformulation of some Bioactive Compounds Used in Modern Pharmaceutical Technology [Contribuții Privind Studiul Preformulării unor Substanțe Bioactive cu Aplicații în Tehnologia Farmaceutică Modernă]. Ph.D. Thesis, University Politehnica Timisoara, Timisoara, Romania, 25 September 2023.
41. Pérez-Maqueda, L.A.; Criado, J.M.; Sánchez-Jiménez, P.E.; Perejón, A. Kinetic studies in solid state reactions by sample-controlled methods and advanced analysis procedures. *J. Therm. Anal. Calorim.* **2013**, *113*, 1447–1453. [CrossRef]
42. Segal, E.; Budrugeac, P.; Carp, O.; Doca, N.; Popescu, C.; Vlase, T. *Analiza Termica: Fundamente si Aplicatii. Analiza Cinetica a Transformarilor Heterogene*; Editura Academiei Române: Bucharest, Romania, 2013.
43. Budrugeac, P.; Segal, E. Some methodological problems concerning nonisothermal kinetic analysis of heterogeneous solid-gas reactions. *Int. J. Chem. Kinet.* **2001**, *33*, 564–573. [CrossRef]

44. Nozela, W.C.; Nozela, C.F.V.; Silva, F.R.; Dias, D.S.; Almeida, S.; Ribeiro, C.A.; Crespi, M.S. Kinetic study of the energetic reuse from torrefied sewage sludge and urban pruning blends. *J. Therm. Anal. Calorim.* **2018**, *134*, 1285–1291. [CrossRef]
45. El-Sadek, M.H.; Ahmed, H.M.; El-Barawy, K.; Morsi, M.B.; El-Didamony, H.; Björkman, B. Non-isothermal carbothermic reduction kinetics of mechanically activated ilmenite containing self-reducing mixtures. *J. Therm. Anal. Calorim.* **2018**, *131*, 2457–2465. [CrossRef]
46. Mothé, C.G.; de Miranda, I.C. Decomposition through pyrolysis process of coconut fiber and rice husk and determination of kinetic parameters according isoconversional methods. *J. Therm. Anal. Calorim.* **2018**, *131*, 601–609. [CrossRef]
47. Antar, K.; Jemal, M. A thermogravimetric study into the effects of additives and water vapor on the reduction of gypsum and Tunisian phosphogypsum with graphite or coke in a nitrogen atmosphere. *J. Therm. Anal. Calorim.* **2018**, *132*, 113–125. [CrossRef]
48. Erceg, M.; Krešić, I.; Vrandečić, N.S.; Jakić, M. Different approaches to the kinetic analysis of thermal degradation of poly(ethylene oxide). *J. Therm. Anal. Calorim.* **2018**, *131*, 325–334. [CrossRef]
49. Verma, R.K.; Szilagyi, I.M.; Pielichowska, K.; Raftopoulos, K.N.; Šimon, P.; Melnikov, A.P.; Ivanov, D.A. Good laboratory practice in thermal analysis and calorimetry. *J. Therm. Anal. Calorim.* **2023**, *148*, 2211–2231. [CrossRef]
50. Vyazovkin, S.; Achilias, D.; Fernandez-Francos, X.; Galukhin, A.; Sbirrazzuoli, N. ICTAC Kinetics Committee recommendations for analysis of thermal polymerization kinetics. *Thermochim. Acta* **2022**, *714*, 179243. [CrossRef]

molecules

MDPI

Article

Rapid Liquid Chromatography–Tandem Mass Spectrometry Method for Determination of Total and Free Testosterone in Human Serum and Its Application to Monitoring Biomarker Response of Elite Athletes

Jianli Zhang [1,*], Hang Yu [2], Yulin Shen [1], Xingya Yang [1] and Yan Wang [2]

1 China Institute of Sport Science, General Administration of Sport of China, Beijing 100061, China; shenyulin@ciss.cn (Y.S.); yangxingya@ciss.cn (X.Y.)
2 State Key Laboratory of Bioactive Substance and Function of Natural Medicines, Institute of Materia Medica, Chinese Academy of Medical Sciences & Peking Union Medical College, Beijing 100050, China; yuhang@imm.ac.cn (H.Y.); wangyan@imm.ac.cn (Y.W.)
* Correspondence: zhangjianli@ciss.cn

Abstract: Total testosterone (TT) and free testosterone (FT) are important biochemical markers for anabolism of the human body, and can also serve as early screening indicators for overtraining syndrome (OTS). Presently, there is no fast and reliable serum TT and FT determination method in the field of sport science that can meet the requirements of sports research. Thus, a rapid and accurate determination method for serum TT and FT to fill the gap is needed urgently in sports training. Herein, a simple and reliable liquid chromatography–tandem mass spectrometry (LC-MS/MS) method for the simultaneous determination of TT and FT in serum was developed and fully validated, followed by the application of professional athletes in training monitoring. Efficient pretreatments based on only one-step liquid–liquid extraction (LLE) for TT and one-step LLE after a 20 min ultrafiltration for FT were adopted in this study, and the isotope internal standard of testosterone-13C3 was used to ensure the reliability of the whole procedure. A linear range of four orders of magnitude with 0.02–100 ng/mL can meet the concentration range requirement between a higher limit for male TT and a lower limit for female FT. The accuracy, precision, stability, and matrix effect were all within the limits of the guidelines. The serum TT and FT levels of 200 professional athletes (98 male athletes and 102 female athletes) were investigated by this method. Serum TT, FT, and FT/TT levels of professional athletes were significantly higher than the general population, and serum TT levels were significantly higher by LC-MS/MS than by a chemiluminescence immunoassay. In conclusion, the LC-MS/MS method for TT and FT measurement developed in this study is time-saving and easy to operate, which can be used as a reliable method for the determination of serum TT and FT in sports training, offering valuable information for monitoring anabolism of athletes and screening OTS in the early stage.

Keywords: total testosterone; free testosterone; LC-MS/MS; biomarker; athlete

1. Introduction

In competitive sports, monitoring biomarker responses to training may inform changes in training load to optimize performance and enhance positive adaptation with the reduction in injury risk for elite athletes. As an anabolic hormone with multiple physiological functions in the human body, testosterone (T) plays an important role in the growth and maintenance of skeletal muscle, bone, and red blood cells [1,2]. Therefore, it can be used as a biomarker to show evidence of some changes in direction, which corresponds to a change in performance, thus indicating the potential for tracking performance [3–6]. In addition, it can also provide some emotional assistance, such as reducing fear-potentiated startle [7].

In the circulation, T is bound tightly to sex hormone-binding globulin (SHBG), approximately 44% in men and 66% in women; a small fraction is bound weakly to human serum albumin (HAS), corticosteroid-binding globulin (CBG), and orosomucoid (ORM); only 1% to 4% of circulating T is unbound or free [8–10]. TT refers to the sum of the combined and uncombined T, and FT refers to the uncombined part of the body circulation, both of which are important biomarkers that need to be monitored in the training supervision and training evaluation of elite athletes [11–16] because they are widely used early indicators of an imbalance between anabolic and catabolic metabolism [12]. FT is referred to as the physiologically active fraction because only the free form can go through the cell membrane to interact with the androgen receptor, which makes it physiological significant. Therefore, FT could improve precision in the assessment of androgenic status compared with TT. Nevertheless, TT and FT present a dynamic equilibrium state in the human body, so the level of TT can not only reflect the reserve of testosterone in the body, but also jointly reflect the athlete's physical state with the ratio of other biomarkers such as cortisol, estrogen, etc. [3]. Thus, specific analysis methods with adequate accuracy and sensitivity for TT and FT need to be developed.

Similar to clinical laboratories, TT was measured by immunoassays in sport scientific research laboratories extensively [17–21]. It is well known that immunoassays lack specificity and thus directly affect measurement accuracy. So far, FT measurement is still a thorny issue whether directly determined by an immunoassay or calculated by using TT, albumin, and SHBG content with algorithms based on incorrect mathematical models of T binding to SHBG [8,22].

LC-MS/MS, as an advanced analytical technique, has been widely used in various fields for quantitative and qualitative measurements due to its high sensitivity and specificity. Recently, more and more LC-MS/MS determination methods for serum TT or FT with different sample preparation techniques have been reported [23–35]. Almost all of these studies can only determine TT [23–33]. There are only a few available articles that can simultaneously measure TT and FT in serum by LC-MS/MS [34,35]. However, these methods involve dialysis and derivatization procedures to prepare FT, which require at least 6 to 12 h for the sample pretreatment. So, these methods are time-consuming and reagent-consuming.

In the field of athlete training monitoring, firstly, the test results need to be fed back to coaches and scientific researchers as soon as possible, so as to adjust the training plan accordingly. This requires the determination method to be both accurate and rapid. In addition, circulating FT levels in women are very low due to significant gender and individual differences in T concentration. Therefore, it is necessary to establish a fast, accurate method with a wide enough linear range to monitor TT and FT. In order to tackle the practical problem, we developed and validated a simple, sensitive, and accurate LC-MS/MS method with an adequately wide linearity range for simultaneously determining TT and FT in this paper. For TT, only one LLE step was required. Compared to TT pretreatment, a protein removal step of a 20 min ultrafiltration prior to LLE for FT was needed. This method could significantly improve work efficiency and reduce sample preprocessing time, satisfying the demands of training monitoring for athletes. Finally, this point was effectively verified by the practical application of using this method on a large-scale serum sample of 200 professional athletes for the first time.

2. Results and Discussion

2.1. Separation of FT from TT

The measurement of FT is still challenging because a simple and fast pretreatment method has not been found. A dialysis method was adopted in previously published articles to extract and purify FT from serum [34–36], which required a significant amount of time, usually several hours or even dozens of hours. The key issue influencing the determination of serum FT is the separation of the protein-binding fraction and unbinding fraction (FT). Thus, removing proteins from the serum matrix is the primary problem to be solved. Ultrafiltration, as an extraction method based on membrane materials, the molecular weight cutoff, and centrifugation conditions that can separate substances of different molecular weights, is well known for its time efficiency. Several published articles have used ultrafiltration to separate free fractions from protein-binding steroids [37,38], proving that ultrafiltration is a simple and reliable method for extracting free steroids. Here, we used an ultrafiltration method in a simple physical manner to achieve the separation of FT and the protein-binding fraction, including SHBG-, HAS-, CBG-, and ORM-bound forms, only by about 20 min. Considering SHBG with a molecular weight of approximately 90 kDa [8], HSA with a molecular weight of approximately 66k Da [39], CBG with a molecular weight of approximately 58 kDa [40], ORM with a molecular weight of approximately 37–54 kDa [41], and β2 microglobulin (BMG) with a molecular weight of about 11 kDa, which is the smallest molecular weight protein in serum, a 10 kDa ultrafiltration tube was selected to achieve a better protein removal effect. Millipore 10 kDa ultrafiltration tubes were adopted in this study due to their vertical membrane structure, which can reduce concentration polarization, accelerate centrifugation speed, and greatly shorten the time required for centrifugation. The whole centrifugation procedure only takes about 20 min with an interception ratio of more than 95%. The interception ratio refers to the proportion of proteins with a molecular weight exceeding 10 kDa that are intercepted by the filter membrane. By using 10 kDa ultrafiltration tubes, almost all the protein-binding T and other interfering proteins in serum could be discarded, and FT was separated into the bottom ultrafiltrate consequently. Compared with previous published articles, the ultrafiltration tube used in this study has a higher molecular weight interception rate and smaller nonspecific adsorption, which can effectively improve the recovery, especially for low-concentration free steroids.

2.2. Method Validation

2.2.1. Ultrafiltration for FT

No difference was found in the measurement results between filtration once and filtration twice (Table 1), indicating that the ultrafiltration membrane used in this study has no adsorption effect on FT. By evaluating the FT determination results of ultrafiltration after three different-level binding proteins (SHBG, HAS, and CBG) were equilibrated with different concentration levels of T, FT increased as the concentrations of the three binding proteins decreased (Table S1), which is consistent with the protein-binding theory of drugs. That is to say, the ultrafiltration method established in this study can accurately determine the amount of FT.

Table 1. Nonspecific adsorption of FT by ultrafiltration membrane.

	Female					Male				
	No. 1	No. 2	No. 3	No. 4	No. 5	No. 1	No. 2	No. 3	No. 4	No. 5
Filtered once (ng/mL)	0.038	0.047	0.098	0.143	0.211	0.121	0.206	0.452	0.559	0.884
Filtered twice (ng/mL)	0.021	0.059	0.071	0.167	0.182	0.139	0.172	0.401	0.522	0.829
RSD (%)	1.2	0.8	1.9	1.7	2.1	1.3	2.4	3.6	2.6	3.9

2.2.2. Specificity

Quantitative ions of T (288.80 → 109.20) and T-C13 (291.90 → 112.10) were investigated by comparing a blank serum matrix spiked with T and T-C13 at a low quality control (LQC, 0.05 ng/mL) level, real human serum sample with a concentration similar to the LQC level, and blank serum matrix to estimate the specificity. No potential interfering substances at retention times of T and T-C13 were found in the above three kinds of matrices (Figure S1) and method specificity was preferred.

2.2.3. Linearity and Lower Limit of Quantitation (LLOQ)

Eight levels of calibrators from 0.02 ng/mL to 100 ng/mL were used to generate the calibration curves. The results showed excellent linearity with correlation coefficients (R2) of greater than 0.99 between 0.02 ng/mL and 100 ng/mL. A representative calibration curve and other details of calibration curves during method validation are presented in Figure S2 and Table S2 in Supplementary Materials, respectively. LLOQ refers to the lowest concentration point in the linear range, which was 0.02 ng/mL and RSD < 20%.

2.2.4. Accuracy and Precision

As shown in Table 2, accuracy ranged from 90.0 to 112.7%; intra-batch and inter-batch precisions (RSD) were between 1.6 and 8.4%. Accuracy and intra-batch and inter-batch precisions for all three level concentrations including low quality control (0.05 ng/mL, LQC), medium quality control (10 ng/mL, MQC), and high quality control (80 ng/mL, HQC) levels were within the limits presented in the published guidelines [42,43]. More details are found in Table S3.

Table 2. Accuracy and precision of testosterone.

Spiked (ng/mL)	Accuracy (%) (n = 15)	Precision (%) Intra-Batch			Inter-Batch (n = 15)
		Batch 1 (n = 5)	Batch 2 (n = 5)	Batch 3 (n = 5)	
0.05	90.0–111.0	6.7	8.4	7.5	7.0
10	94.5–112.7	6.8	1.6	3.4	5.9
80	93.0–105.9	4.8	4.2	5.0	4.9

2.2.5. Recovery

Recoveries at three concentration levels were evaluated with the results of LQC within 85.6–109.8%, MQC within 75.1–93.7%, and HQC within 74.1–81.3% in a five-replicate manner. Results of recovery are presented in Table S4.

2.2.6. Matrix Effect

The matrix effect is one of the challenges encountered in the current application of LC-MS/MS technology. The presence of interfering components in biological sample extracts has a significant impact on the mass spectrometry signal of analytes. Ion suppression and ion enhancement are two forms of the matrix effect, which are manifested as the reduction and enhancement in an analyte signal by matrix components, with ion suppression being more common [44,45].

Serum contains dissolved proteins and other materials, which can interfere with charge transfer during the ionization process in the gas phase between LC and MS [46]. The matrix effect may not always be completely circumventable because a perfectly consistent matrix does not exist, but it can be significantly minimized and largely compensated for by various approaches, such as standard addition, matrix-matched calibration, dilution, and the use of isotopic analogs of the analytes as internal standards [47]. In this study, a 13C-labeled isotope internal standard and matrix-matched calibration were both utilized to minimize the matrix effect. Consequently, a tiny matrix effect was observed at different concentration levels from 96.0 to 97.8% as shown in Table 3.

Table 3. Matrix effect of testosterone in serum.

LQC (0.05 ng/mL) (%)	Avg. (%)	RSD (%)	MQC (10 ng/mL) (%)	Avg. (%)	RSD (%)	HQC (80 ng/mL) (%)	Avg. (%)	RSD (%)
94.0			96.0			95.5		
102.0			92.6			98.2		
102.2	96.7	6.9	98.3	97.8	4.3	103.7	96.0	5.9
86.4			104.2			88.1		
99.0			97.7			94.4		

2.2.7. Stability of Extracts

Considering the large number of batch testing samples, the prepared samples might wait for a certain period of time in the sequence in the automatic sampler for injection. In practical work, automatic sampler temperature of different LC-MS/MS instruments could be controlled to 4 °C or room temperature. Therefore, the stabilities of extracts after being placed at 4 °C and room temperature for 24 h were investigated, respectively. Accuracy ranged from 92.3% to 93.8% after 24 h of storage at 4 °C (Table S5). Accuracy ranged from 95.5% to 103.0% after 24 h of storage at room temperature (Table S6).

2.3. Application for Real Serum Samples of Elite Athletes

In the field of sport science, OTS can cause a decrease in athletic performance, which is an emerging disorder resulting from an excessive training load coupled with inadequate recovery and poor-quality sleep [5]. For competitive athletes, successful training not only involves overload but also avoids the combination of excessive overload with inadequate recovery [48]. Therefore, it is especially important to recognize the OTS as early as possible. One efficient and reliable approach is to monitor and measure biomarkers of athletes. Positive changes in basal T levels are associated with increases in lean mass and strength [49]; it is reasonable to conclude that the alteration of the T level is positively correlated with human sport performance. Moreover, from the previously published literature, decreasing serum TT and FT have been considered as indicators of the overtrained state [4,5,48–51]. Thus, accurately and quickly measuring TT and FT are extremely necessary for sports training.

Utilizing this newly developed simple LC-MS/MS method, both serum TT and FT were evaluated for 200 Chinese national shooting athletes (98 males and 102 females). Serum TT levels of 98 male athletes and 102 female athletes ranged from 2.91 to 11.68 ng/mL and 0.41 to 1.62 ng/mL, respectively (Figure 1). Serum FT levels of 98 male athletes and 102 female athletes ranged from 0.05 to 0.54 ng/mL and 0.02 to 0.06 ng/mL, respectively (Figure 2). The differences between male and female athletes were compared, which are shown in Figure 3a,b. As expected, TT and FT levels of male athletes were significantly higher than the female group. In addition, elite athletes had significantly higher TT and FT upper limit levels compared to the general population ($p < 0.001$). This is mainly because endurably physical exercise could improve FT and TT levels, irrespective of moderate or intense physical activity [49–51]. In fact, endurance exercise increases both FT and TT levels by upregulating FT levels in an SHBG binding affinity-independent manner [51]. As can be seen from Equation (1), the increase in the FT level will directly lead to the increase in the FT/TT ratio. Consequently, FT/TT ratios of professional athletes were conspicuously higher than the upper limit of the general population range, regardless of gender (Figure 4). The FT/TT ratio of female athletes is higher than that of male athletes (Figure 3c), which may be due to women's lower basal testosterone levels, and the increase in testosterone levels caused by exercise is more significant.

$$\frac{\mathrm{FT}}{\mathrm{TT}}(\%) = \frac{\mathrm{FT}}{\mathrm{FT} + \mathrm{CombindedT}} \times 100\% \quad (1)$$

In order to compare the newly established LC-MS/MS method and the original chemiluminescence immunoassay method in our laboratory, both of the two methods were used to measure TT (only serum TT can be measured by a chemiluminescence immunoassay in our laboratory initially) in serum samples from 200 professional athletes. The linear regression plots for the comparison study using Passing and Bablok non-parametric linear regression are displayed in Figure 5a. The resulting regression line was y = 0.0109 (95% CI: −0.0019 to 0.0211) + 1.257 (95% CI: 1.2451 to 1.2699) x. However, the Pearson correlation coefficient (R^2) showed that these methods were highly correlated, $R^2 = 0.9976$, $p < 0.0001$. The methods were also compared using the Bland–Altman plot (Figure 5b). In the percentage plot, the relative bias was 23.5%, and the limits of agreement ranged from 17.2 to 29.9% for the proportional difference. It indicated that the measurement results from these two methods were highly correlated but with systematic bias. The results obtained by the LC-MS/MS method are 17.2–35.4% higher than those of the immunoassay, which is consistent with previously published research [30].

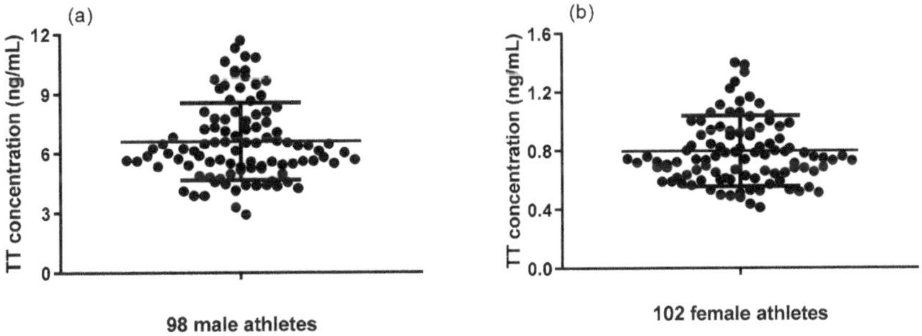

Figure 1. Serum TT levels of (**a**) 98 male athletes and (**b**) 102 female athletes.

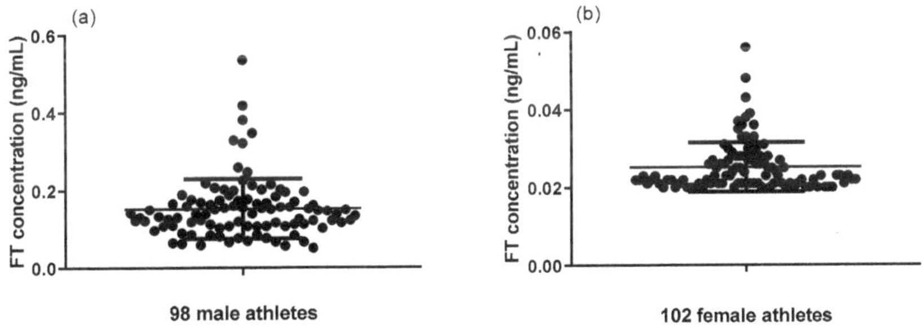

Figure 2. Serum FT levels of (**a**) 98 male athletes and (**b**) 102 female athletes.

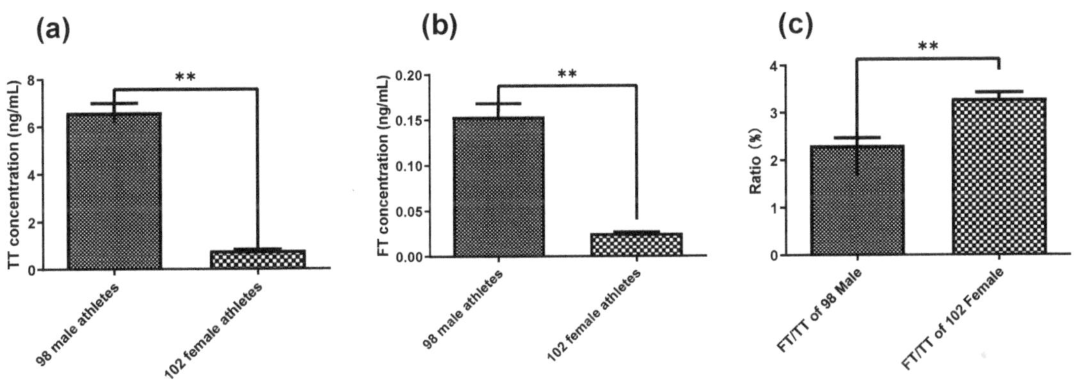

Figure 3. Difference of (**a**) TT between male and female professional athletes and (**b**) FT between male and female professional athletes, and (**c**) FT to TT ratios of 98 male athletes and 102 female athletes (independent two-tailed *t* test, ** $p < 0.001$).

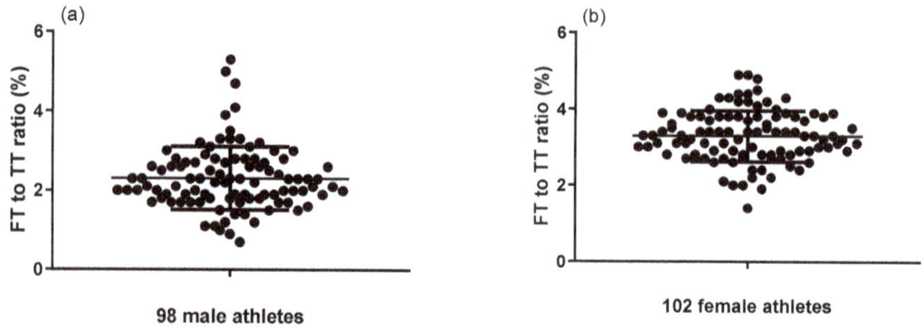

Figure 4. FT-to-TT ratios of (**a**) 98 male athletes and (**b**) 102 female athletes.

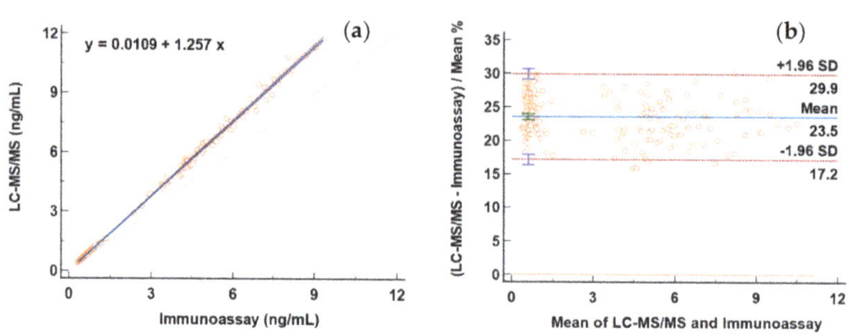

Figure 5. Comparison between LC-MS/MS and immunoassay methods. (**a**) Passing and Bablok regression analysis. (**b**) Bland-Altman plot.

3. Materials and Methods

3.1. Materials and Ethical Approval

T with a purity of 99.3% and T-2,3,4-13C3 (13C 99.4%) with a purity of 99.5% were both purchased from Alta Scientific Co., Ltd. (Tianjin, China). Hexane (HPLC grade), ethyl acetate (HPLC grade), methanol (HPLC grade), and formic acid (HPLC grade) were supplied by Dikma Technologies Inc. (Beijing, China). Ultrapure water was obtained from a Milli-Q ultrapure water system (Millipore, MA, USA). Blank serum for method validation was purchased from Innoreagents (Huzhou, China). The ultrafiltration tube (0.5 mL/10 KD) was supplied by Merck (Millipore, MA, USA).

Athlete serum samples from shooting athletes of the Chinese national team were stored at −80 °C until use. The study was approved by the Research Ethics Committee of the China Institute of Sport Science (20220928).

3.2. Instrumentation and Analytical Conditions

The determination of T and T-13C3 was performed on a SHIMADZU 8060 LC-MS/MS system (Shimadzu, Kyoto, Japan) using a column of Agilent Poroshell 120 EC-C18, 2.1 × 100 mm, 2.7 μm (Agilent, Santa Clara, CA, USA), maintained at 40 °C. Mobile phases consisted of 0.1% formic acid in water (A) and 0.1% formic acid in methanol (B). The injected samples were eluted using a gradient elution at a flow rate of 0.3 mL/min as follows: 0.0–0.1 min, 40% B; 0.1–5.0 min, linear from 40 to 70% B; 5.0–5.1 min, 98% B; 5.1–8.0 min, maintain at 98% B; and 8.0–10.0 min, linear from 98 to 40% B (Table S7). The autosampler was used with an injection volume of 10 μL. Quantitative ion transitions of T were then scanned by multiple reaction monitoring in the positive electrospray ionization mode. Nitrogen was used as nebulizer gas, drying gas, and collision gas. The MS conditions for the analysis were as follows: nebulizer gas, 2.8 L/min; drying gas, 10 L/min; and interface temperature, 300 °C; DL temperature and heat block temperature were maintained at 250 °C and 400 °C, respectively. Quantitative ion transitions were m/z 288.80 → 109.20

(collision energy: 24 eV) for T and m/z 291.90 → 112.10 (collision energy: 27 eV) for T-13C3, respectively.

3.3. Preparation of Standard Solutions

A stock solution of T at 1 mg/mL was prepared in methanol. Working solutions of T with concentrations of 10,000 ng/mL, 8000 ng/mL, 4000 ng/mL, 1000 ng/mL, 100 ng/mL, 10 ng/mL, 5 ng/mL, and 2 ng/mL were prepared by diluting the stock solution in methanol. The purchased commercialized store solution of T-13C3 (internal standard, IS) was in acetonitrile with a concentration of 10 μg/mL. The IS working solution was diluted in a 1:1 (*v:v*) mixed solution of hexane and ethyl acetate at 0.5 ng/mL. All stock and working solutions were stored at −20 °C. Blank serum samples (99 μL) were spiked with a 1 μL T working solution to generate calibration curves with 8 different concentrations at 100 ng/mL, 80 ng/mL, 40 ng/mL, 10 ng/mL, 1 ng/mL, 0.1 ng/mL, 0.05 ng/mL, and 0.02 ng/mL. QC samples including high QC, medium QC, and low QC for T were 80 ng/mL, 10 ng/mL, and 0.05 ng/mL.

3.4. Sample Preparation of TT

One hundred microliters of the serum sample was transferred into a 1.5 mL polypropylene tube. Five hundred microliters of a 1:1 (*v:v*) mixture of hexane/ethyl acetate containing 0.5 ng/mL of T-13C3 was added to the sample tube, vortex-extracted for 30 s, and centrifuged at 10,000 r.p.m for 5 min. The supernatant was transferred to another clean tube and evaporated to dryness by N2. The dry residue was redissolved in a 100 μL initial proportion of the mobile phase, and 10 μL of the solution was injected into the LC–MS/MS system.

3.5. Sample Preparation of FT

Two hundred microliters of the serum sample was transferred into a 0.5 mL/10 KD ultrafiltration tube and centrifuged at 13,000 r.p.m for 20 min. After ultrafiltration, 100 μL separated serum was pretreated by one-step LLE as aforementioned in the "Sample Preparation of TT" section.

3.6. Method Validation

In accordance with widely recognized bioanalytical method validation guidelines [39,40], this method was fully validated by specificity, linearity, LLOQ, accuracy, precision, recovery, the matrix effect, and stability. Additionally, the quantitative collection of FT by the ultrafiltration membrane was also evaluated.

3.6.1. Ultrafiltration for FT

The nonspecific adsorption of FT by the ultrafiltration membrane was investigated by comparing the measurement results of FT in serum samples from ten different athletes (five female and five male) with a once-filtered filtrate and twice-filtered filtrate during which a new membrane was used in the second round of ultrafiltration. In order to assess the accuracy of the ultrafiltration tube for the collection of FT, three different levels of SHBG (20 nmol/L, 80 nmol/L, 150 nmol/L), HAS (20 g/L, 40 g/L, 80 g/L), and CBG (10 mg/L, 20 mg/L, 30 mg/L) were spiked into blank serum matrices without proteins, and then T was spiked into each of them to make three different levels at 2 ng/mL, 10 ng/mL, and 50 ng/mL, respectively. Before ultrafiltration, all the samples were incubated at 37 °C for 1 h.

3.6.2. Specificity

Method specificity was to confirm if there were any interferences from the matrix or other substances present in the sample, which was carried out by analyzing blank matrix samples and blank samples spiked with reference standard solutions of T and IS.

3.6.3. Linearity and LLOQ

Method linearity was investigated by the analysis of calibration curve triplicates. Calibration curves were generated by plotting the peak area ratio of the analyte to IS against concentrations using weighted ($1/\times2$) linear least-squares regression. LLOQ is defined as the concentration at which the quantitative ion signal-to-noise ratio is greater than 10 and should meet a relative standard deviation of less than 20% of at least 5 replicates.

3.6.4. Accuracy and Precision

Accuracy and precision were assessed by the determination of QC samples (n = 5 per level) at three levels (LQC, MQC, and HQC) on three consecutive days. Accuracy was expressed as the percentage of the measured value to true value, which should be within the range of 85–115% (LQC could be within the range of 80–120%). Precision was expressed as relative standard deviation (RSD), which should be no more than 15% (LQC could be no more than 20%).

3.6.5. Recovery

Recovery of T was evaluated by comparing the peak area ratio of the analyte to IS in QC samples at three levels (0.05 ng/mL, 10 ng/mL, and 80 ng/mL) with blank matrices spiked with the analyte prior to extraction with the peak area ratio of the analyte to IS spiked post-extraction of blank matrices.

3.6.6. Matrix Effect

In the bioanalytical LC–MS/MS method, the matrix effect is a very important validation parameter [37–39]. In this study, the effect was evaluated on ion suppression or enhancement with LQC, MQC, and HQC, which was calculated by the ratio of the peak area in the presence of the blank matrix to the peak area in the absence of the matrix by analyzing 5 replicates at each QC concentration level.

3.6.7. Stability of Extracts

The extracts of LQC, MQC, and HQC were retained and analyzed for 24 h at 4 °C and room temperature to estimate stability, respectively.

3.7. Application to Real Athlete Serum Samples

Intravenous blood samples were collected by vacuum tubes with separation gel and coagulation from male and female athletes in fasting states randomly. All the serum samples were measured in replicates. TT and FT of 200 serum samples of shooting athletes from the Chinese national team were analyzed and monitored using this LC-MS/MS method, respectively. Data were processed by Shimadzu LabSolutions software (Version 5.89).

3.8. Statistical Analysis

The statistical analysis of the data from the comparison of LC-MS/MS and chemiluminescence immunoassay methods was performed by using MedCalc Software (Version 23.0.5).

An unpaired 2-tailed t test was performed to analyze the other data by using GraphPad Prism Software (Version 8.0.1).

4. Conclusions

A simple and rapid LC-MS/MS method for simultaneous measurement of TT and FT in human serum was established and fully validated. Compared with the previous reported preparation procedures involving dialysis, derivatization, solid phase extraction, or multi-step LLE, the newly developed method only required one-step LLE for TT and a 20 min ultrafiltration followed by one-step LLE for FT. It is very efficient and time-saving and could effectively meet the needs of physiological monitoring, screening OTS as early as possible in sports training.

Finally, this method was used to study 200 elite national shooters, including 98 male athletes and 102 female athletes. To our knowledge, this is the first large scale evaluation of serum TT and FT in professional athletes. Serum TT and FT from different genders were investigated, respectively. Serum TT, FT, and FT/TT levels of professional athletes were significantly higher than those of the general population, as long-term endurance exercise could increase FT testosterone levels with a mediation by the sympathetic stimulation of T-secreting organs [51]. In addition, the newly developed LC-MS/MS method was compared with the original chemiluminescence immunoassay method in our laboratory by a measurement of serum TT. Similar to the previous report, serum TT levels determined by the LC-MS/MS method were 17.2–35.4% higher than the chemiluminescence immunoassay method. In order to diagnose OTS for a wider range of elite athletes by providing TT and FT levels during sports training as early as possible, it is necessary to investigate TT and FT levels for more athletes in different sports events by using this method.

Supplementary Materials: The following supporting information can be downloaded at: https://www.mdpi.com/article/10.3390/molecules29215007/s1, Figure S1: Specificity of method (A: Blank matrix spiked with testosterone and IS; B: Real human serum spiked with IS; C: Blank matrix). Figure S2: Representative calibration curve of testosterone. Table S1: Nonspecific adsorption of FT to the ultrafiltration membrane. Table S2: Quantitative results of FT with three different levels of SHBG, HAS and CBG. Table S3: Calibration curves and linearity information. Table S4: Accuracy and precision of testosterone. Table S5: Recoveries of testosterone at LQC, MQC and HQC (n = 5). Table S6: Stability of extracts at 4 °C after 24 h of storage. Table S7: Stability of extracts at room temperature after 24 h of storage. Table S8: Gradient elution of testosterone and IS.

Author Contributions: Conceptualization, J.Z.; methodology, J.Z. and H.Y.; validation, J.Z. and H.Y.; formal analysis, J.Z. and H.Y.; investigation, J.Z., H.Y. and X.Y.; resources, X.Y. and Y.W.; data curation, H.Y. and X.Y.; writing—original draft preparation, J.Z.; writing—review and editing, J.Z. and Y.W.; supervision, Y.S. and Y.W.; project administration, J.Z. and Y.S.; funding acquisition, J.Z. and Y.S. All authors have read and agreed to the published version of the manuscript.

Funding: This project was supported by the Fundamental Research Funds for the China Institute of Sport Science (grant numbers 22-28 and 23-18 and 24-09).

Institutional Review Board Statement: The study was approved by the Research Ethics Committee of the China Institute of Sport Science (No. 20220928).

Informed Consent Statement: Informed consent was obtained from all subjects involved in the study.

Data Availability Statement: The data presented in this study are available on request from the corresponding author due to privacy.

Conflicts of Interest: The authors declare no conflicts of interest.

References

1. Brownlee, K.K.; Moore, A.W.; Hackney, A.C. Relationship between circulating cortisol and testosterone: Influence of physical exercise. *J. Sports Sci. Med.* **2005**, *4*, 76–83. [PubMed]
2. Zitzmann, M.; Nieschlag, E. Testosterone levels in healthy men and the relation to behavioural and physical characteristics: Facts and constructs. *Eur. J. Endocrinol.* **2001**, *144*, 183–197. [CrossRef] [PubMed]
3. Greenham, G.; Buckley, J.D.; Garrett, J.; Eston, R.; Norton, K. Biomarkers of physiological responses to periods of intensified, non-resistance-based exercise training in well-trained male athletes: A systematic review and meta-analysis. *Sports Med.* **2018**, *48*, 2517–2548. [CrossRef] [PubMed]
4. Huggins, R.A.; Fortunati, A.R.; Curtis, R.M.; Looney, D.P.; West, C.A.; Lee, E.C.; Fragala, M.S.; Hall, M.L.; Casa, D.J. Monitoring blood biomarkers and training load throughout a collegiate soccer season. *J. Strength Cond. Res.* **2019**, *33*, 3065–3077. [CrossRef]
5. Flavio, A.C.; Claudio, E.K. Basal hormones and biochemical markers as predictors of overtraining syndrome in male athletes: The eros-basal study. *J. Athl. Train.* **2019**, *54*, 906–914.
6. Anderson, T.; Haake, S.; Lane, A.R.; Hackney, A.C. Changes in resting salivary testosterone, cortisol and interleukin-6 as biomarkers of overtraining. *Balt. J. Sport Health Sci.* **2016**, *101*, 2–7. [CrossRef]
7. Hermans, E.J.; Putman, P.; Baas, J.M.; Koppeschaar, H.P.; van Honk, J. A single administration of testosterone reduces fear-potentiated startle in humans. *Biol. Psychiatry* **2006**, *59*, 872–874. [CrossRef]

8. Goldman, A.L.; Bhasin, S.; Wu, F.C.W.; Krishna, M.; Matsumoto, A.M.; Jasuja, R. A Reappraisal of Testosterone's Binding in Circulation: Physiological and Clinical Implications. *Endocr. Rev.* **2017**, *38*, 302–324. [CrossRef]
9. Jasuja, R.; Pencina, K.M.; Peng, L.; Bhasin, S. Accurate Measurement and Harmonized Reference Ranges for Total and Free Testosterone Levels. *Endocrinol. Metab. Clin. N. Am.* **2022**, *51*, 63–75. [CrossRef]
10. Dunn, J.F.; Nisula, B.C.; Rodbard, D. Transport of steroid hormones: Binding of 21 endogenous steroids to both testosterone-binding globulin and corticosteroid-binding globulin in human plasma. *J. Clin. Endocrinol. Metab.* **1981**, *53*, 58–68. [CrossRef]
11. Marinelli, M.; Roi, G.S.; Giacometti, M.; Bonini, P.; Banfi, G. Cortisol, testosterone, and free testosterone in athletes performing a marathon at 4000 m altitude. *Horm. Res.* **1994**, *41*, 225–229. [CrossRef] [PubMed]
12. Lippi, G.; Dipalo, M.; Buonocore, R.; Gnocchi, C.; Aloe, R.; Delsignore, R. Analytical Evaluation of Free Testosterone and Cortisol Immunoassays in Saliva as a Reliable Alternative to Serum in Sports Medicine. *J. Clin. Lab. Anal.* **2016**, *30*, 732–735. [CrossRef] [PubMed]
13. Tsunekawa, K.; Shoho, Y.; Ushiki, K.; Yanagawa, Y.; Matsumoto, R.; Shimoda, N.; Aoki, T.; Yoshida, A.; Nakajima, K.; Kimura, T.; et al. Assessment of exercise-induced stress via automated measurement of salivary cortisol concentrations and the testosterone-to-cortisol ratio: A preliminary study. *Sci. Rep.* **2023**, *13*, 14532–14541. [CrossRef] [PubMed]
14. Cupka, M.; Sedliak, M. Hungry runners-low energy availability in male endurance athletes and its impact on performance and testosterone: Mini-review. *Eur. J. Transl. Myol.* **2023**, *33*, 11104–11112. [CrossRef]
15. Muscella, A.; My, G.; Okba, S.; Zangla, D.; Bianco, A.; Marsigliante, S. Effects of training on plasmatic cortisol and testosterone in football female referees. *Physiol. Rep.* **2022**, *10*, e15291–e15303. [CrossRef]
16. Muscella, A.; Stefàno, E.; Marsigliante, S. The effects of training on hormonal concentrations and physical performance of football referees. *Physiol. Rep.* **2021**, *9*, e14740–e14750. [CrossRef]
17. Handelsman, D.J.; Jones, G.; Kouzios, D.; Desai, R. Evaluation of testosterone, estradiol and progesterone immunoassay calibrators by liquid chromatography mass spectrometry. *Clin. Chem. Lab. Med.* **2023**, *61*, 1612–1618. [CrossRef]
18. Herati, A.S.; Cengiz, C.; Lamb, D.J. Assays of Serum Testosterone. *Urol. Clin. N. Am.* **2016**, *43*, 177–184. [CrossRef]
19. Kanakis, G.A.; Tsametis, C.P.; Goulis, D.G. Measuring testosterone in women and men. *Maturitas* **2019**, *125*, 41–44. [CrossRef]
20. Chen, H.X.; Zhang, X.X. Antibody development to testosterone and its application in capillary electrophoresis-based immunoassay. *Electrophoresis* **2008**, *29*, 3406–3413. [CrossRef]
21. Luque-Ramírez, M.; Jiménez-Mendiguchia, L.; García-Cano, A.; Fernández-Durán, E.; de Dios Rosa, V.; Nattero-Chávez, L.; Ortiz-Flores, A.E.; Escobar-Morreale, H.F. Certified testosterone immunoassays for hyperandrogenaemia. *Eur. J. Clin. Investig.* **2018**, *48*, e13029. [CrossRef] [PubMed]
22. Keevil, B.G.; Adaway, J. Assessment of free testosterone concentration. *J. Steroid Biochem. Mol. Biol.* **2019**, *190*, 207–211. [CrossRef] [PubMed]
23. Chen, Y.; Teo, H.L.; Liu, H.; Loh, T.P.; Liu, Q.; Teo, T.L.; Lee, T.K.; Sethi, S.K. Simple and accurate candidate reference measurement procedure for total testosterone in human serum by one-step liquid-liquid extraction coupled with isotope dilution mass spectrometry. *Anal. Bioanal. Chem.* **2019**, *411*, 7519–7528. [CrossRef]
24. Botelho, J.C.; Shacklady, C.; Cooper, H.C.; Tai, S.S.; Van Uytfanghe, K.; Thienpont, L.M.; Vesper, H.W. Isotope-dilution liquid chromatography-tandem mass spectrometry candidate reference method for total testosterone in human serum. *Clin. Chem.* **2013**, *59*, 372–380. [CrossRef]
25. Sun, G.; Xue, J.; Li, L.; Li, X.; Cui, Y.; Qiao, B.; Wei, D.; Li, H. Quantitative determination of human serum testosterone via isotope dilution ultra-performance liquid chromatography tandem mass spectrometry. *Mol. Med. Rep.* **2020**, *22*, 1576–1582. [CrossRef]
26. Zhou, H.; Wang, Y.; Gatcombe, M.; Farris, J.; Botelho, J.C.; Caudill, S.P.; Vesper, H.W. Simultaneous measurement of total estradiol and testosterone in human serum by isotope dilution liquid chromatography tandem mass spectrometry. *Anal. Bioanal. Chem.* **2017**, *409*, 5943–5954. [CrossRef]
27. Wang, Y.; Gay, G.D.; Botelho, J.C.; Caudill, S.P.; Vesper, H.W. Total testosterone quantitative measurement in serum by LC-MS/MS. *Clin. Chim. Acta* **2014**, *436*, 263–267. [CrossRef]
28. Shi, J.; Bird, R.; Schmeling, M.W.; Hoofnagle, A.N. Using mass spectrometry to overcome the longstanding inaccuracy of a commercially-available clinical testosterone immunoassay. *J. Chromatogr. B. Analyt Technol. Biomed. Life Sci.* **2021**, *1183*, 122969–122983. [CrossRef]
29. Wang, C.; Catlin, D.H.; Demers, L.M.; Starcevic, B.; Swerdloff, R.S. Measurement of total serum testosterone in adult men: Comparison of current laboratory methods versus liquid chromatography-tandem mass spectrometry. *J. Clin. Endocrinol. Metab.* **2004**, *89*, 534–543. [CrossRef]
30. Martínez-Escribano, A.; Maroto-García, J.; Ruiz-Galdón, M.; Barrios-Rodríguez, R.; Álvarez-Millán, J.J.; Cabezas-Sánchez, P.; Plaza-Andrades, I.; Molina-Vega, M.; Tinahones, F.J.; Queipo-Ortuño, M.I.; et al. Measurement of Serum Testosterone in Nondiabetic Young Obese Men: Comparison of Direct Immunoassay to Liquid Chromatography-Tandem Mass Spectrometry. *Biomolecules* **2020**, *10*, 1697. [CrossRef]
31. French, D. Development and validation of a serum total testosterone liquid chromatography-tandem mass spectrometry (LC-MS/MS) assay calibrated to NIST SRM 971. *Clin. Chim. Acta* **2013**, *415*, 109–117. [CrossRef] [PubMed]
32. French, D.; Drees, J.; Stone, J.A.; Holmes, D.T.; van der Gugten, J.G. Comparison of four clinically validated testosterone LC-MS/MS assays: Harmonization is an attainable goal. *Clin. Mass Spectrom.* **2018**, *11*, 12–20. [CrossRef]

33. Star-Weinstock, M.; Dey, S. Development of a CDC-certified total testosterone assay for adult and pediatric samples using LC-MS/MS. *Clin. Mass Spectrom.* **2019**, *13*, 27–35. [CrossRef]
34. Huang, R.; Hong, Y.; Wu, Y.; Li, W.; Liu, W. Simultaneous quantification of total and free testosterone in human serum by LC-MS/MS. *Anal. Bioanal. Chem.* **2023**, *415*, 6851–6861. [CrossRef]
35. Rhea, J.M.; French, D.; Molinaro, R.J. Direct total and free testosterone measurement by liquid chromatography tandem mass spectrometry across two different platforms. *Clin. Biochem.* **2013**, *46*, 656–664. [CrossRef]
36. Fiers, T.; Wu, F.; Moghetti, P.; Vanderschueren, D.; Lapauw, B.; Kaufman, J.M. Reassessing Free-Testosterone Calculation by Liquid Chromatography-Tandem Mass Spectrometry Direct Equilibrium Dialysis. *J. Clin. Endocrinol. Metab.* **2018**, *103*, 2167–2174. [CrossRef]
37. Cibotaru, D.; Celestin, M.N.; Kane, M.P.; Musteata, F.M. Comparison of liquid-liquid extraction, microextraction and ultrafiltration for measuring free concentrations of testosterone and phenytoin. *Bioanalysis* **2022**, *14*, 195–204. [CrossRef]
38. Klock, E.; Kane, M.P.; Musteata, F.M. Measurement of free fraction, total concentration and protein binding for testosterone, triiodothyronine and thyroxine. *Bioanalysis* **2023**, *15*, 1355–1368. [CrossRef]
39. Zhang, Y.; Sun, T.; Jiang, C. Biomacromolecules as carriers in drug delivery and tissue engineering. *Acta. Pharm. Sin. B* **2018**, *8*, 34–50. [CrossRef]
40. Hammond, G.L.; Smith, C.L.; Goping, I.S.; Underhill, D.A.; Harley, M.J.; Reventos, J.; Musto, N.A.; Gunsalus, G.L.; Bardin, C.W. Primary structure of human corticosteroid binding globulin, deduced from hepatic and pulmonary cDNAs, exhibits homology with serine protease inhibitor. *Proc. Natl. Acad. Sci. USA* **1987**, *84*, 5153–5157. [CrossRef]
41. Luo, Z.; Lei, H.; Sun, Y.; Liu, X.; Su, D.F. Orosomucoid, an acute response protein with multiple modulating activities. *J. Physiol. Biochem.* **2015**, *71*, 329–340. [CrossRef]
42. European Medicines Agency. ICH M10 on Bioanalytical Method Validation—Scientific Guideline. Available online: https://www.ema.europa.eu/en/ich-m10-bioanalytical-method-validation-scientific-guideline (accessed on 27 September 2024).
43. U.S. Food & Drug Administration (FDA). Bioanalytical Method Validation Guidance for Industry. Available online: https://www.fda.gov/regulatory-information/search-fda-guidance-documents/bioanalytical-method-validation-guidance-industry (accessed on 27 September 2024).
44. Ryska, M. How to deal with the "matrix effect" as an unavoidable phenomenon. *Eur. J. Mass Spectrom.* **2015**, *21*, 423–432. [CrossRef]
45. Van Eeckhaut, A.; Lanckmans, K.; Sarre, S.; Smolders, I.; Michotte, Y. Validation of bioanalytical LC-MS/MS assays: Evaluation of matrix effects. *J. Chromatogr. B. Analyt. Technol. Biomed. Life Sci.* **2009**, *877*, 2198–2207. [CrossRef]
46. Stahnke, H.; Kittlaus, S.; Kempe, G.; Alder, L. Reduction of matrix effects in liquid chromatography-electrospray ionization-mass spectrometry by dilution of the sample extracts: How much dilution is needed? *Anal. Chem.* **2012**, *84*, 1474–1482. [CrossRef]
47. Furey, A.; Moriarty, M.; Bane, V.; Kinsella, B.; Lehane, M. Ion suppression; a critical review on causes, evaluation, prevention and applications. *Talanta* **2013**, *115*, 104–122. [CrossRef]
48. Meeusen, R.; Duclos, M.; Foster, C.; Fry, A.; Gleeson, M.; Nieman, D.; Raglin, J.; Rietjens, G.; Steinacker, J.; Urhausen, A. Prevention, diagnosis, and treatment of the overtraining syndrome: Joint consensus statement of the European College of Sport Science and the American College of Sports Medicine. *Med. Sci. Sports Exerc.* **2013**, *45*, 186–205. [CrossRef]
49. Hooper, D.R.; Kraemer, W.J.; Focht, B.C.; Volek, J.S.; DuPont, W.H.; Caldwell, L.K.; Maresh, C.M. Endocrinological Roles for Testosterone in Resistance Exercise Responses and Adaptations. *Sports Med.* **2017**, *47*, 1709–1720. [CrossRef]
50. D'Andrea, S.; Spaggiari, G.; Barbonetti, A.; Santi, D. Endogenous transient doping: Physical exercise acutely increases testosterone levels-results from a meta-analysis. *J. Endocrinol. Investig.* **2020**, *43*, 1349–1371. [CrossRef]
51. Fahrner, C.L.; Hackney, A.C. Effects of endurance exercise on free testosterone concentration and the binding affinity of sex hormone binding globulin (SHBG). *Int. J. Sports Med.* **1998**, *19*, 12–15. [CrossRef]

molecules

MDPI

Article

Determination of Cotinine, 3′-Hydroxycotinine and Nicotine 1′-Oxide in Urine of Passive and Active Young Smokers by LC-Orbitrap-MS/MS Technique

Magdalena Szumska [1,2,*], Paweł Mroczek [1], Krystyna Tyrpień-Golder [1], Beata Pastuszka [2] and Beata Janoszka [1,2,*]

1 Department of Chemistry, Faculty of Medical Sciences, Medical University of Silesia, 40-752 Katowice, Poland; d201099@365.sum.edu.pl (P.M.); ktyrpien@sum.edu.pl (K.T.-G.)

2 Research and Implementation Center Silesia LabMed, Medical University of Silesia, 40-752 Katowice, Poland; betty.p2308@gmail.com

* Correspondence: mszumska@sum.edu.pl (M.S.); bjanoszka@sum.edu.pl (B.J.); Tel.: +48-2755188 (M.S.)

Abstract: Tobacco smoke is probably the most significant factor conducing to toxic xenobiotics exposure to humans. The aim of the study was to develop a rapid and sensitive method for the determination of selected nicotine metabolites in urine of tobacco smokers and passive smokers. The method for removing protein and extracting the metabolites involved the centrifugation of urine with acetonitrile. Cotinine, trans-3′-hydroxycotinine, and (2′S)-nicotine 1′-oxide in the supernatant were determined using the LC-Orbitrap-MS/MS technique, with the selected ion monitoring (SIM) and parallel reaction monitoring (PRM) modes used. The recovery of these analytes added to the urine samples ranged from 72% to 101%. Repeatability and reproducibility were less than 3.1% and 10.1%, respectively. The study was carried out among medical students. The group was selected as representatives of young people and who as future physicians should be more aware of the effects of nicotine use. Concentration levels of cotinine and *trans*-3′-hydroxycotinine determined in ng/mL in the urine of cigarette smokers were 70- and 58-fold higher, respectively, compared to passive smokers. Higher concentrations were recorded in the urine of those passively exposed to tobacco smoke than in non-smokers, confirming that passive exposure to tobacco smoke is not harmless to the human body. However, no significant differences were observed in the concentration of (1′S,2′S)-nicotine 1′-oxide in the samples of individuals from various groups.

Keywords: nicotine metabolites; cotinine; passive-smoking; urine; liquid chromatography; mass spectrometry

1. Introduction

Tobacco consumption is one of the most serious public health problems, causing a multitude of chronic diseases such as cancer, cardiovascular disease, and chronic obstructive pulmonary disease. Smoking one cigarette reduces life expectancy by approximately 11 minutes. Cigarette smoke contains over 5000 different chemicals, among which 63 are carcinogenic, as reported by the International Agency for Research on Cancer (IARC); moreover, 11 of these are known human carcinogens, increasing the risk of numerous diseases affecting many organs [1–3].

Nicotine is one of the main addictive constituents of tobacco and the main component of cigarette smoke. However, it is not a carcinogen, but it is still a highly toxic compound with an LD_{50} within the range of 6.5 to 13 mg/kg of body weight. Since nicotine has a relatively short half-life (1–2 h), it is not a useful biomarker for assessing tobacco smoke exposure, especially when estimating the impact of second-hand smoke [4].

Biomarkers of tobacco smoke exposure are considered indicators of the absorbed dose. They can provide objective measures to assess the impact of tobacco products' impact on the general population and their health consequences. In this group of biomarkers, components of tobacco smoke or their biotransformation products and adducts with DNA and proteins, including thiocyanates, carbon monoxide or its adduct with hemoglobin, nicotine and its metabolites, and hydroxy-PAHs (including 1-hydroxypyrene or acrylamide metabolites) can be found [5].

Nicotine metabolites with longer half-lives are most often used as biomarkers of exposure to tobacco products and can be detected in different biological fluids such as plasma, serum, urine, and saliva [1,4,6,7]. The total amount of absorbed nicotine metabolized to cotinine ranges from 70–80%, and approximately 10% of this alkaloid is excreted in urine. Nicotine is mainly metabolized by the P450 enzyme system, which leads to the formation of six primary metabolites: cotinine, *trans*-3′-hydroxycotinine, norcotinine, nornicotine, cotinine oxide, and nicotine oxides [3,8].

Cotinine is the most widely analyzed biomarker of exposure to tobacco smoke and has been successfully used for the study of smoking habits and passive exposure, making it possible to distinguish smokers from non-smokers and passive smokers. Cotinine has a fairly long elimination time from the body and can be detected within days after exposure to tobacco smoke. According to the literature, the biological half-life of cotinine ranges from 8 to even 30 h [2,8,9]. Concentrations of cotinine in urine, saliva, and blood samples are strongly associated with nicotine exposure [10].

Cotinine is the primary metabolite of nicotine. However, it has been estimated that only approximately 10% to 15% of cotinine is found in smokers' urine, as most cotinine is converted into other metabolites in the process of hydroxylation to *trans*-3′-hydroxycotinine. *Trans*-3′-hydroxycotinine has an average half-life of 4 to over 8 h and is another suitable biomarker of exposure to nicotine. Its half-life is similar in both blood and urine samples. *Trans*-3′-hydroxycotinine accounts for approximately up to 40% of the absorbed nicotine dose [2,3]. Measuring the ratio of cotinine to *trans*-3′-hydroxycotinine makes it possible to investigate the metabolism of nicotine. Research has confirmed the usefulness of determining the molar ratio of these two metabolites as a marker of CYP2A6 isoenzyme activity, which determines the rate of metabolization of nicotine [11,12]. Another primary nicotine metabolite is nicotine -N-oxide. It is estimated that approximately up to 7% of nicotine is metabolized to nicotine-N-oxide. This metabolite can be reduced back to nicotine in the digestive tract [2]. Biomarkers of tobacco smoke exposure are considered indicators of the absorbed dose and, according to the current state of knowledge, they can be components of tobacco smoke or their biotransformation products and adducts with DNA and proteins [2,4]. They can be detected and analyzed in different biological matrices such as blood, urine, saliva, and hair [1,7,13,14]. Due to the invasive nature of blood sampling, other non-invasively collected biological samples are often used with great success. For example, urine samples are much easier to collect and contain higher concentrations of nicotine metabolites compared to other biological fluids. This property also makes urine samples the most sensitive matrix for evaluating second-hand tobacco exposure [1,3,13].

Over the years, many analytical methods have been developed for the determination of nicotine metabolites. The most frequently applied methods for the determination of tobacco smoke exposure biomarkers are gas chromatography (GC) and liquid chromatography (LC) coupled with mass spectrometry (MS), but immunological methods can also be found [2,4,6,15]. Immunochemical tests can be used for screening. Despite their high sensitivity, the specificity of these methods is low. For example, cotinine quantification results may be overestimated due to cross-reactions with other nicotine metabolites [16,17].

Chromatographic techniques, especially in combination with mass spectrometry, allow the separation and determination of various metabolites of nicotine due to their high specificity. Before introducing mass spectrometry to chromatographic analysis, different detectors were used for biomonitoring. GC with flame ionization or the more sensitive and specific nitrogen–phosphorus detector (NPD) was applied [16,18,19].

High performance liquid chromatography (HPLC) with an ultraviolet absorption detector has been used to detect nicotine and its metabolites in biological fluids but is now increasingly being replaced by LC methods connected with mass spectrometry [6,16,20].

Biological samples require some pretreatment steps, which include solid-phase extraction (SPE), acidic precipitation, or simple liquid–liquid extraction. The cleanup procedures vary depending on the chosen biological fluid, as these samples have different matrix content that can influence the determination process. However, when it comes to noninvasive samples like urine, the most common method to remove proteins/salts is organic solvent precipitation using acetonitrile or methanol [3,15,21]. Some authors use acetone precipitation pretreatment for urine samples pointing to lower toxicity and faster evaporation of that solvent compared to acetonitrile, water, or methanol. Recovery rates for sample pretreatment procedures vary depending on the procedure used. For solid phase extraction (SPE), they are in the range of 57 to 124%; for liquid–liquid extraction, they range from 53 to 117%, and organic solvent precipitation leads to recoveries in the range of 89 to 105% [3,15,21].

Efficiently examining a large number of samples to study smoke exposure in the general population requires a simple and fast sample preparation and analysis procedure. Although the GC-MS technique is described as a sensitive one, the running time of analysis is much longer (15–30 min) compared to LC-MS (6–14 min). LC-MS/MS is the most promising technique for nicotine metabolite analysis as it allows the determination of a wide range of analytes regardless of their properties. It enables adequate analytical separation and better-quality mass spectral data, leading to enhanced detection limits. For nicotine metabolite analysis, a triple quadrupole MS system is mostly chosen, and the determination of target compounds is performed under multiple reaction monitoring mode with positive electrospray ionization using a C18 or HILIC analytical column [22–24]. In analytical procedures for elution, an isocratic or gradient flow of water and acetonitrile with formic acid (for C18 column methods) or ammonium acetate (for HILIC column methods) is most commonly used. The column temperature between 30 and 40 °C is applied in most LC-MS methods to avoid peak shifting [8,25]. Although the methods used for the determination of nicotine metabolites in biological samples such as urine are sensitive, their quantification can present analytical difficulties. One of the analytical obstacles is the matrix effect. To minimize the matrix effect of biological samples on the results of the analysis, the use of isotopically labeled internal standards is recommended [4,22].

Using LC-MS/MS techniques in monitoring tobacco smoke health hazards makes it possible to detect even very low levels of exposure, giving scientists a tool to verify environmental exposure to tobacco connected to passive smoking. Since urine accumulates higher concentrations of nicotine metabolites, it may be the most sensitive matrix in the assessment of passive smoking [3,16]. Passive smoking is estimated to be the cause of approximately 1.0% of worldwide mortality, and this number is increasing every year. It is of extreme importance to implement the sensitive "golden standard" analytical technique to monitor second-hand smoke exposure [3].

Surveys conducted among medical and emergency medicine students have shown that many young adults in this particular group smoke cigarettes and use e-cigarettes as substitutes for traditional cigarettes. Particularly disturbing is the fact that students do not avoid exposure to tobacco smoke, and willingly as well as frequently stay in rooms where cigarettes are smoked by others. This phenomenon is likely due to the demanding nature of medical studies and the need to solve scientific problems in teams, which is a feature of the profession they are being prepared for. Furthermore, another concern is the results of the questionnaire which show that many medical students have limited knowledge of the harmful compounds formed during smoking. Moreover, staying in one room with tobacco smokers is not considered a significant source of exposure to such compounds. These observations were the reason for undertaking research among medical students to prove to them that nicotine metabolites are present not only in the bodies of tobacco smokers, but also in persons passively exposed to tobacco smoke, leading to further health consequences [26].

The aim of the study was to develop a rapid and sensitive method for the determination of selected nicotine metabolites in urine, which could be applied to assess the level of exposure to harmful compounds in both tobacco smokers and passive smokers.

2. Results and Discussion

2.1. Determination of Nicotine Metabolites by LC-Orbitrap-MS/MS

The determination of the three main nicotine metabolites-Cot, 3′-OH-Cot, and 2′S-Nic-Ox-in urine were chosen to evaluate the study participants' exposure to tobacco smoke. The formulas of these compounds are shown in Table 1.

Table 1. Structures, names, and molecular formulas of the compounds used in this study.

Name (Abbreviation)	Structure	Molecular Formula	Molecular Mass
(-)-Cotinine (Cot)	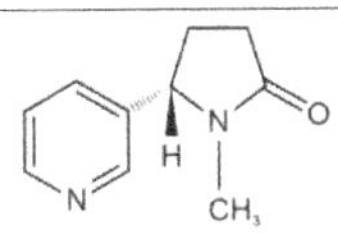	$C_{10}H_{12}N_2O$	176.09496
trans-3′-Hydroxycotinine (3′-OH-Cot)	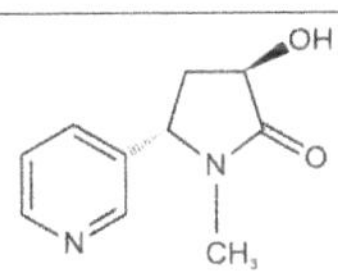	$C_{10}H_{12}N_2O_2$	192.08988
(1′S,2′S)-nicotine 1′-oxide (2′S-Nic-Ox)	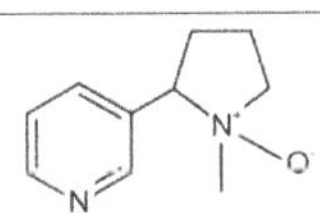	$C_{10}H_{14}N_2O$	178.110161
DL-nicotine-(methyl-D3) (Nic-D3)	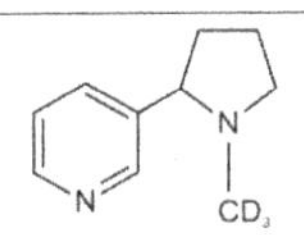	$C_{10}D_3H_{11}N_2$	165.134528

Using an LC-Orbitrap-MS/MS system equipped with a C-18 analytical column and isocratic elution with an acetonitrile-water solution containing 0.1% HCOOH allowed the components of the standard mixture to be separated and detected in less than 3 min.

All of these molecules are nitrogen-containing heterocyclic compounds. In the presence of acid, they undergo protonation and exist as cationic forms, so determinations by LC technique coupled with mass spectrometry can be carried out in positive ionization mode [27].

Initially, the results of chromatographic separation of nicotine metabolites were obtained in selected ion monitoring (SIM) mode, taking into account the molecular weights of analytes increased by 1 (due to the formation of protonated ions). Although the separation results for the standard mixture allowed for the identification and quantification of the analytes, the results for urine sample extracts indicated potential problems with the identification of nicotine metabolites due to the presence of many other components with similar retention times in biological samples. Supplement 1 shows the chromatogram of the standard mixture (Figure S1A), and chromatograms of urine extracts (Figure S1B–D) recorded using the LC-Orbitrap-MS/MS system in SIM mode. Using this method of determining nicotine metabolites could lead to false positive results.

Therefore, another method was developed for the determination of Cot, 3′-OH-Cot, and 2′S-Nic-Ox using parallel reaction monitoring (PRM) mode. In this method, the highly accurate (with a resolution accuracy of 0.000001 m/z) masses of molecular precursor ions and the formed product ions were taken into account. The collision energies for obtaining the appropriate product ions from the protonated precursor molecular ions $[M+H]^+$ for the nicotine metabolites to be determined are shown in Table 2.

Table 2. Parameters for the analysis of nicotine metabolites and internal standard by LC-Orbitrap-MS/MS in parallel reaction monitoring (PRM) mode.

Compound	Retention Time [min]	Precursor Ion $[M+H]^+$ → Product Ion (*m/z*)	Collision Energy (eV)
Cot	1.85	177.10240 → 80.04999	20
3′-OH-Cot	1.72	193.09727 → 80.04999	25
2′S-Nic-Ox	1.68	179.11810 → 84.08128	15
Nic-D3	1.72	166.14189 → 130.06540	25

The chromatograms obtained as a result of the separation of standard mixture components using the LC-Orbitrap-MS/MS system in PRM mode are shown in Figure 1A. Mass spectra recorded during such analysis are presented in Figure 1B. The determination of Cot, 3′-OH-Cot, and 2′S-Nic-Ox was carried out with the addition of DL-nicotine-(methyl-D3) as an internal standard (IS). The retention time of this compound was the same as Cot, but due to its different molecular precursor mass and product ion mass, Nic-D3 could be used as an IS for the determination of the selected three nicotine metabolites in the standard mixture and urine sample extracts.

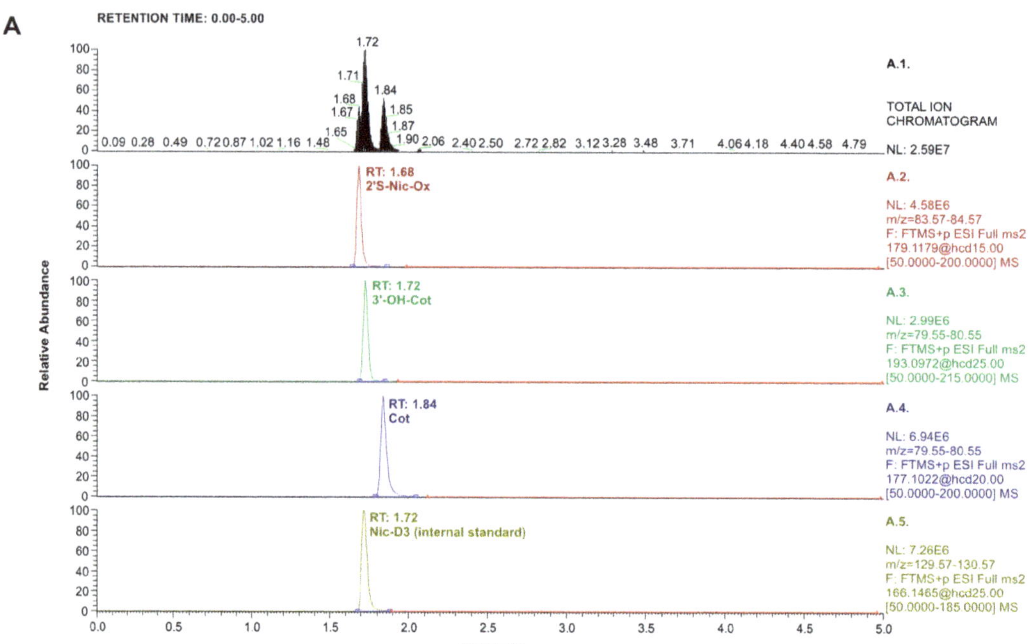

Figure 1. *Cont.*

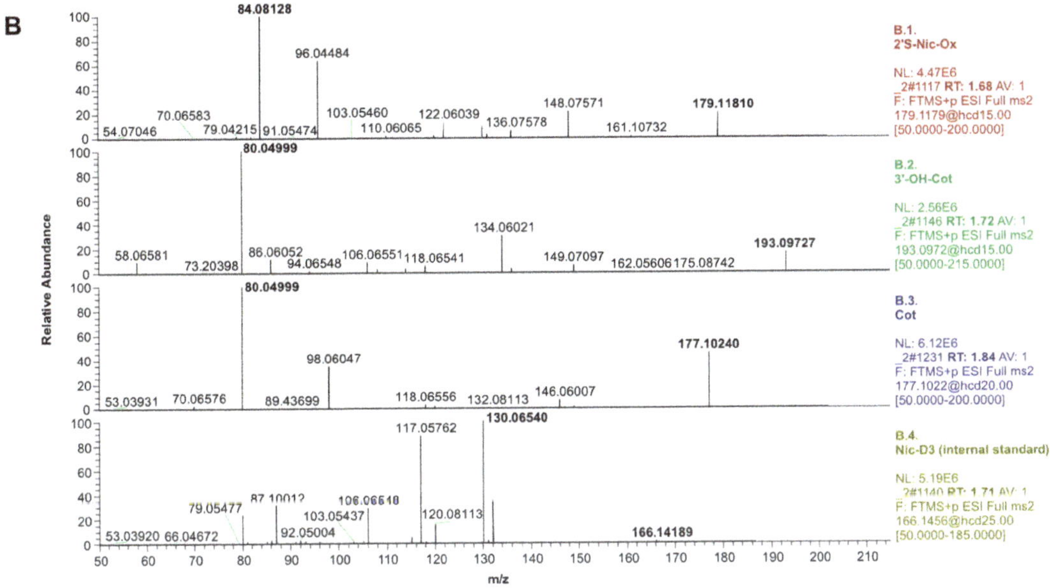

Figure 1. (**A**): LC-Orbitrap-MS/MS chromatograms recorded in total ion mode (top window) and PRM mode (other windows) for nicotine metabolite standards and internal standards (concentration of 0.25 μg/mL; injection: 1 μL). (**B**): Mass spectra recorded in PRM mode of the standards used in the study, shown from the top in order as in Figure 1A, i.e. 2′S-Nic-Ox, 3′-OH-Cot, Cot, and Nic-D3 (internal standard). The appropriate collision energies are presented in Table 2.

Methods for the determination of nicotine metabolites using LC-MS/MS with a tandem or triple quadrupole mass spectrometer often involve monitoring the most intense product ions formed from a protonated molecular ion [4,8,13,21,27]. In such LC-MS/MS systems, this mode of determination is called selected reaction monitoring (SRM) [4,28] or multiple reaction monitoring (MRM) [13,21,25,27,29]. These methods have been used to determine nicotine metabolites, including Cot, 3′-OH-Cot in urine [8,13,25,28,29] and other biological samples, such as plasma [6,27] and saliva [30].

The collision energies of the transitions from precursor ions to product ions in MRM or SRM methods are usually similar to the optimal ones chosen for LC-Orbitrap-MS/MS in our work, although they differ by a few volts in some publications [4,8,13,25,28,29]. The advantage of LC-Orbitrap-MS/MS, however, is that the system allows monitoring of ion masses with high accuracy to 0.000001 m/z.

2.2. Determination of Nicotine Metabolites in Urine Samples

The procedure used to prepare urine samples for the determination of nicotine metabolites included vortexing 0.5 mL of urine with 1 mL of acetonitrile followed by centrifugation of the solution. This stage resulted in protein precipitation. This fast and simple method of sample preparation is commonly used in procedures for determining components of biological samples using sensitive and selective methods such as LC coupled with mass spectrometry [6,7,27,31]. Methanol [27], acetone [4], or aqueous solutions of these solvents are sometimes used instead of acetonitrile [6,7,15,27].

To determine the recovery of the extraction process of nicotine metabolites from urine, a sample of non-smoker urine spiked with standards at three concentration levels (Table 3) was analyzed. This urine sample originally did not contain the analytes selected for testing (Cot, 3′-OH-Cot, and 2′S-Nic-Ox). Chromatograms recorded during the analysis of the non-smoker urine sample without the addition of standards, but with the addition of IS (before the extraction procedure), are shown in Figure 2.

Table 3. Data from the determination of nicotine metabolites by LC-Orbitrap-MS/MS in PRM mode.

Nicotine Metabolite	Calibration Curve Concentration Range (μg/mL)	Regression Coefficients *r*	Recovery (%) and RSD [1] (%) for Spiking Level (μg/mL of Urine); n = 6		
			0.05	0.5	1.0
Cot	0.03–1.00	0.9980	101.4 (1.0)	82.2 (1.0)	81.1 (1.1)
3′-OH-Cot	0.03–1.00	0.9965	100.9 (2.0)	85.6 (1.0)	83.6 (1.0)
2′S-Nic-Ox	0.03–1.00	0.9979	82.8 (1.6)	75.0 (3.1)	72.2 (2.5)

[1] RSD—relative standard deviation.

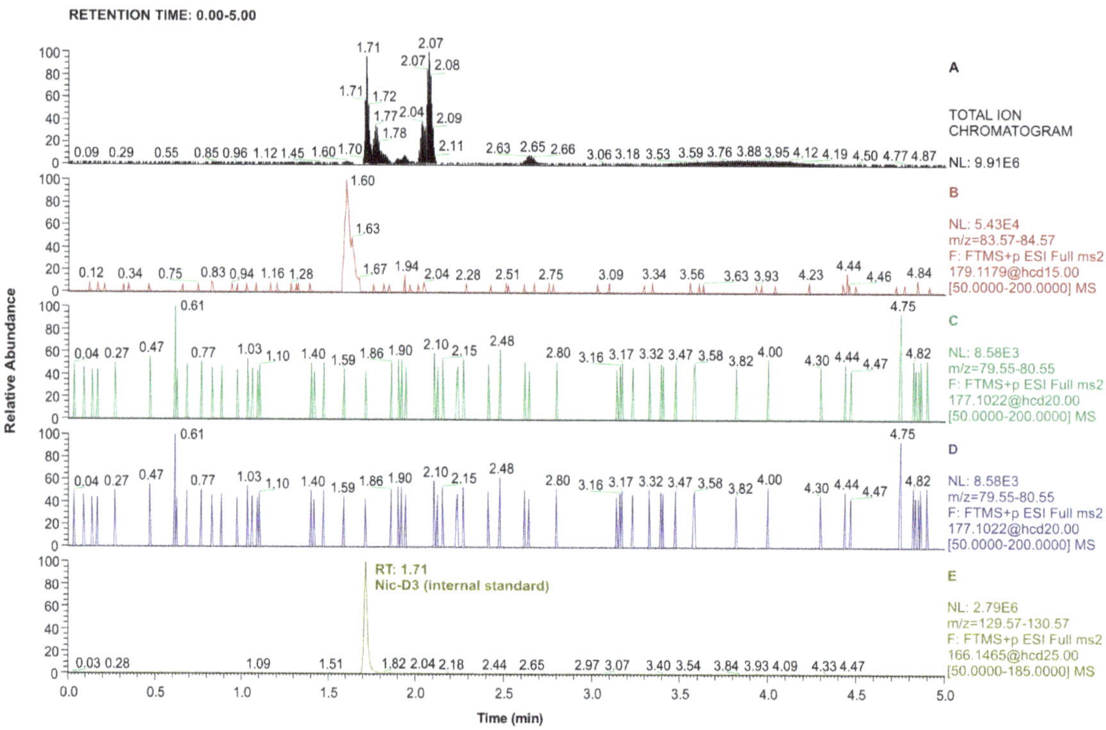

Figure 2. LC-Orbitrap-MS/MS chromatograms recorded in total ion mode (top window) and PRM mode (consecutive windows from the top corresponding to 2′S-Nic-Ox, 3′-OH-Cot, Cot, and Nic-D3) for the urine extract of a non-smoker. Nic-D3 concentration was 0.3 μg/mL (injection volume: 1 μL).

Recovery rates determined from this experiment, shown in Table 3, ranged from 72.2% to 101.4%. These recovery rates are similar to those reported in other studies on the determination of Cot and 2′S-Nic-Ox in urine samples using LC-MS/MS [7,15,22].

The mean concentrations of nicotine metabolites, Cot, 3′-OH-Cot, and 2′S-Nic-Ox, determined in urine samples in our study ranged from: not detected (Cot, non-smokers) to 3819 ng/mL of urine (Cot, smokers). A total of 129 students agreed to take part in the study; 27 of them smoked cigarettes, and 65 admitted to being passive smokers. According to the questionnaire data filled out by the study participants, the remaining individuals (n = 37) did not smoke cigarettes and were not exposed to tobacco smoke either. The results of Cot, 3′-OH-Cot, and 2′S-Nic-Ox determinations in urine samples are shown in Table 4.

The data correspond to the average of three LC-MS/MS analyses conducted for extracts obtained in duplicate for each urine sample.

Table 4. Nicotine metabolites concentration (ng/mL of urine)

Nicotine Metabolite	Status of Smoking		
	Smokers (Min–Max) [1]	**Passive Smokers (Min–Max)**	**Non-Smokers (Min–Max)**
Cot	3819.0 (639.0–12,540.0)	54.6 (n.d.–94.1)	n.d. [2]
3′-OH-Cot	2967.0 (261.0–7080.0)	51.1 (36.0–137.5)	35.4 (0–36.0)
2′S-Nic-Ox	55.4 (n.d.–90.0)	53.4 (n.d.–77.8)	53.7 (0–54.0)
Cot + 3′-OH-Cot	6786.0	105.7	35.4

[1] Concentration range (from minimum to maximum concentration determined); [2] n.d. -not determined (below the limit of quantification (LOQ).

The concentrations of Cot and 3′-OH-Cot determined in the urine of cigarette smokers were 70- and 58-fold higher, respectively, compared to passive smokers. On the other hand, higher concentrations of Cot and 3′-OH-Cot were recorded in the urine of passively exposed individuals than in non-smokers, confirming that passive exposure to tobacco smoke is not harmless to the human body. However, no differences were observed in the averaged concentration of 2′S-Nic-Ox in the urine samples of the different groups of subjects. In numerous samples, even from cigarette smokers, this nicotine metabolite was not detected. Figure 3A shows an example of total and PRM-mass chromatograms recorded during the analysis of a smoker's urine sample. Figure 3B shows the mass spectra of Cot and 3′-OH-Cot determined in this sample. 2′S-Nic-Ox was not detected in this sample because the mass spectrum does not contain the characteristic product ion (m/z = 80.04999) formed from the precursor ion.

The mean urinary cotinine concentration of passive smokers was 55 ng/mL (Table 4). According to Kim, a range of urinary cotinine values of 50–200 ng/mL may serve as a cutoff point for verification of active smokers [32]. A previous study conducted by our team to determine the main metabolites of nicotine by ELISA technique also showed that in the group of active smokers, the concentration of the main metabolites of nicotine was higher than 200 µg/mL of urine. In the group of passive smokers, the concentration of main nicotine metabolites ranged from 20 to 200 µg/mL of urine, and in the group of non-exposed smokers, the concentration of major nicotine metabolites was lower than 20 µg/mL of urine [17].

High concentrations of nicotine metabolites were expected in students who smoked cigarettes. The total content of Cot and 3′-OH-Cot measured in urine was almost 6700 ng/mL in this group of subjects. These data correspond to values determined in urine samples by other authors [4,8,16,22,33–35]. In passive smokers, the mean summed concentration of these two nicotine metabolites determined at almost 106 ng/mL was three times higher than in non-smokers. It is worth noting that Cot was not determined in the urine of some passive smokers, while 3′-OH-Cot was present in these samples. A large-scale Korean National Health and Nutrition Examination Survey confirmed that cotinine may be absent from urine samples, although other nicotine metabolites were detected in them [33]. Previous work by our research group also confirmed that Cot may not be present in the urine of people exposed to tobacco smoke [20].

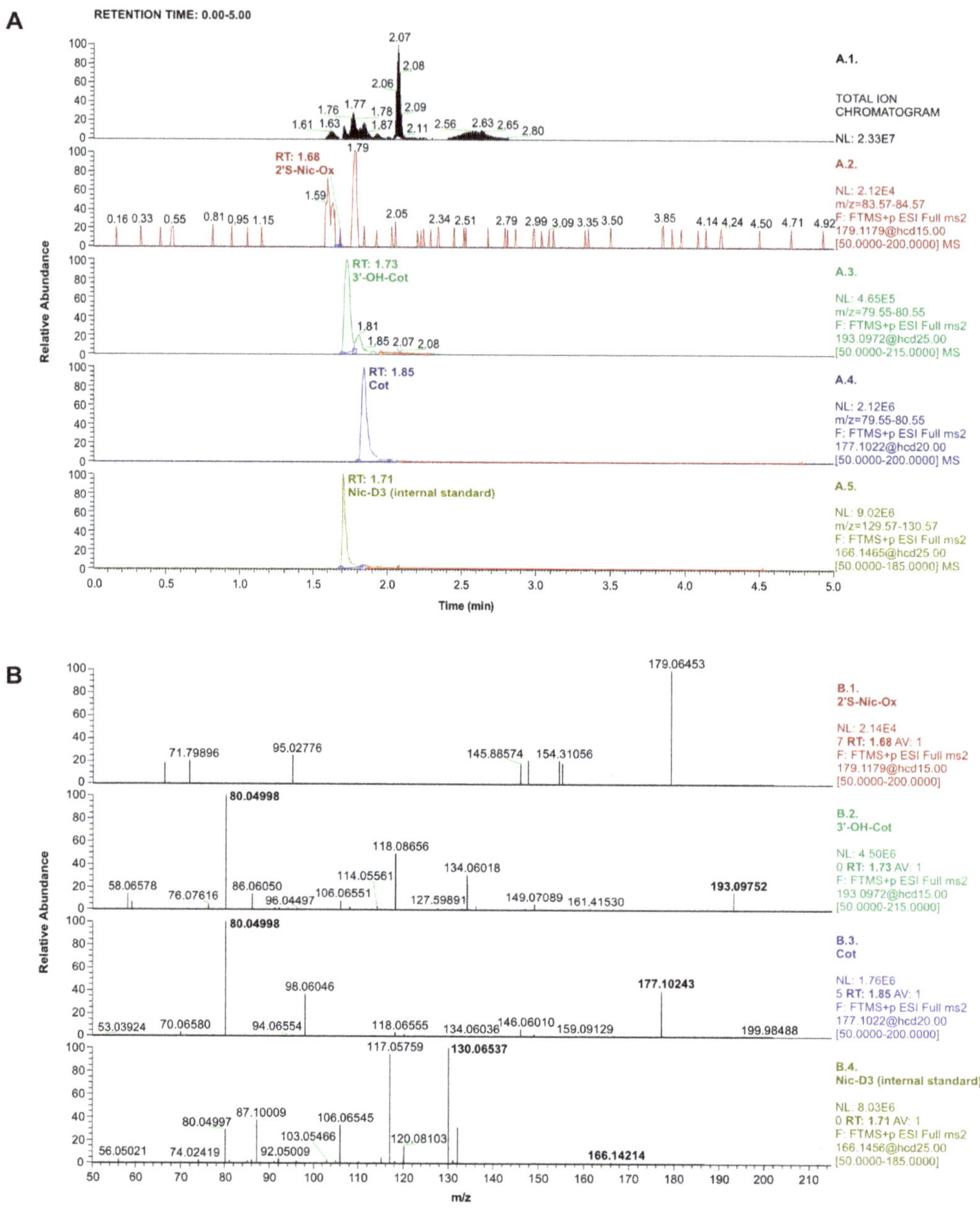

Figure 3. (**A**): LC-orbitrap-MS/MS chromatograms recorded for the smoker's urine extract in total ion mode (top window) and PRM mode (consecutive windows from the top corresponding to 2′S-Nic-Ox, 3′-OH-Cot, Cot, and Nic-D3). The concentration of Nic-D3 was 0.3 μg/mL (injection volume: 1 μL). (**B**): Mass spectra recorded in PRM mode for analytes determined in the smoker's urine extract. Consecutive windows from the top correspond to 2′S-Nic-Ox (not detected in this sample), 3′-OH-Cot, Cot, and Nic-D3 (internal standard). The corresponding collision energies are shown in Table 2.

The opportunity to demonstrate to young people, especially medical students, that smoking cigarettes in the presence of others can lead to the accumulation of harmful compounds in their bodies is perhaps the only way to prevent civilization diseases resulting from smoking in the future. Of even greater concern is the carefree approach to the use of e-cigarettes by increasingly younger people [27,35–37]. The developed easy procedure of nicotine metabolites isolation from biological samples combined with a sensitive, rapid, and reliable determination method of LC-Orbitrap-MS/MS makes it possible to perform such tests during laboratory classes with students.

Few examples are available in the literature of the use of this analytical system for the determination of nicotine metabolites in biological samples, where their concentrations may be low.

The LC-orbitrap-MS/MS method was validated and used (in the SIM mode) for the determination of nicotine metabolites, Cot and 2′S-Nic-Ox, in urine [7] as well as in human plasma, semen, and sperm [31]. The chromatographic separation parameters were analogous to ours. Due to the more complex matrix of plasma and sperm samples, the authors used a methanol solution with trichloroacetic acid for extraction combined with protein precipitation.

Kawasaki et al. used the LC-Orbitrap-MS/MS to determine Cot in the urine of quitting smokers [38] and Cot and 3′-OH-Cot in urine as biomarkers of exposure to secondhand smoke and heated tobacco products [39]. Due to the simultaneous determination of 4-(methyl-nitrosamino)-1-(3-pyridyl)-1-butanol, urine samples were subjected to β-glucuronidase treatment followed by liquid-solid extraction using columns containing diatomaceous earth. The resulting extracts, after evaporation and dissolution in acetonitrile containing 10 mM ammonium acetate, were analyzed by LC-orbitrap-MS/MS technique in PRM mode, to quantify the most intense product ion of each compound [38,39].

Taking advantage of the high selectivity sensitivity of LC-Orbitrap-MS/MS, a validated bioanalytical method was also developed for the simultaneous determination of nicotine and Cot in human blood based on the dry blood spot (DBS) technique after a single extraction step [40]. The calibration range established in this validated method for Cot was linear in the range of 0.010–0.500 μg/mL, which is similar to our work.

3. Materials and Methods

3.1. Reagents and Standards

Pierce™ LTQ Velos ESI Positive Ion Calibration Solution (Thermo Scientific, Rocford, IL, USA) was used to tune and calibrate the mass spectrometer.

Organic solvents: acetonitrile and methanol, both hypergrade for LC-MS, as well as water for LC-MS were purchased from Merck (Darmstadt, Germany). Formic acid for LC-MS LichropurTM used as a mobile phase component was obtained from Sigma-Aldrich (St. Louis, MO, USA).

Four analytical standards were used in the study: DL-nicotine-(methyl-D3) (purity 99%) and (-)-cotinine (purity 98%) which were purchased from Sigma-Aldrich (St. Louis, MO, USA), *trans*-3′-hydroxycotinine (solution, 1 mg/mL methanol) was from Supelco (Round Rock, TX, USA) and (1′S,2′S)-nicotine 1′-oxide bought from Toronto Research Chemicals (Toronto, ON, Canada). Structures and names abbreviation of these compounds are presented in Table 1.

Standard stock solutions of Cot, 3′-OH-Cot, and 2′S-Nic-Ox (each 1.0 mg/mL) were used to prepare a standard mixture of concentration 1 μg/mL in acetonitrile. By diluting this standard mixture, solutions at concentrations of 0.75, 0.50, 0.25, 0.125, 0.060, 0.030, 0.015, and 0.010 μg/mL were prepared and used to create calibration curves and establish detection limits. DL-nicotine-(methyl-D3) in acetonitrile was introduced into each solution as an internal standard, in an amount that referred to a final concentration of 0.30 μg/mL.

3.2. Urine Sample Collection and Extraction

Urine samples were collected from volunteers who were smokers, passively exposed to tobacco smoke, and non-smokers. The samples were centrifuged and stored in a freezer at −20 °C until analysis. The study was conducted among 129 first- and second-year medical and emergency medicine students. The mean age of the study group was 21.1 ± 1.3 years, 21.2 ± 1.4 years for women and 21.1 ± 1.2 years for men. Along with urine samples, survey data were collected using a self-prepared questionnaire designed to assess exposure to tobacco smoke. To conduct such research Approval of the bioethics committee has been obtained (No BNW/NWN/0052/KB1/81/23 from 12.09.2023).

To extract nicotine metabolites, 0.5 mL of urine sample was transferred to an Eppendorf tube and 1 mL of acetonitrile containing 0.45 μg of Nic-D3 (as internal standard) was added. The samples were vortexed for 1 min and then centrifuged for 10 min at 4000 rpm using laboratory centrifuge MPW-260 R (MPW med.Instruments, Warsaw, Poland). Extraction of each sample was carried out in two replicates. Each supernatant was transferred into a 2-mL autosampler vial and analyzed by LC-Orbitrap-MS/MS technique.

3.3. Determination of Nicotine Metabolites by LC-Orbitrap-MS/MS

A Vanquish Thermo Scientific liquid chromatograph assembled with a double split sampler, column thermostat compartment, and double pump (all from Thermo Scientific, Germering, Germany) was used. For the separation of urine extracts components an analytical column Accucore-C18 (150 × 3 mm I.D.; particle size 2.6 μm) from Thermoscientific (Lithuania) was applied. The separations were performed under isocratic conditions by using a mixture of acetonitrile, water, and formic acid (10:90:0.1, *v*/*v*/*v*) as a mobile phase at a constant flow rate of 0.3 mL/min. The temperature of the automatic sampler tray was set to 10 °C, and the temperature of the column thermostat was set to 30 °C. The injection volume of standard solutions and extracts was 1 μL.

The mass spectrometer system connected with the liquid chromatograph consisted of an Orbitrap mass spectrometer (Q Exactive Focus, Thermo Fisher Scientific, Bremen, Germany), equipped with a heated electrospray ionization (HESI) source. The system was controlled by Xcalibur LC Devices 3.2 Robust software.

The mass spectrometer operated in full MS-SIM and secondary mass (MS/MS) monitoring in positive ion mode for the analytes to be determined. The orbitrap was tuned and calibrated using the ESI positive ion calibration solution once a week.

The nitrogen gas for the ion source was produced by the generator Nigen LCMS 40-1 (Claind Brezza, Tremezzina, Italy). The sweep, auxiliary, and sheath gases flow were set at 5, 20, and 45 units, respectively. The ion spray voltage was set at 3.5 kV and the temperature of the ion transfer capillary was set at 310 °C. The MS resolution option was 17.5 at 2 Hz with a mass window of 0.4 Da for each analyte.

The determination of Cot, 3′-OH-Cot, and 2′S-Nic-Ox was performed in parallel reaction monitoring (PRM) mode. The optimal collision energies used to obtain product ions from precursor molecular ions $[M+H]^+$ corresponding to the maximum intensity for the analytes are shown in Table 2. These values were selected based on a series of analyses performed at different collision energies for standard solutions of a concentration of 0.25 μg/mL.

3.4. Calibration Graphs, Limits of Detection and Limits of Quantification, Repeatability and Reproducibility, and Recovery of Extraction

Quantitative analysis of Cot, 3′-OH-Cot, and 2′S-Nic-Ox was performed on the basis of linear calibration graphs recorded in the range of 0.03 to 1.0 μg/mL, for 1 μL of standard mixture injected into the column (Table 3). Each calibration solution contained Nic-D3 as the internal standard at the concentration of 0.3 μg/mL. The graphs were plotted on the ratio of the peak area recorded for the appropriate analyte to the peak area corresponding to IS, relative to the analyte concentration in the standard mixture. The regression coefficients R for the curves were above 0.996.

Limits of detection (LODs) were determined using the stepwise dilution method of standard solutions, taking them as a signal-to-noise (S/N) ratio of 3. A value of 3 times the limit of detection (LOD) was taken as the limit of quantification (LOQ) for an injection volume of 1 μL [41]. They were 0.01 and 0.03 μg/mL, respectively, for all compounds selected for the study, i.e. for Cot, 3′-OH-Cot, and 2′S-Nic-Ox.

The procedure to determine recovery levels was conducted using urine from a non-smoker. The urine was enriched with standards of nicotine metabolites at three concentration levels (i.e., 0.05, 0.5, and 1.0, with six replicates at each level), as well as the internal standard. The samples prepared this way were subjected to extraction, and then the analytes were determined by LC-orbitrap-MS/MS in PRM mode. Recovery values are shown in Table 3. They ranged from 72.2% to 101.4%. The formula $[(C1 - C2)/C] \cdot 100\%$ was used to calculate the recovery, where C1 is the labeled analyte concentration in the urine sample with standards added (μg/mL), C2 is the determined concentration in the urine sample (μg/mL) not enriched with standards, and C is the amount of μg standard added to 1 mL of urine sample. The repeatability (intra-day precision) of the method was assessed by the relative standard deviations (RSD) of the nicotine metabolite determinations obtained on the same day (n = 5) for samples enriched with standards. These ranged from 1.0 % to 3.1%. The inter-day precision (reproducibility) over 5 days ranged from 3.0% (3′-OH-Cot, 0.05 μg/mL) to 10.1% (2′S-Nic-Ox, 1 μg/mL).

4. Conclusions

The procedure for precipitation of proteins with acetonitrile is a very simple and quick technique for preparing biological samples to remove components that can interfere with LC-MS detection. The use of an internal standard allows better control of this process. Combined with LC-orbitrap-MS/MS analysis, in parallel reaction monitoring (PRM) mode, it was possible to determine selected nicotine metabolites (cotinine, *trans*-3′-hydroxycotinine and (1′S,2′S)-nicotine 1′-oxide) in urine samples of smokers and passive smokers using the fast cleanup procedure with acetonitrile. Extraction recoveries of these analytes from urine ranged from 72% (2′S-Nic-Ox) to 101% (Cot and 3′-OH-Cot).

The total determined Cot and 3′-OH-Cot concentration in the urine samples of cigarette smokers was 6786 ng/mL, whereas in samples of participants passively exposed to tobacco smoke, the concentration was only 106 ng/mL, but still 3 times higher than in non-smokers not exposed to cigarettes.

Although metabolite concentrations in passively exposed individuals were much lower than in smokers, it still raises the question of concern of neglecting, by young adults and additionally medicine students, the potential sources of exposure to harmful chemicals and tolerating smoking in the company of non-smokers. It also shows how important it is to implement healthy habits at the early stage of adult life, especially in the population of future physicians who are supposed to promote pro-health attitudes.

While many countries ban the use of tobacco products in public places (restaurants, public transport, public beaches, outdoor swimming pools), newly introduced tobacco products such as e-cigarettes may pose other health risks. In recent years, the use of e-cigarettes has become fashionable among young people because of their perceived safety and the widespread belief that they can help smokers quit traditional cigarettes and minimize withdrawal symptoms. It is also believed that e-cigarettes do not contain many harmful substances [8,26]. Our future research goal is to implement the method described in this paper to analyze urine samples of young adults and medical students who use e-cigarettes, and individuals passively exposed to e-cigarette smoke.

Supplementary Materials: The following supporting information can be downloaded at: https://www.mdpi.com/article/10.3390/molecules29153643/s1, Figure S1A. Total and selected ion chromatograms (SIM) of nicotine metabolites in standard mixture and of internal standard; Figure S1B. Total and selected ion chromatograms (SIM) of nicotine metabolites in urine from a cigarette smoker; Figure S1C. Total and selected ion chromatograms (SIM) of nicotine metabolites in urine from a passive smoker; Figure S1D. Total and selected ion chromatograms (SIM) of nicotine metabolites in urine from a non-smoker.

Author Contributions: Conceptualization, M.S., P.M. and B.J.; Data curation, M.S. and B.P.; Investigation, M.S., P.M. and B.P.; Methodology, M.S., P.M. and B.P.; Supervision, K.T.-G. and B.J.; Validation, M.S. and B.P.; Visualization, M.S. and B.J.; Writing—original draft, M.S. and B.J.; Writing—review & editing, K.T.-G. and B.J. All authors have read and agreed to the published version of the manuscript.

Funding: This work was supported by Medical University of Silesia, grant number BNW-2-096/N/3/I and BNW-1-125/N/3/O.

Institutional Review Board Statement: This study was conducted in accordance with the Declaration of Helsinki, and approved by the Ethics Committee of Medical University of Silesia (No BNW/NWN/0052/KB1/81/23 from 12 November 2023) for studies involving humans.

Informed Consent Statement: Informed consent was obtained from all subjects involved in the study.

Data Availability Statement: The raw data supporting the conclusions of this article will be made available by the authors on request.

Conflicts of Interest: The authors declare no conflicts of interest.

References

1. Chen, C.-Y.; Jhou, Y.-T.; Lee, H.-L.; Lin, Y.-W. Simultaneous, Rapid, and Sensitive Quantification of 8-Hydroxy-2′-Deoxyguanosine and Cotinine in Human Urine by on-Line Solid-Phase Extraction LC-MS/MS: Correlation with Tobacco Exposure Biomarkers NNAL. *Anal. Bioanal. Chem.* **2016**, *408*, 6295–6306. [CrossRef]
2. Habibagahi, A.; Alderman, N.; Kubwabo, C. A Review of the Analysis of Biomarkers of Exposure to Tobacco and Vaping Products. *Anal. Methods* **2020**, *12*, 4276–4302. [CrossRef] [PubMed]
3. Torres, S.; Merino, C.; Paton, B.; Correig, X.; Ramírez, N. Biomarkers of Exposure to Secondhand and Thirdhand Tobacco Smoke: Recent Advances and Future Perspectives. *Int. J. Environ. Res. Public Health* **2018**, *15*, 2693. [CrossRef] [PubMed]
4. Habibagahi, A.; Siddique, S.; Harris, S.A.; Alderman, N.; Aranda-Rodriguez, R.; Farhat, I.; Chevrier, J.; Kubwabo, C. Challenges Associated with Quantification of Selected Urinary Biomarkers of Exposure to Tobacco Products. *J. Chromatogr. B* **2021**, *1162*, 122490. [CrossRef] [PubMed]
5. Dai, H.; Benowitz, N.L.; Achutan, C.; Farazi, P.A.; Degarege, A.; Khan, A.S. Exposure to Toxicants Associated with Use and Transitions Between Cigarettes, e-Cigarettes, and No Tobacco. *JAMA* **2022**, *5*, e2147891. [CrossRef]
6. Baj, J.; Flieger, W.; Przygodzka, D.; Buszewicz, G.; Teresiński, G.; Pizoń, M.; Maciejewski, R.; Flieger, J. Application of HPLC-QQQ-MS/MS and New RP-HPLC-DAD System Utilizing the Chaotropic Effect for Determination of Nicotine and Its Major Metabolites Cotinine, and Trans-3′-Hydroxycotinine in Human Plasma Samples. *Molecules* **2022**, *27*, 682. [CrossRef] [PubMed]
7. Abu-awwad, A.; Arafat, T.; Schmitz, O.J. Study the Influence of Licorice and Pomegranate Drinks on Nicotine Metabolism in Human Urine by LC-Orbitrap MS. *J. Pharm. Biomed. Anal.* **2017**, *132*, 60–65. [CrossRef] [PubMed]
8. Gallart-Mateu, D.; Dualde, P.; Coscollà, C.; Soriano, J.M.; Garrigues, S.; de la Guardia, M. Biomarkers of Exposure in Urine of Active Smokers, Non-Smokers, and Vapers. *Anal. Bioanal. Chem.* **2023**, *415*, 6677–6688. [CrossRef]
9. Murphy, S.E. Biochemistry of Nicotine Metabolism and Its Relevance to Lung Cancer. *J. Biol. Chem.* **2021**, *296*, 100722. [CrossRef]
10. Esther, C.R.; O'Neal, W.K.; Alexis, N.E.; Koch, A.L.; Cooper, C.B.; Barjaktarevic, I.; Raffield, L.M.; Bowler, R.P.; Comellas, A.P.; Peters, S.P.; et al. Prolonged, Physiologically Relevant Nicotine Concentrations in the Airways of Smokers. *Am. J. Physiol. Lung Cell. Mol. Physiol.* **2023**, *324*, L32–L37. [CrossRef]
11. Tomaz, P.R.X.; Gonçalves, T.S.; Santos, J.R.; Scholz, J.; Abe, T.O.; Gaya, P.V.; Figueiredo, E.C.; de Faria, H.D.; Martins, I.; Pego, A.M.F.; et al. Evaluation of the Nicotine Metabolite Ratio in Smoking Patients Treated with Varenicline and Bupropion. *Front. Pharmacol.* **2022**, *13*, 1–7. [CrossRef] [PubMed]
12. Siegel, S.D.; Lerman, C.; Flitter, A.; Schnoll, R.A. The Use of the Nicotine Metabolite Ratio as a Biomarker to Personalize Smoking Cessation Treatment: Current Evidence and Future Directions. *Cancer Prevent. Res.* **2020**, *13*, 261–272. [CrossRef] [PubMed]
13. Piller, M.; Gilch, G.; Scherer, G.; Scherer, M. Simple, Fast and Sensitive LC–MS/MS Analysis for the Simultaneous Quantification of Nicotine and 10 of Its Major Metabolites. *J. Chromatogr. B* **2014**, *951–952*, 7–15. [CrossRef] [PubMed]
14. Kataoka, H.; Kaji, S.; Moai, M. Risk Assessment of Passive Smoking Based on Analysis of Hair Nicotine and Cotinine as Exposure Biomarkers by In-Tube Solid-Phase Microextraction Coupled On-Line to LC-MS/MS. *Molecules* **2021**, *26*, 7356. [CrossRef] [PubMed]

15. McGuffey, J.E.; Wei, B.; Bernert, J.T.; Morrow, J.C.; Xia, B.; Wang, L.; Blount, B.C. Validation of a LC-MS/MS Method for Quantifying Urinary Nicotine, Six Nicotine Metabolites and the Minor Tobacco Alkaloids—Anatabine and Anabasine—in Smokers' Urine. *PLoS ONE* **2014**, *9*, e101816. [CrossRef] [PubMed]
16. Fernandes, A.; Santos, L.; Pinheiro, G.; Vasconcellos, D.S.; Telles de Oliva, S.; Fernandes, B.; Couto, R. Urinary Cotinine as a Biomarker of Cigarette Smoke Exposure: A Method to Differentiate Among Active, Second-Hand, and Non-Smoker Circumstances. *Open Biomark. J.* **2020**, *10*, 60–68. [CrossRef]
17. Wielkoszynski, T.; Tyrpien, K.; Szumska, M. The enzyme-linked immunosorbent assay (ELISA) method for nicotine metabolites determination in biological fluids. *J. Pharmaceut. Biomed.* **2009**, *49*, 1256–1260. [CrossRef] [PubMed]
18. Malafatti, L.; Maia, P.; Martins, M.; Bastos de Siqueira, M.; Martins, I. Single gas chromatography method with nitrogen phosphorus detector for urinary cotinine determination in passive and active smokers. *Brazilian J. Pharm. Sci.* **2010**, *46*, 769–776. [CrossRef]
19. Cunha, L.; Oliveira, F.; Santos, L.; Pucci, L.; Neto, J.; Rahal, R.; Freitas, R.J. A simplified method for the analysis of urinary cotinine by GC-MS. *Rev. Ciênc. Farm. Básica Apl.* **2013**, *34*, 177–182.
20. Tyrpień, K.; Wielkoszynski, T.; Janoszka, B.; Dobosz, C.; Bodzek, D.; Stęplewski, Z. Application of liquid separation techniques to the determination of the main urinary nicotine metabolites. *J. Chromatogr. A* **2000**, *870*, 29–38. [CrossRef]
21. Oh, J.; Park, M.-S.; Chun, M.-R.; Hwang, J.H.; Lee, J.-Y.; Jee, J.H.; Lee, S.-Y. A Simple and High-Throughput LC–MS-MS Method for Simultaneous Measurement of Nicotine, Cotinine, 3-OH Cotinine, Nornicotine and Anabasine in Urine and Its Application in the General Korean Population. *J. Anal. Toxicol.* **2022**, *46*, 25–36. [CrossRef] [PubMed]
22. Jin, S.; Pang, W.; Zhao, L.; Zhao, Z.; Mei, S. Review of HPLC–MS Methods for the Analysis of Nicotine and Its Active Metabolite Cotinine in Various Biological Matrices. *Biomed. Chromatogr.* **2022**, *36*, e5351. [CrossRef] [PubMed]
23. El Mubarak, M.A.; Danika, C.; Cachon, C.; Korovila, C.; Atsopardi, K.; Panagopoulos, N.; Margarity, M.; Poulas, K.; Sivolapenko, G.B. In Vivo Quantification and Pharmacokinetic Studies of Cotinine in Mice after Smoke Exposure by LC–MS/MS. *Biomed. Chromatogr.* **2020**, *34*, e4752. [CrossRef] [PubMed]
24. Kaisar, M.A.; Kallem, R.R.; Sajja, R.K.; Sifat, A.E.; Cucullo, L. A Convenient UHPLC-MS/MS Method for Routine Monitoring of Plasma and Brain Levels of Nicotine and Cotinine as a Tool to Validate Newly Developed Preclinical Smoking Model in Mouse. *BMC Neurosci.* **2017**, *18*, 71. [CrossRef]
25. Scheidweiler, K.; Shakleya, D.; Huestis, M. Simultaneous Quantification of Nicotine, Cotinine, trans-3′-Hydroxycotinine, Norcotinine and Mecamylamine in Human Urine by Liquid Chromatography-Tandem Mass Spectrometry. *Clin. Chim. Acta* **2012**, *413*, 978–984. [CrossRef] [PubMed]
26. Mroczek, P.J.; Tyrpień-Golder, K.M.; Janoszka, B.M.; Szumska, M. Medical Students' Awareness of e-Cigarette Use Addiction. *Environ. Med.* **2024**, *26*, 60–66. [CrossRef]
27. Mallock, N.; Andrea Rabenstein, A.; Laux, P.; Rüther, T.; Hutzler, C.; Parr, M.K.; Luch, A. Rapid, sensitive, and reliable quantitation of nicotine and its main metabolites cotinine and trans-3′-hydroxycotinine by LC-MS/MS: Method development and validation for human plasma, *J. Chromatogr. B* **2021**, *1179*, 122736. [CrossRef]
28. Weymarn, L.; Lu, X.; Thomson, N.; LeMarchand, L.; Park, S.; Murphy, S. Quantitation of Ten Urinary Nicotine Metabolites, Including 4-Hydroxy-4-(3-pyridyl) Butanoic Acid, a Product of Nicotine 2′- Oxidation, and CYP2A6 Activity in Japanese Americans, Native Hawaiians, and Whites. *Chem. Res. Toxicol.* **2023**, *36*, 313–321. [CrossRef] [PubMed]
29. Rangiah, K.; Hwang, W.; Mesaros, C.; Vachani, A.; Blair, A. Nicotine exposure and metabolizer phenotypes from analysis of urinary nicotine and its 15 metabolites by LC–MS. *Bioanalysis* **2011**, *3*, 745–761. [CrossRef]
30. Martínez-Sánchez, J.M.; González-Marrón, A.; Sánchez, J.; Sureda, X.; Fu, M.; Ortuno, R.; Moyano, C.; Galán, I.; Pascual, J.; Fernández, E. Validity of self-reported intensity of exposure to second-hand smoke at home against environmental and personal markers. *Gac. Sanit.* **2017**, *32*, 393–395. [CrossRef]
31. Abu-awwad, A.; Arafat, T.; Schmitz, O.J. Simultaneous determination of nicotine, cotinine, and nicotine N-oxide in human plasma, semen, and sperm by LC-Orbitrap MS. *Anal. Bioanal. Chem.* **2016**, *408*, 6473–6481. [CrossRef] [PubMed]
32. Kim, S. Overview of Cotinine Cutoff Values for Smoking Status Classification. *Neurosci. Nicotine* **2019**, *13*, 419–431. [CrossRef]
33. Lee, H.S.; Chun, M.R.; Lee, S.Y. Simultaneous Measurement and Distribution Analysis of Urinary Nicotine, Cotinine, Trans-3′-Hydroxycotinine, Nornicotine, Anabasine, and Total Nicotine Equivalents in a Large Korean Population. *Molecules* **2023**, *28*, 7685. [CrossRef] [PubMed]
34. Kim, J.; Lee, S. Daily Cigarette Consumption and Urine Cotinine Level between Dual Users of Electronic and Conventional Cigarettes, and Cigarette-Only Users. *J. Psychoact. Drugs* **2020**, *52*, 20–26. [CrossRef]
35. Feng, J.; Sosnoff, C.; Bernert, J.; Blount, B.; Li, Y.; Del Valle-Pinero, A.; Kimmel, H.; Bemmel, D.; Rutt, S.; Barreto, J.; et al. Urinary Nicotine Metabolites and Self-Reported Tobacco Use Among Adults in the Population Assessment of Tobacco and Health (PATH) Study, 2013–2014. *Nicotine Tob. Res.* **2022**, *24*, 768–777. [CrossRef] [PubMed]
36. Górna, I.; Napierala, M.; Florek, E. Electronic Cigarette Use and Metabolic Syndrome Development. Critical Review. *Toxics* **2020**, *8*, 105. [CrossRef] [PubMed]
37. Visser, W.F.; Klerx, W.N.; Cremers, H.; Ramlal, R.; Schwillens, P.; Talhout, R. The health risks of electronic cigarette use to bystanders. *Int. J. Environ. Res. Public Health* **2019**, *16*, 1525. [CrossRef]
38. Kawasaki, Y.; Li, Y.; Ootsuyama, Y.; Nagata, K.; Yamato, H.; Kawai, K. Effects of smoking cessation on biological monitoring markers in urine. *Genes Environ.* **2020**, *42*, 26. [CrossRef] [PubMed]

39. Kawasaki, Y.; Li, Y.; Watanabe, S.; Ootsuyama, Y.; Kawai, K. Urinary biomarkers for secondhand smoke and heated tobacco products exposure. *J. Clin. Biochem. Nutr.* **2021**, *69*, 37–43. [CrossRef]
40. Awwad, A.; Schmitz, O.; Arafat, T. Determination of Nicotine and Cotinine in Human Blood by Dried Blood Spot-LC-Orbitrap MS Technique. *J. Anal. Pharm. Res.* **2016**, *3*, 00070. [CrossRef]
41. Konieczka, P.; Namieśnik, J. *Evaluation and Quality Control of Analytical Measurement Results*, 1st ed.; PWN: Warsaw, Poland, 2019; pp. 225–300.

molecules

MDPI

Article

Protein Tyrosine Phosphatase 1B Inhibitors of *Pueraria lobata* Based on the Spectrum–Effect Relationship by Q-Marker Selection

Yong Zhang [1], Haipeng Liu [2], Tianci Lv [2], Mengqian Xiao [2] and Guihua Gao [1,*]

1 School of Pharmacy, Jining Medical University, Rizhao 276826, China; zhang_2211@163.com
2 School of Pharmacy, Shandong University of Traditional Chinese Medicine, Jinan 250355, China; liuhaipeng1005@163.com (H.L.); ays51314@163.com (T.L.); 15954112102@163.com (M.X.)
* Correspondence: guihua526@163.com; Fax: +86-633-2983690

Abstract: *Pueraria lobata* (*P. lobata*), a traditional anti-diabetic medicine mainly composed of flavonoids and isoflavones, has a long history in diabetes treatment in China. However, the anti-diabetic active component is still unclear. Recently, protein tyrosine phosphatase 1B (PTP1B) has been a hot therapeutic target by negatively regulating insulin signaling pathways. In this study, the spectrum–effect relationship analysis method was first used to identify the active components of *P. lobata* that inhibit PTP1B. The fingerprints of 12 batches of samples were established using high-performance liquid chromatography (HPLC), and sixty common peaks were identified. Meanwhile, twelve components were identified by a comparison with the standards. The inhibition of PTP1B activity was studied in vitro by using the p-nitrophenol method, and the partial least squares discriminant analysis, grey relational analysis, bivariate correlation analysis, and cluster analysis were used to analyze the bioactive compounds in *P. lobata*. Peaks 6, 9 (glycitin), 11 (genistin), 12 (4′-methoxypuerarin), 25, 34, 35, 36, 53, and 59 were considered as potentially active substances that inhibit PTP1B. The in vitro PTP1B inhibitory activity was confirmed by glycitin, genistin, and 4′-methoxypuerarin. The IC_{50}s of the three compounds were 10.56 ± 0.42 μg/mL, 16.46 ± 0.29 μg/mL, and 9.336 ± 0.56 μg/mL, respectively, indicating the obvious PTP1B inhibitory activity. In brief, we established an effective method to identify PTP1B enzyme inhibitors in *P. lobata*, which is helpful in clarifying the material basis of *P. lobata* on diabetes. Additionally, it is evident that the spectrum–effect relationship method serves as an efficient approach for identifying active compounds, and this study can also serve as a reference for screening bioactive constituents in traditional Chinese medicine.

Keywords: PTP1B; inhibitory activity; *Pueraria lobata*; spectrum–effect relationship

1. Introduction

Diabetes mellitus (DM) is the collective name for a group of metabolic diseases characterized by hyperglycemia, which is a major challenge to public health due to the high incidence and low cure rate [1]. According to the report of the International Diabetes Federation (IDF), the global prevalence of diabetes among adults aged 20–79 increased from 4.6% to 10.5% of the worldwide population from 2000 to 2021. More seriously, global morbidity will sharply increase to 11.3% by 2030 and 12.2% by 2045, respectively. The prevalence of diabetes has also imposed a substantial economic burden on nations and families, with the global healthcare expenditure for individuals aged 20–79 escalating from USD 232 billion in 2007 to USD 966 billion in 2021 [2]. Type 2 DM (T2DM) is the most common type of DM, accounting for approximately 90% of cases [3]. The main pathogeneses of T2DM are insulin secretion defects and insulin resistance. Insulin is a protein hormone synthesized by pancreatic β-cells and secreted after stimulation. It is the only hormone that can promote the synthesis of glycogen, fat, and protein while reducing blood glucose in the body. Its main physiological role is to act on tissues, such as the liver, muscle, and fat, to regulate metabolism [4].

As shown in Figure 1, the insulin signaling cascade is highly regulated by two key classes of enzymes: protein tyrosine kinases (PTKs) and protein tyrosine phosphatases (PTPs). Insulin binds to the α-subunit of the insulin receptor, thereby activating the tyrosine kinase activity of the β-subunit located in the cell and the phosphorylation of specific tyrosine residues catalyzed by PTK. Then, the phosphorylation of the tyrosine residues of downstream signaling proteins (such as insulin receptor substrates) is initiated, thereby initiating intracellular signal transduction [5] and activating the transduction of the PI3K-Akt pathway. Thus, glucose is taken up by the cells from the blood [6], resulting in the inhibition of glycogen synthesis and, thus, lowering the blood glucose.

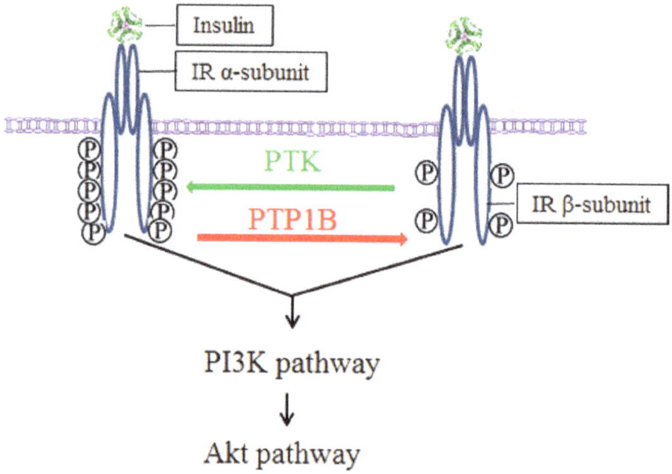

Figure 1. Mechanisms of TPK and PTP1B regulating insulin signaling. Green arrow represents phosphorylation; red arrow represents dephosphorylation.

Protein tyrosine phosphatase 1B (PTP1B), an intracellular PTP and member of the NT1 subfamily, has a total length of approximately 50 kDa and is located in the endoplasmic reticulum. The PTP domain is oriented in the cytoplasm due to hydrophobic stretching at the C-terminal. PTP1B dephosphorylates activated insulin receptors and insulin receptor substrates and is, therefore, a key enzyme in the cellular regulation of insulin signaling. Only the phosphorylated protein can exert the biological effect of lowering blood glucose. Phosphorylation usually occurs at the tyrosine (Tyr), serine (Ser), and threonine (Thr) residues of proteins [7,8]. The maintenance of a dynamic balance between phosphorylation and dephosphorylation is essential for the regulation of signaling pathways and the preservation of glucose homeostasis [9].

The phosphorylation level is key to glucose metabolism in the insulin signaling pathway. Insulin stimulates Tyr66 phosphorylation in the PTP1B catalytic region, which, in turn, activates PTP1B to dephosphorylate insulin receptors and their substrates, thereby negatively regulating insulin-promoted glucose metabolism. Physiologically, PTKs and PTPs collaborate to maintain the homeostasis of the insulin signaling pathway [10]. However, when PTP1B is overexpressed in vivo, phosphorylated insulin receptors are restored to their non-phosphorylated form within minutes [11]. Long-term preservation of low phosphorylation levels of insulin receptors is one of the causes of insulin resistance [6]. In T2DM individuals, PTP1B is overexpressed [12], resulting in the insufficient phosphorylation of insulin receptors, thereby impairing the normal regulatory function of the insulin signaling pathway. The change in the biological activity of insulin in the tissue prompts an increase in insulin secretion by pancreas islet β-cells, counteracting the lack of insulin action and resulting in hyperinsulinemia. As the disease progresses, pancreas islet β-cells begin to fail, and a vicious cycle ensues, leading to further a decline in insulin secretion and the worsening of hyperglycemia. Knockout of PTP1B promotes the phosphorylation of insulin receptors in mice, leading to the enhancement of glucose metabolism [13] and the decrease of glucose and insulin in the blood [14]. Compared to PTP1B+/+ mice, PTP1B−/− mice show increased insulin sensitivity in the liver and muscle and insulin resistance induced by a high-fat diet [15,16]. Therefore, PTP1B is an important target for the treatment of insulin resistance, and PTP1B inhibitors have potential for the treatment of T2DM.

In a clinic, T2DM is generally treated by administering hypoglycemic and lipid-lowering medications and supplemented by controlling dietary regimen and engaging in moderate physical activity [17,18]. In recent decades, natural products have emerged as valuable bioactive resources for the development of anti-T2DM medications. Traditional Chinese medicines (TCM) have been long-term used for the treatment of T2DM. For example, Astragalus, Bupleurum, *Pueraria lobata* (*P. lobata*), Poria cocos, Atractylodes, Codonopsis pilosula, and Scutellaria baicalensis are commonly used in clinics for the therapy of T2DM [19]. Gegen Qinlian Decoction, composed of *P. lobata*, Scutellaria baicalensis, Coptis, and Licorice, is beneficial for the regulation of pancreas islet conduction, inflammatory signaling transduction, oxidative stress, and intestinal flora, thereby improving insulin resistance [20].

In TCM, *P. lobata* is one of the most used herbal drugs for the treatment of DM. Approximately 35 species of *P. lobata* are known worldwide, mainly distributed in the tropical and subtropical regions of East and Southeast Asia [21], north to China, Japan, south to Malaysia, west to India, Sri Lanka, and east to Indonesia. There are eight species and two varieties in China. Both *Puerariae lobatae radix* and *Puerariae thomsonii radix* are included in the Pharmacopoeia of the People's Republic of China.

P. lobatae, the dried root of *P. lobata* (Willd.) Ohwi, is a kind of TCM used as both medicine and food. Its components include flavonoids [22], isoflavones [23], pentacyclic triterpenoids, triterpenoid saponins, coumarins, Pueraria glycosides, and polysaccharides [24,25]. The modern pharmacological studies found various bioactivities, such as anti-inflammation [26], anti-atherosclerosis [27], anti-hyperlipidemia [28], anti-hypertension [29] and liver protection [30]. More and more research has indicated that *P. lobata* exerts anti-diabetic effects by reducing insulin resistance, increasing insulin release, and improving insulin sensitivity, enhancing glucose uptake and metabolism [31,32]. In detail, puerarin, the main component of *P. lobata*, exerted an anti-diabetic effect by inhibiting the TGF-β1/Smad2 pathway in streptozotocin-induced mice [33]. A 75% butanol fraction of *P. lobata* extract inhibited PTP1B activity to enhance the glucose uptake capacity in insulin-resistant HepG2 cells and improve glucose control and tolerance in diabetic mice [34]. Moreover, Genistein, another of the main components of *P. lobata*, has a reversible effect on insulin release [15]. However, the effective components of *P. lobata* for the therapy of T2DM have not been fully demonstrated.

Spectrum–effect relationship analysis uses powerful mathematical and statistical methods, such as grey relational analysis (GRA), correlation analysis (CA), and partial least squares discrimination analysis (PLS-DA), to study the relationship between fingerprint data and pharmacodynamic indicators, to clarify the material basis of the pharmacological action of TCM, and to analyze the bioactive substances in fingerprints [35]. However, there is a lack of reports regarding the spectrum–effect relationship analysis between *P. lobata* and its inhibition of PTP1B. In the present study, the spectrum–effect relationship method was proposed to study the components of *P. lobata* on inhibiting PTP1B activities and aimed to explore the material basis of the hypoglycemic mechanism of *P. lobata*.

2. Results and Discussion

2.1. Establishment of Fingerprint of *P. lobata* Extract

The HPLC column plays a crucial role in HPLC analysis, and it is also very important to the establishment of TCM chromatogram fingerprints. Therefore, following a comparison of the Tskgel ODS-100V C18 (Tosoh, Tokyo, Japan) (4.6 mm× 250 mm, 3 μm) and the ZORBAX SBC18 RP (Agilent Technologies, Santa Clara, CA, USA) (4.6 mm × 250 mm, 5 μm) columns, the Tskgel ODS-100V column was selected for the separation of compounds from the *P. lobata* extract, as it had a better peak-to-peak resolution. Moreover, the number of peaks in the chromatogram obtained using this column was the highest, at 254 nm.

Under optimized chromatographic conditions, the HPLC fingerprints of the *P. lobata* samples were arranged neatly, and the reference fingerprints were generated using the Similarity Evaluation System for Chromatographic Fingerprint of TCM, shown in Figure 2. In the fingerprints, the peak for glycitein (peak 20) was selected as the reference peak for calculating the relative retention time and relative peak area of each common peak because of its moderate size and good separation from other chromatographic peaks. As a result, 60 characteristic peaks were identified, which accounted for over 80% of the total peak areas in the fingerprint. A similarity analysis of the chromatographic profiles of all the samples revealed that all the similarity values between the fingerprints of the different batches of *P. lobata* and the reference fingerprints exceeded 0.80, with a range of 0.888 to 0.967 (Table 1). These results demonstrated minimal chemical variation among samples in the major producing regions, thereby confirming the reliability of the established fingerprint. However, the similarity values between S11 and S1, S2, S3, S5, S6, S7, S9, and S10 were 0.735, 0.727, 0.810, 0.856, 0.757, 0.842, 0.796, and 0.823, respectively, indicating that the contents of the chemical components of S11 were different from those of the other samples. The results demonstrated that the production area had a significant influence on the chemical composition and content of *P. lobata*.

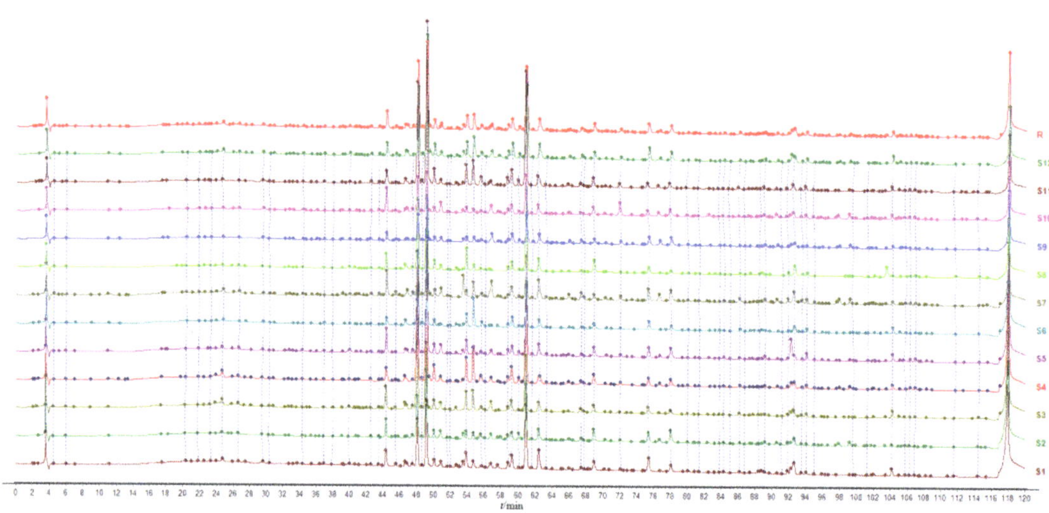

Figure 2. HPLC fingerprint chromatograms of *P. lobata*.

Table 1. The origin and similarity of 12 batches of *P. lobata*.

Sample No.	Origin	Longitude and Latitude	Similarity
S1	Liuan, Anhui Province, China	116.53 east, 32.01 north	0.959
S2	Anqing, Anhui Province, China	116.50 east, 30.20 north	0.955
S3	Tongling, Anhui Province, China	117.61 east, 30.89 north	0.969
S4	Nanyang, Henan Province, China	112.39 east, 32.96 north	0.981
S5	Pingdingshan, Henan Province, China	113.24 east, 33.56 north	0.987
S6	Liuan, Anhui Province, China	116.28 east, 32.06 north	0.966
S7	Luohe, Henan Province, China	113.77 east, 33.61 north	0.962
S8	Huaibei, Anhui Province, China	116.71 east, 33.91 north	0.963
S9	Fuyang, Anhui Province, China	116.13 east, 32.75 north	0.978
S10	Anqing, Anhui Province, China	116.26 east, 30.31 north	0.961
S11	Xinyang, Henan Province, China	114.50 east, 31.83 north	0.888
S12	Nanyang, Henan Province, China	112.77 east, 32.90 north	0.946

Moreover, 12 reference standards, including 3′-hydroxypuerarin, puerarin 6″-*O*-xyloside, 3′-methoxypuerarin, glycitin, genistin, 4′-methoxypuerarin, ononin, daidzein, glycitein, genistein, formononetin, and biochanin A, were dissolved in methanol to prepare a pure reference standard solution and a mixed reference standard solution. They were then injected into the HPLC system, and the retention time of each reference standard was recorded for qualitative analysis. Twelve peaks of fingerprints were identified by comparing their retention times. Chromatograms of the reference fingerprints and the mixed reference standard solution are shown in Figure S1.

To validate the HPLC method, one sample (S4) was randomly selected for precision, repeatability, and stability validation tests. Precision was evaluated using six injections of the same sample, while six sample solutions were analyzed to determine repeatability. Stability studies on the sample solution were performed at 0, 2, 4, 6, 12, 18, and 24 h. The RSDs of precision, repeatability, and sample stability were calculated in terms of the compound relative peak area and were less than 3%, indicating that the established method was reliable and reproducible and that the sample solution could be stable at room temperature for 24 h.

2.2. Analysis of Inhibitory Activity of *P. lobata* Extract on PTP1B

The inhibition ratio and IC_{50} were employed to evaluate the inhibitory activity of *P. lobata* extract on PTP1B. The dose–effect curve was drawn using the logarithm value of sample concentration as the abscissa and the inhibition ratio as the ordinate. The results in Figure 3A demonstrated a clear dose-dependent relationship between the inhibition ratio and the concentration, with higher concentrations resulting in greater inhibition activity. As shown in Figure 3B, the IC_{50} ranged between 5.318 ± 0.17 μg/mL and 11.68 ± 0.22 μg/mL, indicating that the *P. lobata* samples exhibited inhibitory activity against PTP1B. Since the IC_{50} was negatively correlated with the inhibitory activity of PTP1B, and S4 and S7 exhibited higher inhibitory activity than other batch samples, with IC_{50} values of 5.318 ± 0.17 μg/mL and 5.690 ± 0.044 μg/mL, respectively. The IC_{50} of S1 was 2.2 times that of S4, showing a significant difference in inhibitory activities against PTP1B between different batches of samples.

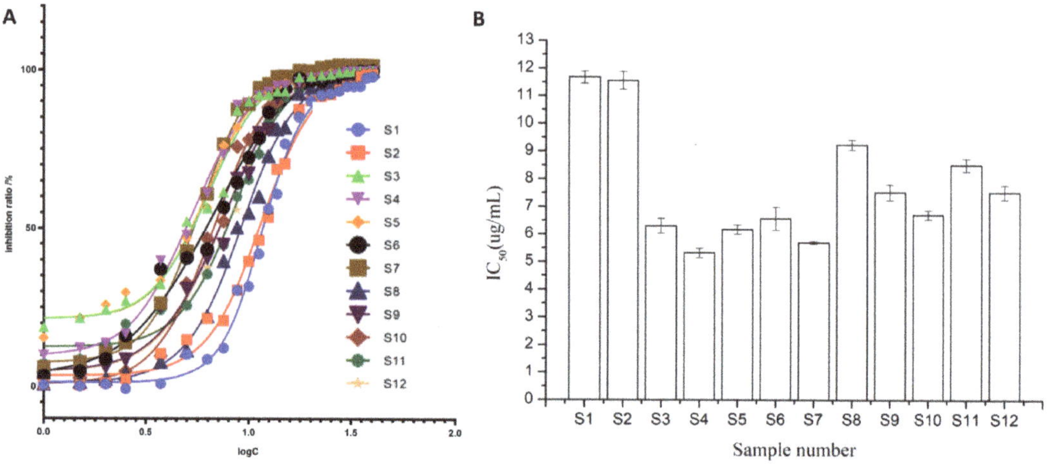

Figure 3. The inhibitory activity of 12 batches of *P. lobata* samples on PTP1B. (**A**), Dose–effect curve; (**B**), IC_{50}.

2.3. *Multivariate Analysis Results*

2.3.1. Grey Relational Analysis (GRA)

Grey relational theory, a method used widely in many fields, was first proposed by Deng in his 1982 paper and was designed to analyze uncertain and incomplete information systems with limited collected data. The grey relational grade (GRD) indicates the relational grade between each compared series and the reference series. In this study, IC_{50} was chosen as the reference series, and the peak areas of the 60 common peaks were regarded as the compared series. The grey relational coefficient between the compared and reference series was calculated at a resolution rate of 0.5, and GRD reflected the inhibitory activity of PTP1B.

The results of GRA are shown in Figure 4A, and the contributions of the common peaks to the inhibitory activity against PTP1B are shown in Table S2. The GRD between the relative contents of the 60 common peaks and the inhibitory activity of PTP1B were in the range of 0.758–0.921. The GRD values of all the peaks were greater than 0.6, suggesting that the common peak areas were correlated with IC_{50}; thus, all peaks demonstrated an important effect on the inhibitory activity of *P. lobata*. Moreover, the GRD values of the peaks, except for 2, 4, 6, 8, 11, and 25, were greater than 0.8, demonstrating the strong correlation between common peak area and inhibitory activity.

2.3.2. Bivariate Correlation Analysis (BCA)

A BCA was performed using Pearson's correlation analysis. The peak areas of the common peaks were considered as a group of variables, and the drug efficacy index, IC_{50}, was taken as another group of variables. Pearson's correlation coefficient was used as an index to analyze the correlation size, significance, and direction of change between the two variables. The results (Figure 4B) showed that the peak areas of 48 among the 60 common peaks were negatively correlated with IC_{50}.; That is, with an increase in peak area, the inhibitory activity of the extracts on PTP1B increased. The absolute values of correlation coefficients of peaks 5 (puerarin 6″-*O*-xyloside), 6, 8, 9 (glycitin), 11 (genistin), 12 (4′-methoxypuerarin), 23, 25, 34, 35, 36, 45, 53, 54, 55, 58, and 59 were all greater than 0.4, indicating a good correlation. As shown in Table S3, peaks 5 and 6 showed a significant negative correlation ($p \leq 0.05$).

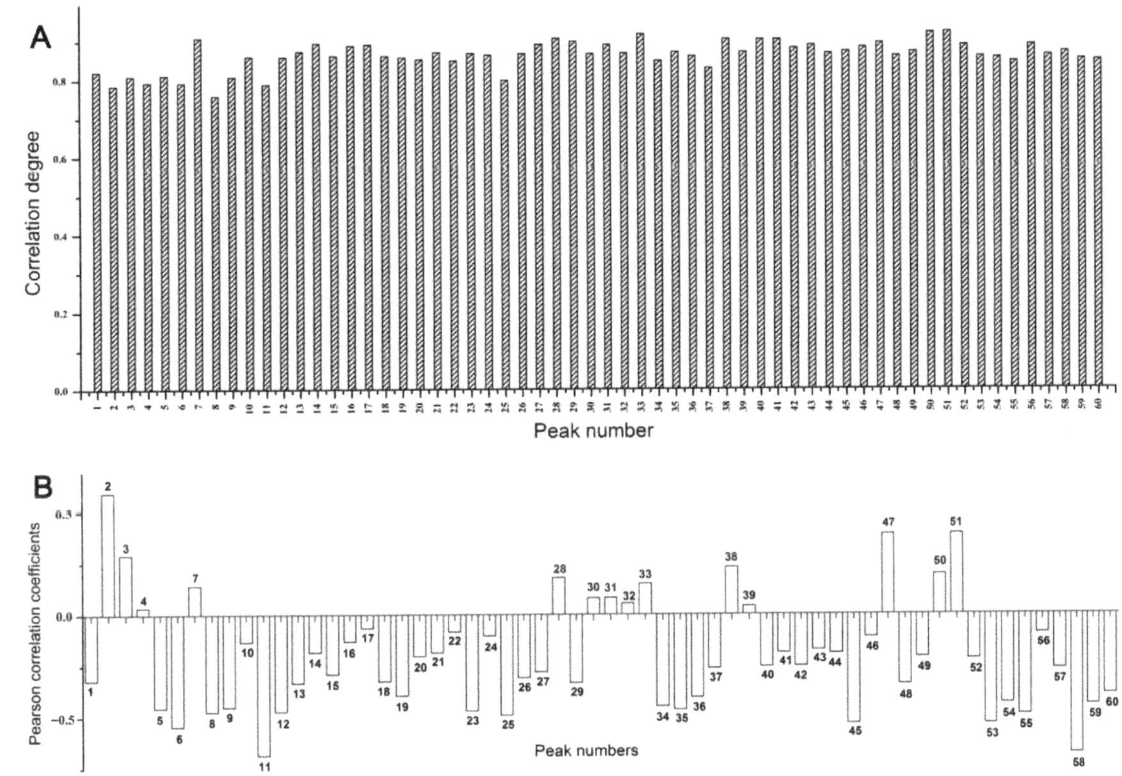

Figure 4. The grey relational grade (**A**) and the Pearson correlation coefficients (**B**) between the peak area and inhibitory activity of PTP1B (IC_{50}).

2.3.3. Hierarchical Cluster Analysis (HCA)

Cluster analysis is a multivariate statistical method used to classify research subjects into relatively homogeneous categories. In our HCA, the peak areas of the 60 common peaks were used as variables. The variables were standardized by Z-score. The distance was measured by cosine, and the within-groups linkage method was used for cluster analysis. A dendrogram is shown in Figure 5. When the classification distance was 19, 12 batches of *P. lobata* were divided into two categories. Samples S1, S2, S3, S6, S8, S9, and S10, collected from Anhui Province, were clustered into one category, and S4, S5, S7, S11, and S12, collected from Hebei Province, were clustered into another. The results of cluster analysis were consistent with those of the origin of the samples.

2.3.4. Partial Least Squares Discriminant Analysis (PLS-DA)

To reflect the differences between the groups, PLS-DA in the supervised mode was used to identify and screen the components with a large contribution rate to the differences between groups. The PLS-DA model was used to identify the potential active compounds related to the inhibitory activity on PTP1B by correlating the fingerprint chromatographic data and the IC_{50} values. An X matrix of dimensions (12 × 60) from the common peak areas and a Y matrix of IC_{50} values were used. The R^2X, R^2Y, and Q^2 values of the model were 0.587, 0.99, and 0.867, respectively, which were greater than 0.5, indicating good fitting and prediction abilities. This model can be used for the pattern recognition of *P. lobata* from different origins. For these *P. lobata* samples, all peaks, except peaks 2, 8, 14, 15, 16, 17, 23, 37, 42, 49, 50, 51, and 54, were negatively correlated with the IC_{50} values (Figure 6A).

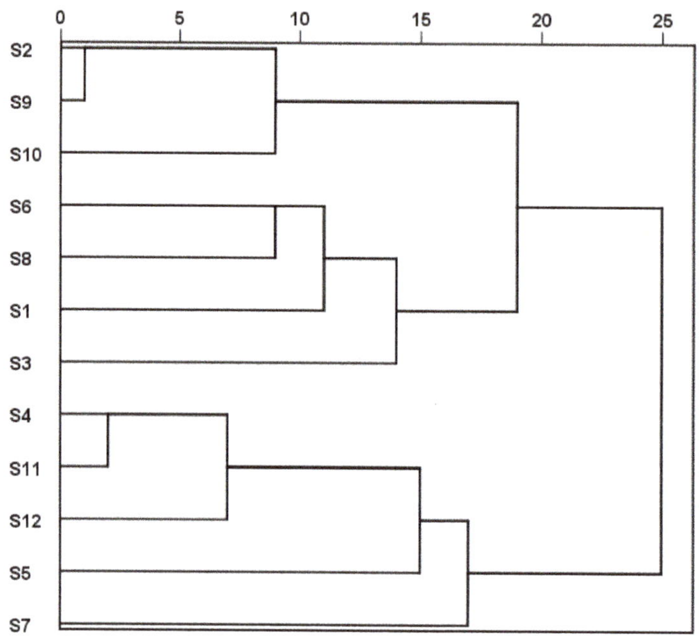

Figure 5. The clustering analysis tree diagram of 12 batches of *P. lobata*.

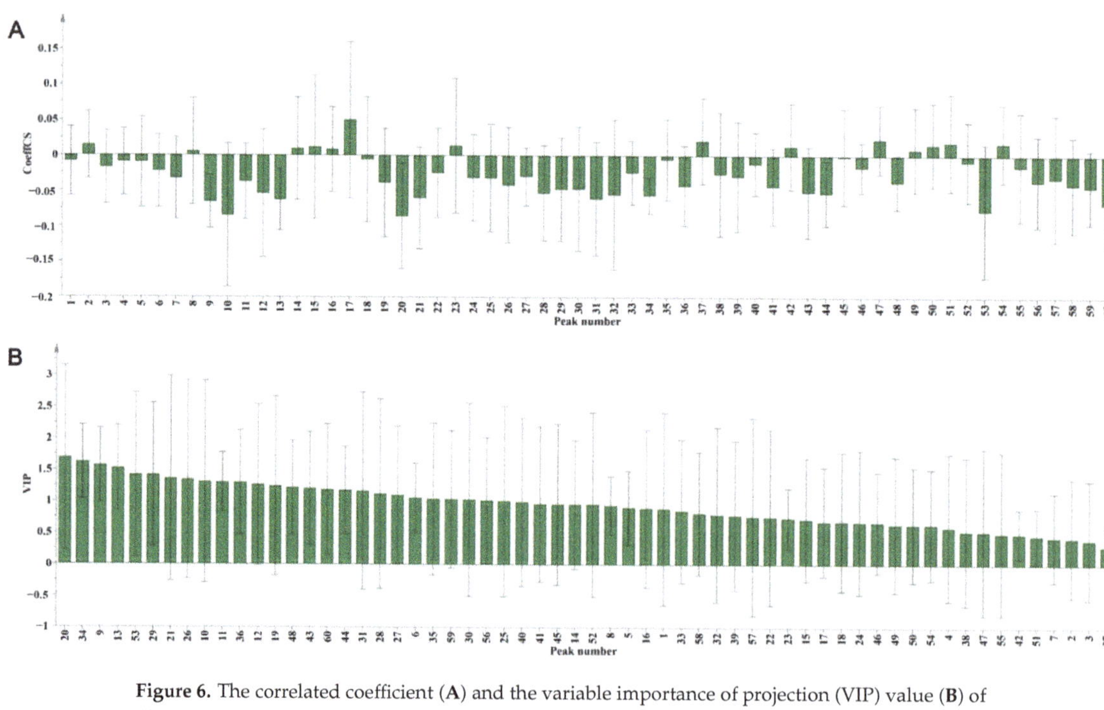

Figure 6. The correlated coefficient (**A**) and the variable importance of projection (VIP) value (**B**) of 60 common peaks of *P. lobata* extracts analysis by PLS-DA.

Furthermore, we used the variable importance of projection (VIP) parameter to screen for variables responsible for the inhibitory activity on PTP1B. Compounds with variables above the VIP-value threshold of 1.0 were filtered out as candidate bioactive compounds. There were 26 peaks with VIP values greater than 1.0, which were considered candidate active substances (Figure 6B, Table S4).

2.3.5. Integration of Analytical Results

In the comprehensive analysis, 10 common peaks satisfied the following criteria: negative correlation and a VIP value greater than one in the PLS-DA, negative correlation and a correlation coefficient greater than 0.4 in the BCA, and a correlation degree greater than 0.6 in the GCA. Based on this, peaks 6, 9 (glycitin), 11 (genistin), 12 (4′-methoxypuerarin), 25, 34, 35, 36, 53, and 59 were preliminarily identified as candidates with PTP1B inhibitory activity.

2.4. Verification Experiment of PTP1B Inhibitory Activity

According to the results of the spectrum activity relationship analysis, glycitin, genistin, and 4′-methoxypuerarin were identified as the active substances in *P. lobata* inhibiting PTP1B, and the activity was verified using standard samples. Furthermore, chlorogenic acid has been reported to have inhibitory activity against PTP1B [36], so it was selected as a positive control in this study. In addition, the inhibition of oleanolic acid activity against PTP1B was determined. Glycitin, genistin, 4′-methoxypuerarin, chlorogenic acid, and oleanolic acid were diluted in 10% dimethyl sulfoxide (DMSO) Tris-HCL to determine the inhibitory activity of single compounds on PTP1B. The IC_{50} values of glycitin, genistin, 4′-methoxypuerarin, and chlorogenic acid were 10.56 ± 0.42 μg/mL, 16.46 ± 0.29 μg/mL, 9.336 ± 0.56 μg/mL, and 15.28 ± 0.35 μg/mL, respectively, while oleanolic acid had no inhibitory activity against PTP1B. Therefore, oleanolic acid was employed as a negative control for the subsequent experiment.

2.5. Molecular Docking

The crystal structure of PTP1B (1XBO) was downloaded from the RCSB Protein Data Bank (http://www.pdb.org/ (accessed on 1 July 2023)) and was modified using a protein preparation procedure. The receptor grid was generated by removing ligands and water, adding hydrogen, optimizing amino acids, and refining. The ligand structures of glycitin (PubChem CID: 187808), genistin (PubChem CID: 5281377), 4′-methoxypuerarin (PubChem CID: 5319486), chlorogenic acid (positive control) (PubChem CID: 1794427), and oleanolic acid (negative control) (PubChem CID: 10494) were downloaded from the National Library of Medicine (https://pubchem.ncbi.nlm.nih.gov/ (accessed on 8 January 2024)). The ligPrep procedure was used to create the 3D chemical structures with minimal energy. The ligand and receptor grids were then imported into the ligand-docking procedure for calculation and visualization.

As shown in Figure 7, four ligands (glycitin, genistin, 4′-methoxypuerarin, and chlorogenic acid) and the PTP1B protein showed good binding activity, and the molecules were connected by hydrogen bond and π bond, whereas oleanolic acid (negative control) failed to dock. This validates the reliability of the molecular docking model. Moreover, glycitin formed hydrogen bonds with the amino acid residues GLY259, SER28, ASP29, and ARG24 of the protein. Genistin formed hydrogen bonds with the amino acid residues ASP48, CYS215, and ARG221 of the protein, and π bonds with TYR46. 4′-methoxypuerarin formed hydrogen bonds with the amino acid residues ASP29, ARG24, and CYS32 and a salt bridge with ARG24 of the protein. Chlorogenic acid formed hydrogen bonds with the amino acid residues ASP181, ASP48, and ARG47 of the protein. The docking scores were −3.908, −5.414, −4.018, and −5.369, respectively.

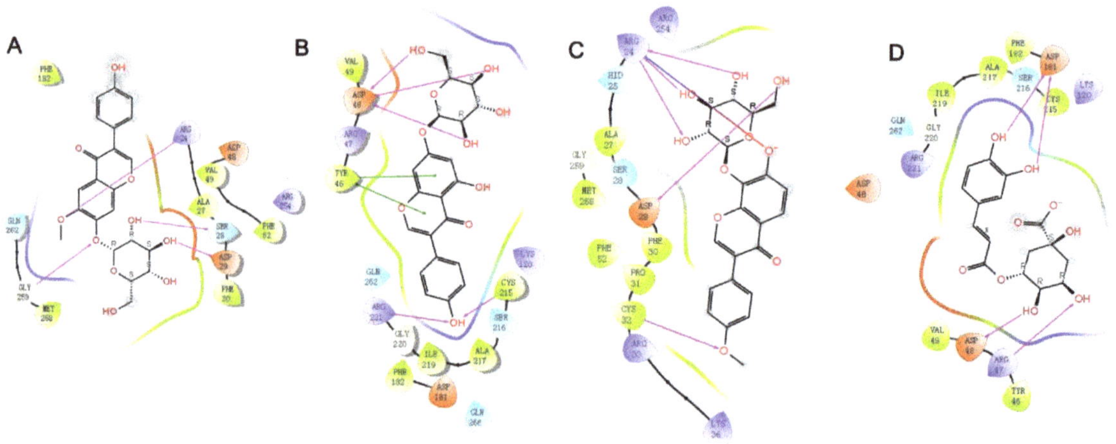

Figure 7. Ligand–protein interaction diagrams. (**A**), Glycitin; (**B**), genistin; (**C**), 4′-Methoxypuerarin; (**D**), chlorogenic acid; ➔ represent hydrogen bonds; — represent salt bridge; ●–● represent π–π stacking.

3. Materials and Methods

3.1. Reagents and Materials

Reference standards of 3′-hydroxypuerarin (DST220316-075), puerarin 6″-*O*-xyloside (DST200912-054), 3′-methoxypuerarin (DST211203-224), glycitin (DST201129-008), genistin (DST 210319-003), 4′-methoxypuerarin (DST200425-310), ononin (DST211218-044), glycitein (DSTDH000901), genistein (DSTDR000201), formononetin (DSTDM001102), and biochanin A (DSTDY010901) were obtained from Chengdu Desite Bio-Technology Co., Ltd. (Chengdu, China). Daidzein (20211031) was obtained from Shandong Xiya Chemical Co., Ltd. (Linyi, China). PTP1B was obtained from Zoonbio Biotechnology Co., Ltd. (Nanjing, China). HPLC-grade methanol was obtained from Shanghai CINC High Purity Solvent Co., Ltd. (Shanghai, China), and formic acid was obtained from Tianjin Kemiou Chemical Reagent Co., Ltd. (Tianjin, China). Analytical grade reagents, including Tris, anhydrous sodium carbonate, ethanol, DMSO, and chloroform, were obtained from Sinopharm Chemical Reagent Co., Ltd. (Shanghai, China). 4-Nitrophenyl phosphate disodium salt hexahydrate (PNPP) was obtained from Shanghai Macklin Biochemical Technology Co., Ltd. (Shanghai, China).

Twelve batches of samples were collected from different regions of China. The details of the plant samples are presented in Table 1.

3.2. Sample Preparation

P. lobata samples were crushed and sifted through a 60-mesh sieve. The powders (10 g) were added to flasks containing 100 mL of chloroform, and reflux was performed three times for 2 h at a time. The resulting solution was evaporated under reduced pressure. *P. lobata* extract was obtained by vacuum drying.

Approximately 10 mg of *P. lobata* extract was diluted with DMSO to obtain a 10 mg/mL stock solution. On one hand, 1 mL of the stock solution was then diluted to 10 mL with methanol for HPLC analysis. On the other hand, the sample solution for activity analyses was prepared by diluting the stock solution to a concentration of 1–40 μg/mL with a 10% DMSO Tris-HCl (10 mM, pH 7.35) solution. All solutions were filtered through a 0.22 μm syringe filter before testing.

3.3. HPLC Analysis

The fingerprints of the *P. lobata* extract were performed using the Shimadzu LC-20A ultra-fast liquid chromatography (UFLC) system equipped with a UV detector (Shimadzu,

Kyoto, Japan). A Tskgel ODS-100V C18 column (4.6 mm× 250 mm, 3 μm) (Tosoh) was employed to separate the extract at 40 °C. As mobile phase at the flow rate of 1.0 mL/min, 0.1% formic acid aqueous (A) and methanol (B) were used, and the optimized gradient elution procedure was as follows: 0–8 min, 22% B; 30–50 min, 45%–60% B; 70–100 min, 70–90% B; 110–115 min, 90–22% B. The detection wavelength was set at 254 nm, and 20 μL of the sample were injected into the HPLC system.

3.4. Determination of Inhibitory Activity of *P. lobata* on PTP1B In Vitro

The inhibitory activity on PTP1B was evaluated using a modified procedure. Briefly, a 12.5 μg/mL PTP1B solution was prepared by diluting 50 μL PTP1B stock solution (0.5 mg/mL) with Tris-HCl (10 mM, pH 7.35) to 2 mL. Simultaneously, a PNPP solution was prepared by dissolving 10 mg of PNPP in 20 mL water, resulting in a concentration of 0.5mg/mL as the substrate solution. Subsequently, a test solution was prepared by combining 100 μL of the sample solution (1–40 μg/mL), 100 μL of the PNPP solution, 180 μL of 10% DMSO Tris-HCl (10 mM, pH 7.35), and 20 μL of the PTP1B solution. The test solution was incubated at 37 °C for 30 min, followed by the immediate addition of a 200 μL sodium carbonate solution (1M) to terminate the reaction. Subsequently, 200 μL of the resulting solution were transferred to a well in a 96-well plate, and the absorbance (A1) was measured at 405 nm using a SyncrgyH1 Hybrid Multi-Mode Microplate Reader (BioTek, Winoski, VT, USA).

By employing the same methodology, the absorbance (A2) of the blank solution was determined when 100 μL of 10% DMSO Tris-HCl (10 mM, pH 7.35) was used to replace the sample solution, and another 100 μL of Tris-HCl (10 mM, pH 7.35) was used instead of the PTP1B solution. Additionally, the absorbance (A0) of the normal solution was obtained by replacing the sample solution with 100 μL of 10% DMSO Tris-HCl (10 mM, pH 7.35). The inhibition ratio (%) was calculated by the following formula:

$$\text{Inhibition}(\%) = \frac{\text{A0} - \text{A1}}{\text{A0} - \text{A2}} \times 100\%$$

3.5. Multivariate Analysis

The HPLC fingerprints generated using 12 batches of *P. lobata* samples were assessed using the Similarity Evaluation System for Chromatographic Fingerprint of TCM (2013.130723 Version, Committee of Chinese Pharmacopoeia). Specifically, the data files of 12 batches of *P. lobata* generated by HPLC analysis were initially exported into AIA format files, which were then imported into the Similarity Evaluation System for Chromatographic Fingerprint of TCM software (2013.130723 Version). In the software, the median method was adopted, the time window width was set at 0.2, multi-point correction was carried out, and the full chromatogram peak matching was performed. Finally, the reference chromatogram was generated using sample S4 as the reference.

GraphPad Prism 9 (GraphPad Software, San Diego, CA, USA) was used to draw the dose–effect curve and calculate the IC_{50} value. The dose–effect curve was plotted with the logarithm value of the sample concentration as the *X*-axis and the inhibition rate against PTP1B as the *Y*-axis. The IC_{50} value was then calculated using the dose-response-inhibition equation in the non-linear regression model.

SPSS 25.0 statistical software (SPSS Inc, Chicago, IL, USA) was used to perform BCA and HCA. The peak areas of the common peaks and IC_{50} values for each sample were input into SPSS 25.0 software for analysis. For BCA, the Pearson correlation coefficient was utilized to examine the relationship between the peak area and the IC_{50} value. When HCA was carried out on the data, the peak area was taken as the variable, and the Z-score was used to standardize the data. Subsequently, cosine values were employed to measure distances, and a tree graph was generated through the within-groups clustering method.

SIMCA 14.1 (Umetrics, Umea, Sweden) was used for PLS-DA. The peak areas and IC_{50} values of the common peaks in each sample were inputted into SIMCA 14.1 software. Homologous samples were grouped together, and the confidence level was set at 95%. The coefficients were calculated using the scaled and centered method, and the UV method was used for scaling.

SPSS Pro software (www.spsspro.com (accessed on 8 July 2023)) was used for GRA. The feature sequence was determined as the peak area of the common peaks in each sample, while the parent sequence was defined as IC_{50}. The distinguishing coefficient ρ was set to 0.5. The data were standardized prior to analysis.

4. Conclusions

In this study, an HPLC fingerprint of *P. lobata* extract was established, revealing 60 common peaks. The inhibitory activity of *P. lobata* extract on PTP1B was investigated in vitro, considering the crucial role of PTP1B in insulin regulation. The results of the pharmacological activity tests showed that the chloroform extract of *P. lobata* had a significant inhibitory effect on PTP1B. Based on multivariate analyses, 10 common peaks met the set requirements simultaneously and were selected as substances with PTP1B inhibitory activity. Three compounds, genistein, glycine, and 4′ -methoxypuerarin, were identified by comparison with reference standards, and they bound to the PTP1B protein through hydrogen, π bonds, and salt bridge. The IC_{50}s of these three compounds against PTP1B in vitro were 10.56 ± 0.42 μg/mL, 16.46 ± 0.29 μg/mL, and 9.336 ± 0.56 μg/mL, respectively, demonstrating that the spectrum–effect relationship method is of great significance in screening the active components of TCMs. This study provides a method for screening PTP1B enzyme inhibitors from *P. lobata*, which helps to clarify the material basis of *P. lobata* in the treatment of diabetes. It also provides a reference for TCM to treat diabetes based on new targets and also promotes the discovery of the active components in TCMs that act on new targets.

Supplementary Materials: The following supporting information can be downloaded at https://www.mdpi.com/article/10.3390/molecules29122731/s1: Figure S1: HPLC chromatograms of reference fingerprint and reference standards; Figure S2: Molecular docking diagram; Table S1: Retention time and peak area of common peaks of 12 batches of *P. lobata*; Table S2: GRA results of common peaks on PTP1B inhibitory activity; Table S3: BCA results of common peaks on PTP1B inhibitory activity; Table S4: VIP value of the common peaks in PLS-DA.

Author Contributions: Conceptualization, Y.Z. and G.G.; methodology, Y.Z. and H.L.; validation, T.L. and M.X.; investigation, Y.Z. and H.L.; data curation, Y.Z. and G.G.; writing—original draft preparation, Y.Z.; writing—review and editing, G.G.; supervision, G.G.; project administration, Y.Z. and G.G.; funding acquisition, Y.Z. and G.G. All authors have read and agreed to the published version of the manuscript.

Funding: The National Natural Science Foundation of China (82003706) and Science and Technology Project of Traditional Chinese Medicine of Shandong Province (M-2023193).

Institutional Review Board Statement: Not applicable.

Informed Consent Statement: Not applicable.

Data Availability Statement: The data presented in this study are available on request from the corresponding author.

Acknowledgments: The authors would like to thank all the reviewers who participated in the review during the preparation of this manuscript.

Conflicts of Interest: The authors declare no conflicts of interest.

References

1. Harding, J.L.; Pavkov, M.E.; Magliano, D.J.; Shaw, J.E.; Gregg, E.W. Global trends in diabetes complications: A review of current evidence. *Diabetologia* **2018**, *62*, 3–16. [CrossRef] [PubMed]
2. IDF. *IDF Diabetes Atlas*, 10th ed.; IDF: Brussels, Belgium, 2021; pp. 1–141.
3. Sacks, D.B.; Arnold, M.; Bakris, G.L.; Bruns, D.E.; Horvath, A.R.; Lernmark, Å.; Metzger, B.E.; Nathan, D.M.; Kirkman, M.S. Guidelines and Recommendations for Laboratory Analysis in the Diagnosis and Management of Diabetes Mellitus. *Diabetes Care* **2023**, *46*, e151–e199. [CrossRef] [PubMed]

4. Genovese, M.; Nesi, I.; Caselli, A.; Paoli, P. Natural α-Glucosidase and Protein Tyrosine Phosphatase 1B Inhibitors: A Source of Scaffold Molecules for Synthesis of New Multitarget Antidiabetic Drugs. *Molecules* **2021**, *26*, 4818. [CrossRef] [PubMed]
5. White, M.F. Insulin Signaling in Health and Disease. *Science* **2003**, *302*, 1710–1711. [CrossRef] [PubMed]
6. Rath, P.; Ranjan, A.; Chauhan, A.; Verma, N.K.; Bhargava, A.; Prasad, R.; Jindal, T. A Critical Review on Role of Available Synthetic Drugs and Phytochemicals in Insulin Resistance Treatment by Targeting PTP1B. *Appl. Biochem. Biotechnol.* **2022**, *194*, 4683–4701. [CrossRef] [PubMed]
7. Chen, X.; Yu, J.; Shi, J. Management of Diabetes Mellitus with Puerarin, a Natural Isoflavone FromPueraria lobata. *Am. J. Chin. Med.* **2019**, *46*, 1771–1789. [CrossRef]
8. Wang, S.; Zhang, S.; Wang, S.; Gao, P.; Dai, L. A comprehensive review on Pueraria: Insights on its chemistry and medicinal value. *Biomed. Pharmacother.* **2020**, *131*, 110734. [CrossRef]
9. Wang, C.; Xu, N.; Cui, S. Comparative transcriptome analysis of roots, stems, and leaves of Pueraria lobata (Willd.) Ohwi: Identification of genes involved in isoflavonoid biosynthesis. *PeerJ* **2021**, *9*, e10885. [CrossRef] [PubMed]
10. Combs, A.P. Recent Advances in the Discovery of Competitive Protein Tyrosine Phosphatase 1B Inhibitors for the Treatment of Diabetes, Obesity, and Cancer. *J. Med. Chem.* **2009**, *53*, 2333–2344. [CrossRef]
11. Jin, J.; He, Y.; Guo, J.; Pan, Q.; Wei, X.; Xu, C.; Qi, Z.; Li, Q.; Ma, S.; Lin, J.; et al. BACH1 controls hepatic insulin signaling and glucose homeostasis in mice. *Nat. Commun.* **2023**, *14*, 8428. [CrossRef]
12. Xiong, Y.; Ma, P.; Huang, L.; Li, Y.; Wang, X. Widely targeted metabolomics analysis of different parts of kudzu. *Biomed. Chromatogr.* **2022**, *37*, e5545. [CrossRef] [PubMed]
13. Fan, Y.; Huang, G. Preparation, Structural Analysis and Antioxidant Activity of Polysaccharides and Their Derivatives from Pueraria lobata. *Chem. Biodivers.* **2023**, *20*, e202201253. [CrossRef] [PubMed]
14. Sun, Y.-J.; Cao, S.-J.; Liang, F.-N.; Li, J.-Y.; Zhang, X.-Y.; Li, W.; Ding, L.-Q.; Qiu, F. Puerol and pueroside derivatives from Pueraria lobata and their anti-inflammatory activity. *Phytochemistry* **2023**, *205*, 113507. [CrossRef] [PubMed]
15. Elchebly, M.; Payette, P.; Michaliszyn, E.; Cromlish, W.; Collins, S.; Loy, A.L.; Normandin, D.; Cheng, A.; Himms-Hagen, J.; Chan, C.-C.; et al. Increased Insulin Sensitivity and Obesity Resistance in Mice Lacking the Protein Tyrosine Phosphatase-1B Gene. *Science* **1999**, *283*, 1544–1548. [CrossRef] [PubMed]
16. Ding, H.; Zhang, Y.; Xu, C.; Hou, D.; Li, J.; Zhang, Y.; Peng, W.; Zen, K.; Zhang, C.-Y.; Jiang, X. Norathyriol reverses obesity- and high-fat-diet-induced insulin resistance in mice through inhibition of PTP1B. *Diabetologia* **2014**, *57*, 2145–2154. [CrossRef] [PubMed]
17. Xu, L.; Li, Y.; Dai, Y.; Peng, J. Natural products for the treatment of type 2 diabetes mellitus: Pharmacology and mechanisms. *Pharmacol. Res.* **2018**, *130*, 451–465. [CrossRef] [PubMed]
18. van Smoorenburg, A.N.; Hertroijs, D.F.L.; Dekkers, T.; Elissen, A.M.J.; Melles, M. Patients' perspective on self-management: Type 2 diabetes in daily life. *BMC Health Serv. Res.* **2019**, *19*, 605. [CrossRef] [PubMed]
19. Dou, Z.; Xia, Y.; Zhang, J.; Li, Y.; Zhang, Y.; Zhao, L.; Huang, Z.; Sun, H.; Wu, L.; Han, D.; et al. Syndrome Differentiation and Treatment Regularity in Traditional Chinese Medicine for Type 2 Diabetes: A Text Mining Analysis. *Front. Endocrinol.* **2021**, *12*, 728032. [CrossRef] [PubMed]
20. Yimin, L.; Yanhua, J.; Mengjie, C.; Xiaojun, Y.; Hongning, L.; Zhijun, Z. Research progress of Gegen Qinlian Decoction and its modified prescriptions in improving insulin resistance in type 2 diabetes. *Chin. J. Exp. Tradit. Med. Formulae* **2024**, *30*, 1–11. [CrossRef]
21. Li, J.; Li, Y.; Yuan, X.; Yao, D.; Gao, Z.; Niu, Z.; Wang, Z.; Zhang, Y. The effective constituent puerarin, from Pueraria lobata, inhibits the proliferation and inflammation of vascular smooth muscle in atherosclerosis through the miR-29b-3p/IGF1 pathway. *Pharm. Biol.* **2022**, *61*, 1–11. [CrossRef]
22. Su, B.; Kan, Y.; Xie, J.; Hu, J.; Pang, W. Relevance of the Pharmacokinetic and Pharmacodynamic Profiles of Puerariae lobatae Radix to Aggregation of Multi-Component Molecules in Aqueous Decoctions. *Molecules* **2016**, *21*, 845. [CrossRef] [PubMed]
23. Wu, W.; Yang, S.; Liu, P.; Yin, L.; Gong, Q.; Zhu, W. Systems Pharmacology-Based Strategy to Investigate Pharmacological Mechanisms of Radix Puerariae for Treatment of Hypertension. *Front. Pharmacol.* **2020**, *11*, 11345. [CrossRef]
24. He, H.; Peng, S.; Song, X.; Jia, R.; Zou, Y.; Li, L.; Yin, Z. Protective effect of isoflavones and triterpenoid saponins from pueraria lobata on liver diseases: A review. *Food Sci. Nutr.* **2021**, *10*, 272–285. [CrossRef]
25. Zhang, Z.-T.; Guo, N.; Zhuang, G.-D.; Deng, S.-M.; He, W.-J.; Chen, Z.-Q.; Xu, Y.-H.; Tang, D.; Wang, S.-M. Metabolic Profiling of Carbonyl Compounds for Unveiling Protective Mechanisms of Pueraria lobata against Diabetic Nephropathy by UPLC-Q-Orbitrap HRMS/MS Analysis. *J. Agric. Food Chem.* **2021**, *69*, 10943–10951. [CrossRef]
26. Yang, L.; Chen, J.; Lu, H.; Lai, J.; He, Y.; Liu, S.; Guo, X. Pueraria lobatafor Diabetes Mellitus: Past, Present and Future. *Am. J. Chin. Med.* **2019**, *47*, 1419–1444. [CrossRef]
27. She, S.; Liu, W.; Li, T.; Hong, Y. Effects of puerarin in STZ-induced diabetic rats by oxidative stress and the TGF-β1/Smad2 pathway. *Food Funct.* **2014**, *5*, 944–950. [CrossRef]
28. Sun, R.; Deng, X.; Zhang, D.; Xie, F.; Wang, D.; Wang, J.; Tavallaie, M.S.; Jiang, F.; Fu, L. Anti-diabetic potential of Pueraria lobata root extract through promoting insulin signaling by PTP1B inhibition. *Bioorg. Chem.* **2019**, *87*, 12–15. [CrossRef] [PubMed]
29. Jonas, J.C.; Plant, T.D.; Gilon, P.; Detimary, P.; Nenquin, M.; Henquin, J.C. Multiple effects and stimulation of insulin secretion by the tyrosine kinase inhibitor genistein in normal mouse islets. *Br. J. Pharmacol.* **2012**, *114*, 872–880. [CrossRef] [PubMed]

30. Klaman, L.D.; Boss, O.; Peroni, O.D.; Kim, J.K.; Martino, J.L.; Zabolotny, J.M.; Moghal, N.; Lubkin, M.; Kim, Y.B.; Sharpe, A.H.; et al. Increased energy expenditure, decreased adiposity, and tissue-specific insulin sensitivity in protein-tyrosine phosphatase 1B-deficient mice. *Mol. Cell. Biol.* **2020**, *20*, 5479–5489. [CrossRef]
31. Goldstein, B.J.; Bittner-Kowalczyk, A.; White, M.F.; Harbeck, M. Tyrosine dephosphorylation and deactivation of insulin receptor substrate-1 by protein-tyrosine phosphatase 1B. Possible facilitation by the formation of a ternary complex with the Grb2 adaptor protein. *J. Biol. Chem.* **2020**, *275*, 4283–4289. [CrossRef]
32. Agouni, A.; Tual-Chalot, S.; Chalopin, M.; Duluc, L.; Mody, N.; Martinez, M.C.; Andriantsitohaina, R.; Delibegović, M. Hepatic protein tyrosine phosphatase 1B (PTP1B) deficiency protects against obesity-induced endothelial dysfunction. *Biochem. Pharmacol.* **2014**, *92*, 607–617. [CrossRef] [PubMed]
33. Vakili, S.; Ebrahimi, S.S.S.; Sadeghi, A.; Gorgani-Firuzjaee, S.; Beigy, M.; Pasalar, P.; Meshkani, R. Hydrodynamic-based delivery of PTP1B shRNA reduces plasma glucose levels in diabetic mice. *Mol. Med. Rep.* **2013**, *7*, 211–216. [CrossRef] [PubMed]
34. Liu, R.; Mathieu, C.; Berthelet, J.; Zhang, W.; Dupret, J.-M.; Rodrigues Lima, F. Human Protein Tyrosine Phosphatase 1B (PTP1B): From Structure to Clinical Inhibitor Perspectives. *Int. J. Mol. Sci.* **2022**, *23*, 7027. [CrossRef] [PubMed]
35. Zhang, H.; Wei, Z.; Tong, Y.; Song, X.; Li, S.; Sun, Y.; Liu, C.; Han, F.; Yu, J. Spectrum-effect relationship study to reveal the pharmacodynamic substances in Flos Puerariae-Semen Hoveniae medicine pair for the treatment of alcohol-induced liver damage. *J. Ethnopharmacol.* **2023**, *314*, 116628. [CrossRef]
36. Baskaran, S.K.; Goswami, N.; Selvaraj, S.; Muthusamy, V.S.; Lakshmi, B.S. Molecular Dynamics Approach to Probe the Allosteric Inhibition of PTP1B by Chlorogenic and Cichoric Acid. *J. Chem. Inf. Model.* **2012**, *52*, 2004–2012. [CrossRef]

molecules

Article

Study of Bitespiramycin Distribution in Rats and Cerebrospinal Fluid of Patients by a Sensitive LC-MS/MS Method with Rapid Sample Preparation

Yujie Zhang [1,†], Jingjie Cao [1,†], Jiahan Su [1], Tingting He [1], Qianru Wang [1], Feng Wei [2], Xin Guo [3], Qibing Mei [1,3,*] and Jing Zeng [1,*]

1 School of Pharmacy, Southwest Medical University, Luzhou 646000, China; yujiezhang_995@163.com (Y.Z.)
2 Shanghai Tonglian Pharmaceutical Co., Ltd., Shanghai 201611, China
3 Luzhou New Drug Safety Evaluation Research Center, Luzhou 646000, China
* Correspondence: qbmei@swmu.edu.cn (Q.M.); zengjing@swmu.edu.cn (J.Z.)
† These authors contributed equally to this work.

Abstract: Bitespiramycin, has been shown to have a therapeutic effect against respiratory tract inflammation, including a potential effect against COVID-19. A current clinical trial in China showed that bitespiramycin was an effective treatment for severe pneumonia and intracranial infection. However, there is lack of an analytical method to elucidate the distribution of bitespiramycin. In this study, a highly sensitive, rapid and reliable UPLC–MS/MS method was developed to comprehensively characterize the bitespiramycin distribution in various bio-samples, which is significantly improved upon the published work. A rapid sample preparation method was developed by using *n*-butanol as the solvent to extract bitespiramycin from different bio-samples. The extract was then directly analyzed by UPLC–MS/MS coupled with an alkaline-resistant column after centrifugation which avoids the time-consuming concentration process under nitrogen and redissolution. The method was employed to accurately quantify bitespiramycin and its metabolites in rat plasma, tissues, and human cerebrospinal fluid. Notably, the presence of bitespiramycin and its metabolites was identified for the first time in various rat organs including brain, testis, bladder and prostate as well as in human cerebrospinal fluid. This newly developed approach shows great promise for drug distribution assays including other antibiotics and can help elucidate the ADME of bitespiramycin.

Keywords: bitespiramycin; distribution; LC–MS/MS; rat; human cerebrospinal fluid

1. Introduction

Bitespiramycin, a new 16-membered macrolide antibiotic, has been shown to have excellent therapeutic effect against respiratory tract inflammation, including the potential effect against COVID-19 [1]. It has been listed in a clinical trial (ClinicalTrials.gov Identifier: NCT04672564) to validate the efficacy against COVID-19 by National Institutes of Health (NIH) [2]. In addition, isovaleryl spiramycins I (ISV-SPM I), the main component of bitespiramycin, has been proved to be an active ingredient to suppress cancer cell growth and tumor metastases by targeting selenoprotein H [3]. Different from other drugs with a single ingredient, bitespiramycin is mainly composed of ISV-SPMs I, II, and III [4], which could interact with multiple targets in microenvironments. The three pharmacologically active ingredients can rapidly undergo in vivo de-esterification, resulting in the formation of spiramycins (SPMs) I, II, and III as metabolites (Figure 1) [5]. These three metabolites exhibit a strong positive correlation with therapeutic efficacy.

Recently, a new clinical trial in China has shown that bitespiramycin has excellent therapeutic effect against severe pneumonia, prostatitis, and intracranial infection (unpublished data). However, comprehensive distribution of bitespiramycin in tissues and body fluids, especially in lung, prostate, and cerebrospinal fluid needs to be further elucidated.

An applicable analytical method with simplicity, high sensitivity and specificity is urgently needed, since the lowest concentration of bitespiramycin in some tissues would be in the range of ppb to ppt.

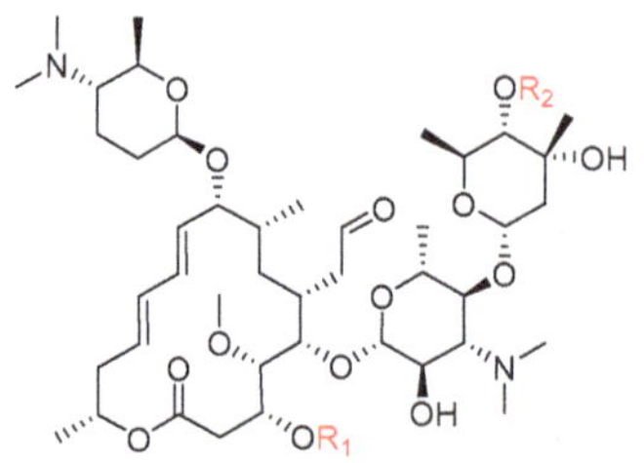

Component	R_1	R_2	Formula
Isovaleryl spiramycins I (ISV-SPM I)	H	$COCH_2CH(CH_3)_2$	$C_{48}H_{82}N_2O_{15}$
Isovaleryl spiramycins II (ISV-SPM II)	$COCH_3$	$COCH_2CH(CH_3)_2$	$C_{50}H_{84}N_2O_{16}$
Isovaleryl spiramycins III (ISV-SPM III)	$COCH_2CH_3$	$COCH_2CH(CH_3)_2$	$C_{51}H_{86}N_2O_{16}$
Spiramycins I (SPM I)	H	H	$C_{43}H_{74}N_2O_{14}$
Spiramycins II (SPM II)	$COCH_3$	H	$C_{45}H_{76}N_2O_{15}$
Spiramycins III (SPM III)	$COCH_2CH_3$	H	$C_{46}H_{78}N_2O_{15}$

Figure 1. The structures and formulas of bitespiramycin and its metabolites.

LC–MS/MS-based quantitation methods were usually used to determine metabolic profiles in vivo, tissue distribution, and pharmacokinetics of bitespiramycin in recent years, while sample preparation was very time-consuming [6–9]. Reversed phase (C_{18})—based solid phase extraction (SPE) combined with being concentrated under a stream of nitrogen and redissolution was used to prepare body fluids, including plasma, urea, and bile. In addition, a mixed solvent, ethyl acetate–isopropanol (95:5, *v*/*v*) based liquid–liquid extraction (LLE) method at pH 9~10 was developed to extract bitespiramycin and its metabolites in tissues, followed by being concentrated and redissolved before LC–MS/MS measurement. In another study, acetonitrile and ethyl acetate were employed as the protein precipitating and extracting agent respectively in the study of bitespiramycin pharmacokinetics in beagle dogs [10]. The extract was then concentrated and redissolved before UHPLC–MS/MS analysis, which was similar to the previous studies [6,7].

In this study, a rapid, sensitive and reliable UHPLC–MS/MS method, specifically to simplify the sample preparation, detect bitespiramycin and its metabolites simultaneously in rat plasma, tissues, and human cerebrospinal fluid, was developed. It revealed the underlying mechanism of the clinical therapeutic effect on severe pneumonia, prostatitis, and intracranial infection. The newly developed approach could provide a promising strategy to analyze macrolide antibiotics in vivo.

2. Results and Discussion

2.1. Method Development

Sample extraction, purification, and enrichment are often the rate-limiting steps of an analytical method, making them critical aspects of method development. As for the extraction solvent selection, acetonitrile, ethyl acetate, chloroform, dichloromethane, and *n*-butanol are commonly used solvents for sample extraction. Acetonitrile is a well-known solvent for protein precipitation, but it is not functional for the isolation of liposoluble compounds (phospholipids, etc.), resulting in rapid decrease of efficiency and increase of back pressure of columns. Chloroform and dichloromethane are good choices for the LLE, but not compatible with the aqueous mobile phase in the UHPLC separation. Another option for sample extraction is a mixed solvent composed of acetonitrile and ethyl acetate, followed by several steps including vortexing, centrifugation, nitrogen concentration, and redissolution [9,10]. N-butanol is a popular solvent to extract polar or medium-polar components in the area of phytochemistry, which would be a potential proper solvent for the extraction of bitespiramycin and its metabolites. In this study, the solubility of bitespiramycin in *n*-butanol was found to be about 2 times higher than that in ethyl acetate. The boiling point of *n*-butanol (117 °C) is much higher than other solvents and less volatile, resulting in the concentration of *n*-butanol extract being more stable. Moreover, a microliter of *n*-butanol extract is soluble in the aqueous mobile phase, facilitating direct analysis of the extract. Therefore, *n*-butanol was selected as the extraction solvent in this study. Bio-samples extracted by *n*-butanol could be directly injected to the UHPLC–MS/MS system without being further concentrated under nitrogen and redissolution, which reduced the whole process time and significantly improved the analytical throughput and stability.

The pH of the mobile phase is widely recognized to have a significant impact on separation and detection processes. Bitespiramycin, being a weak alkaline compound due to its amino sugar moiety, typically exhibits low efficiency or peak tailing when subjected to acidic mobile phase conditions owing to the strong ion interaction between dissociated silanol (anions) and bitespiramycin (cations) [7–10]. Under the condition of alkaline mobile phase ($pH_{mobile\ phase} > pI_{bitespiramycin}$), bitespiramycin is a kind of neutral molecules, and the strong ion interactions were greatly reduced, improving peak shape and column efficiency. Therefore, the alkaline mobile phase (pH 8.8) coupled with Waters Acquity BEH–C_{18} column with the widest pH range (pH 1–12) was used. The peak shapes of bitespiramycin and internal standard azithromycin were significantly improved, which increases the sensitivity of this method. Additionally, UHPLC gradient, parameters of ion source, MRM transitions and collision energy were carefully optimized. As a result, a sensitive LC-MS/MS method with rapid sample preparation was successfully developed.

2.2. Method Validation

2.2.1. Specificity and Selectivity

The MRM chromatograms of a mixture containing six standards, plasma samples after administration of bitespiramycin and spiked internal standard azithromycin were overlaid, as depicted in Figures 2a and 2b, respectively. Each MRM chromatogram exhibited a single peak in its molecular ion−product ion transition without interference from endogenous impurities due to the high resolution and specificity of the triple quadrupole mass analyzer. However, it should be noted that the six components exhibited baseline separation with a minimum resolution of approximately 1.5. Furthermore, Figure 2c displays the MRM chromatograms of blank plasma without any interfering peaks observed, indicating exceptional specificity and selectivity.

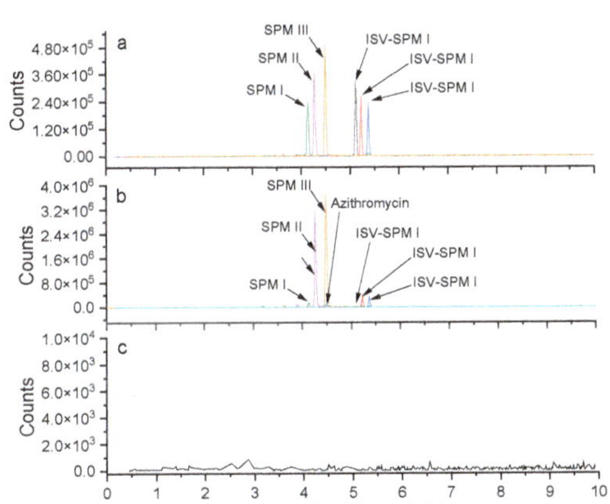

Figure 2. The MRM chromatograms of mixtures of six standards (**a**), plasma after bitespiramycin administration and spiked internal standard azithromycin (**b**), and blank plasma (**c**).

2.2.2. Linearity and LLOQ

Six standards spiked in rat plasma, tissues and cerebrospinal fluid of patients with different concentrations were measured, and the representative linear curve of cerebrospinal fluid was plotted in Figure 3. The linear ranges were shown in Table 1. The concentration (X) is linearly related to the ratio of analyte to internal standard (Y). The deviation between the actual concentrations with standards at LLOQ (signal to noise = 10) was within 20%, and the deviations of other concentration were all within ±15%. The LLOQ was significantly improved in this study compared to the LLOQ published in the literature [9,10], as shown in Table 2.

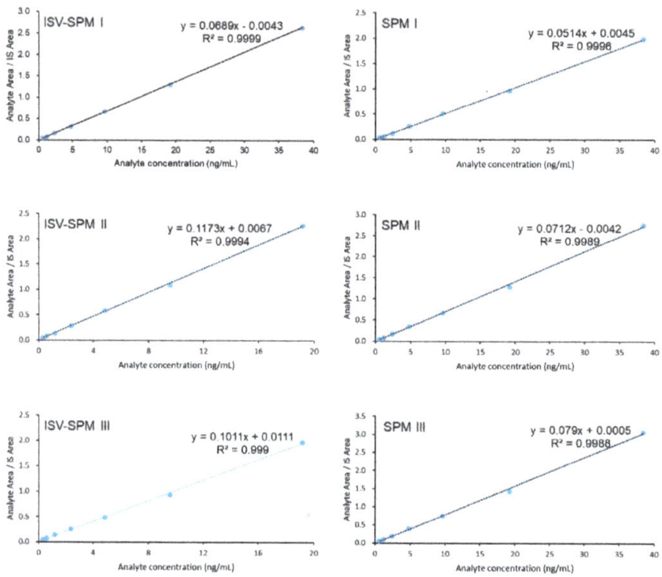

Figure 3. The Linearity of SPMs I, II, III, and ISV-SPM I, II, III in cerebrospinal fluid of patients.

Table 1. Linear ranges of ISV-SPM I, II, III and SPM I, II, III in tissues and body fluid.

Tissues/Body Fluid	Linear Ranges	
	ISV-SPM I, SPM I, II, III	**ISV-SPM II, III**
Brain	1.2–72.8 ng/g	0.6–38.4 ng/g
Testis	0.6–800 ng/g	0.3–400 ng/g
Prostate	1.2–8000 ng/g	0.6–4000 ng/g
Uterus and ovary	1.8–12,000 ng/g	0.9–6000 ng/g
Bladder	2.4–16,000 ng/g	1.2–8000 ng/g
Lung	3.6–24,000 ng/g	1.8–12,000 ng/g
Plasma	0.6–4000 ng/mL	0.3–2000 ng/mL
Cerebrospinal fluid	0.6–38.4 ng/mL	0.3–19.2 ng/mL

Table 2. Comparison of LLOQs of SPMs and ISV-SPMs (ng/mL).

	ISV-SPM I	**ISV-SPM I**	**ISV-SPM I**	**SPM I**	**SPM II**	**SPM III**
LLOQ reported in [9]	4	12	18	4	12	18
LLOQ reported in [10]	1	1	1	1	1	1
LLOQ of this study	0.3	0.6	0.6	0.6	0.3	0.3

2.2.3. Accuracy and Precision

The accuracy fell within the range of 89.90% to 111.10%. The intra-day and inter-day precision errors were below 13.62% and 13.71%, respectively, indicating compliance with the specifications outlined in the Chinese Pharmacopoeia (Guidelines for Validation of Quantitative Analysis Methods for Biological Samples).

2.2.4. Extraction Recovery Rate and Matrix Effect

The extraction recovery rate of bitespiramycin and internal standard varied from 85% to 115%. The ion suppression/enhancement was less than 15% for QC samples, indicating good accuracy and reproducibility.

2.2.5. Stability

RSD of ISV-SPMs I, II, III and SPMs I, II, III under four storage conditions were all less than ±15%, indicating bitespiramycin and its metabolites were stable.

2.3. Distribution of Bitespiramycin in Rats

2.3.1. Distribution in Rat Plasma

Using the validated UHPLC–MS/MS method, bitespiramycin and its metabolites concentration in rat plasma was measured at 1, 3, 6, 12, 24, and 48 h post intragastric administration of 100 mg/kg of bitespiramycin. As shown in Figure 4a and Table S1, ISV-SPMs I, II, and III were rapidly metabolized to SPMs I, II, and III, respectively. The concentration of ISV-SPMs I (5.08 ng/mL), II (25.05 ng/mL), and III (13.15 ng/mL) after 6 h administration remained at lower levels, while that of SPMs I (33.65 ng/mL), II (338.67 ng/mL), and III (970.00 ng/mL) remained at significantly higher levels, indicating that the three metabolites of bitespiramycin would be the pharmacologically active metabolites. After 48 h post-administration, the ISV-SPM I in plasma cannot be detected, while the concentration of SPM III remained at 47.12 ng/mL.

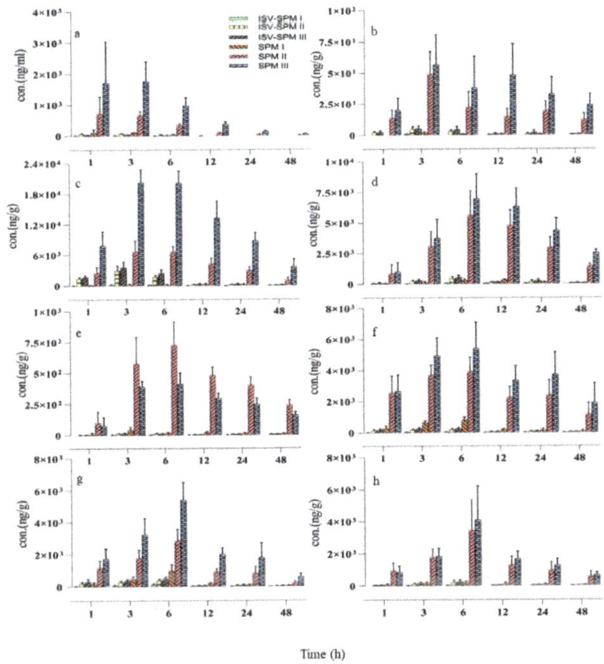

Figure 4. The distribution of SPMs I, II, III and ISV-SPMs I, II, III in the plasma (**a**), brain (**b**), lung (**c**), prostate (**d**), testis (**e**), womb (**f**), ovary (**g**), and bladder (**h**) of rats after intragastric administration of 100 mg/kg of bitespiramycin (Mean ± SD, n = 6).

2.3.2. Distribution in Rat Tissues

The concentrations of six analytes in tissues were also measured at 1, 3, 6, 12, 24, and 48 h after intragastric administration of 100 mg/kg of bitespiramycin, shown in Figure 4b–h and Tables S2–S7. ISV-SPMs I, II, III and SPMs I, II, III were absorbed, distributed, accumulated and eliminated in various tissues, including lung, uterus, ovary, testis, prostate, bladder and brain. Their concentrations in these tissues were significantly different. Concentrations of ISV-SPMs I, II, III and SPMs I, II, III in lung were the highest, which was about 200 times more than that in the lowest tissue, the brain. The highest concentration of ISV-SPMs I, II, III and SPMs I, II, III in these tissues was observed at 6 h, except in the brain which was at 3 h.

These results indicated that ISV-SPMs I, II, III and SPMs I, II, III tend to distribute in lung, followed by prostate, bladder, womb, ovary, testis, brain and cerebrospinal fluid. Bladder, prostate and cerebrospinal fluid of patients were examined for the first time. It was also the first time that bitespiramycin was successfully detected in brain and testis due to the significantly improved sensitivity of this method. Furthermore, after comparing the concentrations of SPMs and ISV-SPMs, it can be concluded that the concentrations of SPMs in tissues were about 10 times higher than that of ISV-SPMs, indicating that parent drugs, ISV-SPMs I, II, and III were rapidly metabolized to SPMs I, II, III, respectively, similar to the results in plasma. Moreover, SPMs I, II, III were eliminated slowly, which cannot completely be cleared from the rat's body, even after 48 h. It was interesting that the concentration of SPM III in lungs was as high as 3700.33 ng/g after 48 h, which would directly contribute to the therapeutic efficacy against pneumonia. The highest concentration of ISV-SPMs I, II, III and SPMs I, II, III in these tissues was observed at 6 h, except in the brain which was at 3 h.

The tissue-to-plasma concentration ratio (C_t/C_p) at 1, 3, 6, 12, 24, and 48 h was statistically analyzed in Table 3 and Tables S9–S13. At the 24 h post-dose time point, ISV-SPMs II, III and SPMs I, II, III were detected except for ISV-SPM I. There were significant differences in the concentrations of six components in various tissues. The average C_t/C_p value in the brain was found to be 0.51, whereas it was significantly higher in the lung (104.86), followed by the prostate (82.06), womb (46.55), ovary (30.11), bladder (22.11), and testis (4.88). These findings indicate that bitespiramycin exhibits extensive tissue distribution with the exception of in the brain.

Table 3. The tissue-to-plasma concentration ratio (Ct/Cp) determined 24 h following oral administration of bitespiramycin at a dosage of 100 mg/kg.

	Brain	Lung	Prostate	Testis	Womb	Ovary	Bladder
ISV-SPM I	/	/	/	/	/	/	/
ISV-SPM II	0.31	83.10	78.81	2.70	30.66	26.28	12.35
ISV-SPM III	1.34	221.04	182.44	5.82	60.86	57.28	37.17
SPM I	0.22	89.29	53.50	5.60	62.18	35.75	31.27
SPM II	0.40	61.84	61.11	8.34	49.47	17.06	19.91
SPM III	0.26	69.01	34.46	1.95	29.59	14.19	9.87

2.4. Distribution in Cerebrospinal Fluid of Patients

To further identify if bitespiramycin could cross the blood-brain barrier, bitespiramycin and its metabolites in cerebrospinal fluid of patients were measured. As shown in Table 4, the concentration of the six analytes collected on the first day was in the range of 0–1.36 ng/mL and on the fifth day in the range of 0–3.40 ng/mL, respectively, indicating a trend of drug accumulation. Moreover, the concentrations of the six analytes in another sample collected on nineteenth and twentieth days increased to the range of 0–13.00 ng/mL, which was about 5 times higher than that on the fifth day. Bitespiramycin and its metabolites were definitively accumulated in the brain to a relatively high concentration, due to the properties of slow elimination.

Table 4. The concentrations of SPMs I, II, III and ISV-SPMs I, II, III in cerebrospinal fluid of patients.

No.	Sex	Age	Sampling Time *	Concentration (ng/mL)					
				ISV-SPM I	ISV-SPM II	ISV-SPM III	SPM I	SPM II	SPM III
1	Male	77	1	ND	0.47	0.60	ND	ND	ND
			5	ND	1.54	1.76	ND	ND	1.27
2	Male	44	1	ND	ND	ND	ND	ND	ND
			5	ND	0.31	0.42	ND	ND	0.75
3	Male	57	1	ND	ND	ND	ND	ND	ND
			5	ND	ND	ND	ND	0.99	0.67
4	Female	53	1	ND	ND	0.33	ND	1.36	0.83
			5	ND	1.73	1.83	ND	2.77	3.40
5	Male	43	1	ND	ND	ND	ND	0.74	ND
			5	ND	ND	ND	ND	1.13	0.86
6	Male	72	19	0.72	5.21	9.14	ND	4.69	13.00
			20	1.05	6.55	9.82	ND	5.61	11.90

* The days after continuous oral administration. ND: Not Detected.

3. Material and Methods

3.1. Reagents

Bitespiramycin Tablets (Shanghai Tonglian Pharmaceutical Co., Ltd., Shanghai, China, Lot No. 0220200204); SPMs I, II, III and ISV-SPMs I, II, III standards (Shanghai Tonglian Pharmaceutical Co., Ltd., purity 98%); Acetonitrile (chromatographic grade; Fisher Chemical, Waltham, MA, USA); *n*-butanol (HPLC; Macklin, Shanghai, China); Methanol (chromatographic grade; Aladdin, Shanghai, China); Ammonium acetate (LC-MS; Aladdin, Shanghai, China); Ammonia (LC-MS; Aladdin, Shanghai, China); Azithromycin in methanol (internal standard, 100 μg/mL, purity 98%, Ehrenstorfer, London, UK); Sodium carboxymethyl cellulose (Jiangsu Anxin Food Ingredients Mall, Nanjing, China); Sodium heparin (Leagene, Beijing, China); Endothelial Cell Medium (ECM, Sciencell, Carlsbad, CA, USA); Pancreatin (Gibco, Waltham, MA, USA); DPBS 1X (Gibco, Waltham, MA, USA); Matrigel matrix (Corning, New York, NY, USA).

3.2. Animals

Eighty-four Sprague Dawley (SD) rats of Specific Pathogen Free (SPF) grade were purchased from the Laboratory Animal Center of Southwest Medical University (Luzhou, Sichuan, China). The rats were kept in an animal room with a barrier system at temperature (20 ± 2 °C), humidity (50 ± 5%), and a 12 h light/dark cycle and were fasted for 18 h with free drinking water before administration. Experimental animal production license is No: SCXK (Chuan) 2018-17 and experimental animal use license number is No: SYXK (Chuan) 2018-065. All animal procedures were approved (No. SYXK202112–01) by the animal ethics committee of Luzhou New Drug Evaluation and Research Center (Luzhou, Sichuan, China).

The rats selected for the study weighed an average of 210 ± 23 g for males and 193 ± 14 g for females. They were randomly divided into seven groups, with each group consisting of twelve rats (six males and six females) sacrificed after administration periods of 0 (control group), 1, 3, 6, 12, 24, and 48 h. The rats were then given bitespiramycin by gavage (100 mg/kg). After 1, 3, 6, 12, 24 and 48 h post administration, rats were sacrificed by CO_2 and the blood was collected from the inferior vena cava for further testing. Rat tissues, including lung, prostate, testis, bladder, ovary, uterus and brain were collected and frozen at −20 °C.

3.3. Patient Cerebrospinal Fluid Collection

Clinical trials were approved (No. KY20212078–C–1) by the Medical Ethics Committee of the Xijing Hospital of Fourth Military Medical University. Limited cerebrospinal fluid samples of six inpatients in the period of September 2021–January 2022 were collected at the 0 (control sample), 1, 5, 19 or 20 days after bitespiramycin treatment for further biochemical assay.

3.4. Standards and Sample Preparation for UHPLC–MS/MS

SPMs I, II, III and ISV-SPMs I, II, III standard stock solution (1 mg/mL) was prepared with acetonitrile-water (1:1, *v:v*) and stored at 4 °C, which was used for the calibration curves and the low, medium and high quality controls. ISV-SPMs II, III: 0.5 ng/mL (low), 9 ng/mL (medium) and 15 ng/mL (high), ISV-SPM I, SPMs I, II, III: 1 ng/mL (low), 18 ng/mL (medium) and 30 ng/mL (high) with acetonitrile-water (1:1, *v:v*). The concentrations of the low, medium, and high quality controls was determined based on the linear ranges of each component. The low quality controls (QCL) were set at approximately twice the lowest concentration within the linear range. The medium quality controls (QCM) were set at around 50% of the concentration within the linear range. The high quality controls (QCH) were set at about 70% of the concentration within the linear range.

The accurately weighed tissues were cut, fully dispersed, and homogenized using a glass grinder with the addition of pure water in different proportions: 1 mL/mg for brain, testis, and prostate; 2 mL/mg for uterus and ovary; 3 mL/mg for bladder; and 5 mL/mg for lung. The resulting tissue homogenate was then prepared along with plasma and cerebrospinal fluid for UHPLC-MS/MS analysis.

40 µL of each biological sample (rat's plasma or tissue homogenate, patients' cerebrospinal fluid) was mixed with 50 µL of pure water, 5 µL of internal standard (Azithromycin, 20 ng/mL), 5 µL of 50% acetonitrile-water, and 200 µL of *n*-butanol. The mixture was vortexed for 1 min, sonicated for 2 min, and centrifuged at 13,000 rpm at 4 °C for 10 min. The supernatant was collected for UHPLC-MS/MS analysis (Figure 5).

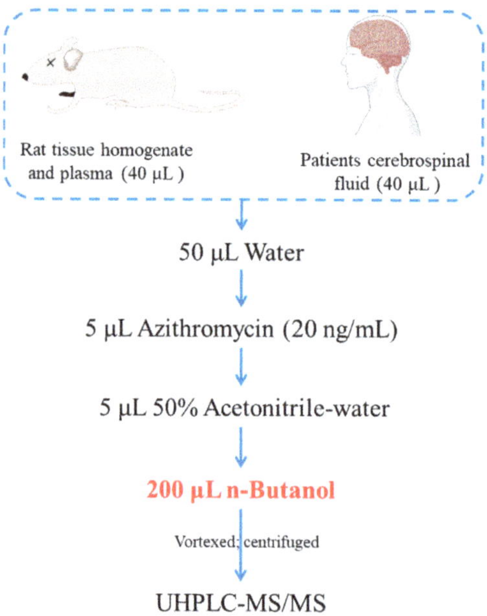

Figure 5. The sample preparation diagram.

3.5. UHPLC–MS/MS

UHPLC-MS/MS system was equipped with an UHPLC separation module (Acquity UPLC I-Class PLUS, Waters, Milford, MA, USA) and a triple Quad mass spectrometer. The Acquity BEH–C_{18} column (50 mm × 2.1 mm, 1.7 μm) combined with 5 mM ammonium acetate with 0.05% ammonia (pH 8.8) (A) and acetonitrile (B) was used for the separation with gradient elution (0–0.5 min, 5% B; 0.5–5 min, 5%–95% B; 5–9 min, 95% B). The flow rate was set at 0.3 mL/min, the injection volume was 2 μL, and the column temperature was set at 45 °C.

The mass spectrometry was performed on triple Quad 6500 plus mass spectrometer (Sciex, Boston, MA, USA). The parameters of the ES ion source in positive mode were set as follows: voltage 5500 V, curtain gas (CUR) 35 psi, spray gas (GS1) 50 psi, auxiliary heater (GS2) 50 psi, ionization temperature (TEM) 550 °C, the dwell-time values 0.2 s. The MS data recordings were carried out in multiple reaction monitoring (MRM) mode and the scanning time was 0.5 s. The declustering potential (DP), collision energy (CE) and transitions for each of the analytes and for internal standard were shown in Table 5.

Table 5. The declustering potential, collision energy and transitions for MS/MS detection.

Component	m/z	CE/V	DP/V
SPM I	843.7 → 174.2	43	80
SPM II	885.7 → 174.3	44	80
SPM III	899.7 → 174.4	46	80
ISV-SPM I	927.7 → 174.0	51	80
ISV-SPM II	969.8 → 174.1	48	80
ISV-SPM III	983.8 → 174.1	48	80
Azithromycin	749.7 → 591.6	42	90

3.6. Method Validation

Specificity and selectivity were validated by comparing blank samples, spiked samples, and samples after administration. Six standards spiked into blank bio-samples with gradient concentrations were determined. The concentrations of the analytes were subjected to linear regression with the ratio of the peak area of analytes to that of internal standard. LLOQ is defined as the lowest point of the linear regression curve with a precision less than 20%. Intra-day accuracy and precision were evaluated by continuously analyzing of six replicates of LLOQ, low, medium and high QC samples in a day. Inter-day accuracy and precision were assessed by analyzing of three batches of samples with the same material in three days. The precision should not be greater than 15% and the mean accuracy should be within ±15%, except for the LLOQ which can be within ±20% of the standard concentration.

Recovery and matrix effect were measured at low, medium, and high QC samples in six replicates. Matrix effect was assessed by comparing the concentrations of standards spiked bio-samples with that of solution spiked standards. Extraction recovery rate was evaluated by comparing the concentrations of pre-extraction samples with that of post-extraction samples.

The stability of bitespiramycin was investigated by comparing the concentrations of low, medium and high QC samples under the following four storage conditions: short-term stability of body fluid and tissues at 25 °C for 2 h (all the replicate samples were prepared within 2 h); long-term stability at –20 °C for 7 days; freeze-thaw stability for three cycles; autosampler stability at 4 °C for 24 h. The analytes were considered stable when the deviation was within ±15%.

4. Conclusions

An improved UHPLC–MS/MS method was developed and carefully validated, which simplified the sample preparation and improved the peak shape, resulting in significantly enhanced sensitivity and stability. This validated method was then successfully applied to the study of bitespiramycin distribution in rat plasma, tissues, and in human cerebrospinal fluid, in which bladder, prostate and cerebrospinal fluid of patients were examined for the first time. Moreover, bitespiramycin and its metabolites were detected for the very first time in rat brain, testis, bladder, and prostate, and in human cerebrospinal fluid.

Bitespiramycin was first proved to cross the blood-brain barrier, providing evidence of its therapeutic efficacy against intracranial infection. Distribution in lung and prostate with a higher concentration would directly contribute to its anti-pneumonia and anti-prostatitis efficacy. A promising strategy has been established for the treatment drug monitoring and elucidation of the underlying ADME of bitespiramycin, as well as other macrolide antibiotics.

Supplementary Materials: The following supporting information can be downloaded at: https://www.mdpi.com/article/10.3390/molecules29051037/s1.

Author Contributions: Conceptualization, J.Z.; Methodology, J.C., J.S., Q.M. and J.Z.; Formal analysis, J.S. and T.H.; Investigation, Y.Z. and J.C.; Resources, F.W. and J.Z.; Data curation, Y.Z., J.C. and Q.W.; Writing—original draft, Y.Z.; Writing—review & editing, J.C., X.G. and J.Z.; Supervision, Q.M.; Project administration, Q.M. and J.Z. All authors have read and agreed to the published version of the manuscript.

Funding: This study was funded by the National Natural Science Foundation of China (82273889), the Science and Technology Department of Sichuan Province (2023NSFSC0657), and the Cooperative Project of Sichuan University and Luzhou (2022CDLZ-17).

Institutional Review Board Statement: The study was conducted in accordance with the Declaration of Helsinki, and approved by the Ethics Committee of the Xijing Hospital of Fourth Military Medical University (protocol code KY20212078–C–1) at 5 January 2021. The animal study protocol was approved by the Ethics Committee of Luzhou New Drug Evaluation and Research Center (protocol code SYXK202112–01) at 8 August 2021.

Informed Consent Statement: Informed consent was obtained from all subjects involved in the study.

Data Availability Statement: The data presented in this study are available in article and Supplementary Materials.

Acknowledgments: The authors would like to express their gratitude to Xiaofan Jiang, Shijie Hu, Peng Luo, Xiaopeng Shi, and Shan Miao from Xijing Hospital, Fourth Military Medical University in Xian, China for their invaluable assistance in sample collection.

Conflicts of Interest: Author Feng Wei was employed by the company Shanghai Tonglian Pharmaceutical Co., Ltd. The remaining authors declare that the research was conducted in the absence of any commercial or financial relationships that could be construed as a potential conflict of interest.

References

1. Yan, H.Y.; Sun, J.; Wang, K.; Wang, H.; Wu, S.; Bao, L.; He, W.; Wang, D.; Zhu, A.; Zhang, T.; et al. Repurposing CFDA-approved drug carrimycin as an antiviral agent against human coronaviruses, including the currently pandemic SARS-CoV-2. *Acta Pharm. Sin. B* **2021**, *11*, 2850–2858. [CrossRef] [PubMed]
2. U. S. National Library of Medicine. Study to Evaluate Safety and Efficacy of Carrimycin for Treatment of Severe COVID-19 in Hospitalized Patients. Available online: https://clinicaltrials.gov/study/NCT04672564 (accessed on 19 February 2024).
3. Cui, J.; Zhou, J.; He, W.; Ye, J.; Westlake, T.; Medina, R.; Wang, H.; Thakur, B.L.; Liu, J.; Xia, M.; et al. Targeting selenoprotein H in the nucleolus suppresses tumors and metastases by Isovalerylspiramycin I. *J. Exp. Clin. Cancer Res.* **2022**, *41*, 126. [CrossRef]
4. Wang, M.J.; Xue, J.; Zou, W.-B.; Wang, Y.; Hu, C.-Q.; Hoogmartens, J.; Adams, E. Identification of multi components in bitespiramycin by liquid chromatography-mass spectrometry. *J. Pharm. Biomed. Anal.* **2012**, *66*, 402–410. [CrossRef] [PubMed]
5. Yang, Y.L.; Yang, J.N.; Hu, M.; Hu, C.Q. Analysis of components of bitespiramycin by HPLC. *Acta Pharm. Sin. B* **2009**, *44*, 1183–1186. [CrossRef]
6. Shi, X.G.; Fawcett, J.P.; Chen, X.Y.; Zhong, D.F. Structural identification of bitespiramycin metabolites in rat: A single oral dose study. *Xenobiotica* **2005**, *35*, 343–358. [CrossRef] [PubMed]
7. Shi, X.G.; Zhong, D.F.; Sun, Y.M.; Zhang, Y.F. Metabolites of A Novel Antibiotic Bitespiramycin in Rat Urine and Bile. *Chin. Chem. Lett.* **2004**, *4*, 431–434.
8. Zhong, D.; Shi, X.; Sun, L.; Chen, X. Determination of three major components of bitespiramycin and their major active metabolites in rat plasma by liquid chromatography-ion trap mass spectrometry. *J. Chromatogr. B* **2003**, *791*, 45–53. [CrossRef] [PubMed]

9. Shi, X.G.; Sun, Y.M.; Zhang, Y.F.; Zhong, D.F. Tissue distribution of bitespiramycin and spiramycin in rats. *Acta Pharmacol. Sin.* **2004**, *25*, 1396–1401. [PubMed]
10. Gao, H.L.; Liu, Q.; Liu, H.Z.; He, Z.G. Pharmacokinetics of kelimycin tablets in beagle dogs. *J. Shenyang Pharma Univ.* **2019**, *36*, 973–980. [CrossRef]

molecules

MDPI

Article

Ultrasound-Assisted Extraction, Followed by Gas Chromatography–Mass Spectrometry for the Simultaneous Quantification of Ethinyl Estradiol and Drospirenone in Contraceptive Formulations

Javier Peña , Iria González-Mariño * and José L. Pérez Pavón

Department of Analytical Chemistry, Nutrition and Bromatology, Faculty of Chemical Sciences, 37008 Salamanca, Spain; javierpena@usal.es (J.P.); jlpp@usal.es (J.L.P.P.)

* Correspondence: iriagonzalez@usal.es

Abstract: Contraceptive tablets typically contain a combination of two synthetic versions of an estrogen and a progestogen, which work together to inhibit the ovulation process. An accurate and precise quantification of these components is essential for contraceptive producers. In this study, we have developed the first gas chromatography–mass spectrometry (GC–MS) method for the simultaneous quantification of 17α-ethinyl estradiol (EE) and drospirenone (DP) in contraceptive formulations. Under the final working conditions, analytes were extracted from the solid by ultrasound-assisted extraction (15 min) in methanol. The resulting suspension was diluted in ethyl acetate, subjected to centrifugation and, finally, the supernatant was directly injected into the GC–MS system. No derivatization reagents were utilized. To correct for instrumental variations, calibration was performed using the internal standard method, with cholesterol as the internal standard. A good linearity was achieved throughout the calibration range for both EE (3–12 μg mL^{-1}) and DP (300–1200 μg mL^{-1}), with R^2 values exceeding 0.99. Trueness, assessed in terms of percentages of recovery, was also found to be satisfactory for both analytes, with recovery rates of 106 ± 8% for EE and 93 ± 9% for DP. Furthermore, intra-day and inter-day precision studies yielded relative standard deviation values below 6% for both analytes. In terms of sensitivity, the instrumental limits of detection were 0.25 μg mL^{-1} for EE and 6.6 μg mL^{-1} for DP, and the instrumental limits of quantification 0.82 μg mL^{-1} for EE and 22 μg mL^{-1} for DP. The method was successfully applied to the analysis of contraceptive tablets from three different pharmaceutical companies. No differences were observed between the measured and the declared amount of active principle per tablet, demonstrating the applicability of the procedure. In addition, a stability study conducted on both the standards and sample extracts demonstrated that they can be stored at room temperature for a minimum period of seven days.

Keywords: ethinyl estradiol; drospirenone; contraceptive tablets; gas chromatography; mass spectrometry

1. Introduction

One of the most commonly used contraception methods nowadays consists of the administration of an estrogen and a progestogen, which together inhibit the ovulation process. Two synthetic versions of these hormones that are typically combined in contraceptive tablets are 17α-ethinyl estradiol (EE) and drospirenone (DP, 6β,7β:15β,16β-dimethylene-3-oxo-17α-pregn-4-ene-21,17-carbolactone).

To minimize their known adverse side effects, such as hypercoagulability, and ensure an appropriate balance between efficacy and safety, the hormone dosage per tablet is carefully optimized. Currently, DP and EE can be administered to women in two different formulations. One contains 0.02 mg of EE and 3 mg of DP (DP/20EE), while the other contains 0.03 mg of EE and 3 mg of DP (DP/30EE) per tablet. Another difference between these treatments relates to the dosage strategy: DP/30EE is administered for 21 consecutive days, followed by 7 days of rest, whereas DP/20EE is administered for 24 consecutive days, followed by 4 days of rest [1–3].

Due to the small doses used and the long duration of the hormonal treatments (years, even decades), accurately quantifying the amount of the active principles is crucial in the preparation of contraceptive pills [4]. Most of the analytical methods published up to date for quantifying EE and DP in tablets are based on liquid chromatography (LC) with spectrophotometric detection, either with a single UV detector [5,6], or with a UV detector for DP and a fluorescence detector (FD) in series for EE [4]. In fact, LC–UV is the preferred technique for DP quantification [7], while both LC–UV [8–11] and LC–FD [12] have been applied to contraceptive formulations combining EE with progestins other than DP.

The use of mass spectrometry (MS) has been gaining attention as a complementary detection technique, also for hormones. However, MS-based methods have primarily been applied to determine EE and DP in complex biological and environmental samples, in line with the need of an increased selectivity and sensitivity. For instance, liquid chromatography coupled to mass spectrometry (LC–MS) has been used to determine DP in plasma [13], blood [14], and water [15], whereas gas chromatography coupled to mass spectrometry (GC–MS) has been applied to separate and quantify EE and other related hormones in urine [16], cattle hair [17], water [18–21], and sediments [22]. When determined via GC–MS, hormones are usually derivatized via silylation to increase their volatility and reduce their polarity, thereby improving the physicochemical properties for their subsequent GC analysis. However, this step involves longer analysis times, the use of organic (sometimes toxic) reagents, and a higher sample manipulation that may increase the uncertainty of the obtained results. There are only a few examples where EE has been determined via GC–MS without derivatization [23–25], but neither in combination with DP nor in pharmaceutical formulations.

In this line, and since GC–MS has never been applied to neither the analysis of EE-containing tablets nor to the determination of DP, herein we report the first GC–MS method for the simultaneous and accurate quantification of both active principles in contraceptive tablets. Unlike other methods, our procedure does not require the use of derivatizing agents. Instead, analytes are extracted by ultrasound-assisted extraction, the extract is diluted and centrifuged, and the supernatant is directly injected into the GC–MS system with a split mode regime. To correct for potential instrumental variations, calibration is performed by the internal standard method, using cholesterol as the internal standard (IS). The aim of this work is to provide a simple, accurate, and reliable method based on GC–MS as an alternative technique to LC (which has sometimes required the use of two different detectors) for the determination of EE and DP in contraceptive formulations.

2. Results and Discussion

2.1. GC–MS Optimization

The main aim of the study was to optimize and validate a GC–MS-based method, as an alternative to those based on LC–UV and/or LC–FD, for the simultaneous determination of EE and DP in contraceptive pills. With the aim of prioritizing simplicity and sustainability, no derivatization step was performed, which is a real challenge considering the polarity and boiling points of both active principles: predicted values of log K_{OW} = 3.67 (EE) and 4.02 (DP); boiling points = 411 °C (EE) and 552 °C (DP) [26].

MS detection was performed in SIM mode. Different dwell times were evaluated, and the selected optimal values were 20 ms for EE, ES, and the IS (100 ms was the optimal value observed for DP).

Chromatographic separation was performed on an HP-5MS UI capillary column (5% phenyl-methylpolysiloxane). Isothermal conditions (325 °C) were initially explored to achieve the shortest possible chromatographic run time. In addition, different programs were tested, considering different initial oven temperatures (from 200 °C to 300 °C), applying the maximum allowed ramps by the system, and maintaining the final temperature at 325 °C (the maximum recommended by the column manufacturer). No difference was observed, but in terms of precision, the program starting at 250 °C and ramping up to 325 °C at a rate of 45 °C min^{-1} yielded the best results, with less than 6% of RSD for both analytes. Under these conditions, the retention times of EE, the IS, and DP were 2.32 min, 3.19 min, and 6.15 min, respectively.

Initial studies were conducted using an injector filled with an empty straight glass liner, with the temperature set to the maximum recommended value of 275 °C. However, this setup resulted in very low injection precision, with RSD values exceeding 40%. The replacement of this liner by another one filled with deactivated glass wool improved the % RSD values considerably (less 6%). Consequently, this packing material was selected to continue with the study. The injection volume was set at 1 µL. Different split ratios (1:2, 1:10, 1:50) were tested in order to obtain a non-saturated peak for DP and, at the same time, a satisfactory peak height for EE (in contraceptive tablets, and so in our sample extracts, DP is 100 or 150 times more concentrated than EE). The best compromise was achieved with a split ratio of 1:50, which was the ratio finally selected to continue with the study.

2.2. Extraction Optimization

Initially, all standard solutions used in this study were prepared in methanol due to its suitability as an organic solvent for the subsequent solid–liquid extraction of the active principles, considering their polarities. However, when these methanolic solutions were injected into the GC–MS system, we observed a partial degradation of EE to form ES (confirmed by its MS spectrum). This could be related to the reactivity of methanol at high temperatures (in the injector), which does not happen with ethyl acetate and which leads to the formation of ES from EE by oxidation of the hydroxyl moiety and liberation of the ethinyl group. In contrast, when standard solutions were prepared in ethyl acetate, no traces of ES were detected at any of the concentration levels studied during the linearity tests (Figure 1). Hence, all standard solutions were prepared in ethyl acetate hereinafter.

The solid–liquid extraction procedure was based on other procedures described in literature for similar formulations [4,8,27,28]. Both methanol and ethyl acetate were evaluated as extractant solvents. To this end, the mass powder equivalent to the weight of one tablet (determined by weighing 20 individual tablets, see Section 2.2) was subjected to ultrasound-assisted extraction (15 min) in 1.0 mL of either methanol or ethyl acetate. After centrifugation, the extracts in ethyl acetate (n = 3) were made up to a final volume of 5.0 mL with the same solvent. For the extracts in methanol, two approaches were considered: either diluting them to 5.0 mL with ethyl acetate (n = 3) or evaporating them to dryness and reconstituting them in 5.0 mL of ethyl acetate (n = 3). As shown in Figure 2, ethyl acetate was not able to extract EE from the solid pill. Conversely, this compound was completely recovered from the solid when using methanol as an extractant solvent. No formation of ES was observed neither when the extract was evaporated to dryness and reconstituted in 5.0 mL of ethyl acetate, nor when it was directly diluted to a final volume of 5.0 mL with ethyl acetate. Hence, the latter (and simpler) protocol was selected: ultrasound assisted extraction with 1.0 mL of methanol for 15 min, and dilution to 5.0 mL with ethyl acetate. It is worth noting that standard solutions prepared in this solvent composition (20% of methanol in ethyl acetate) still showed the formation of ES from EE (Figure 1), unlike what was observed in sample extracts. This difference could be attributed to other ingredients present in the pharmaceutical formulation, which might have prevented the oxidation from EE to ES. Thus, we kept the preparation of the standard solutions in pure ethyl acetate.

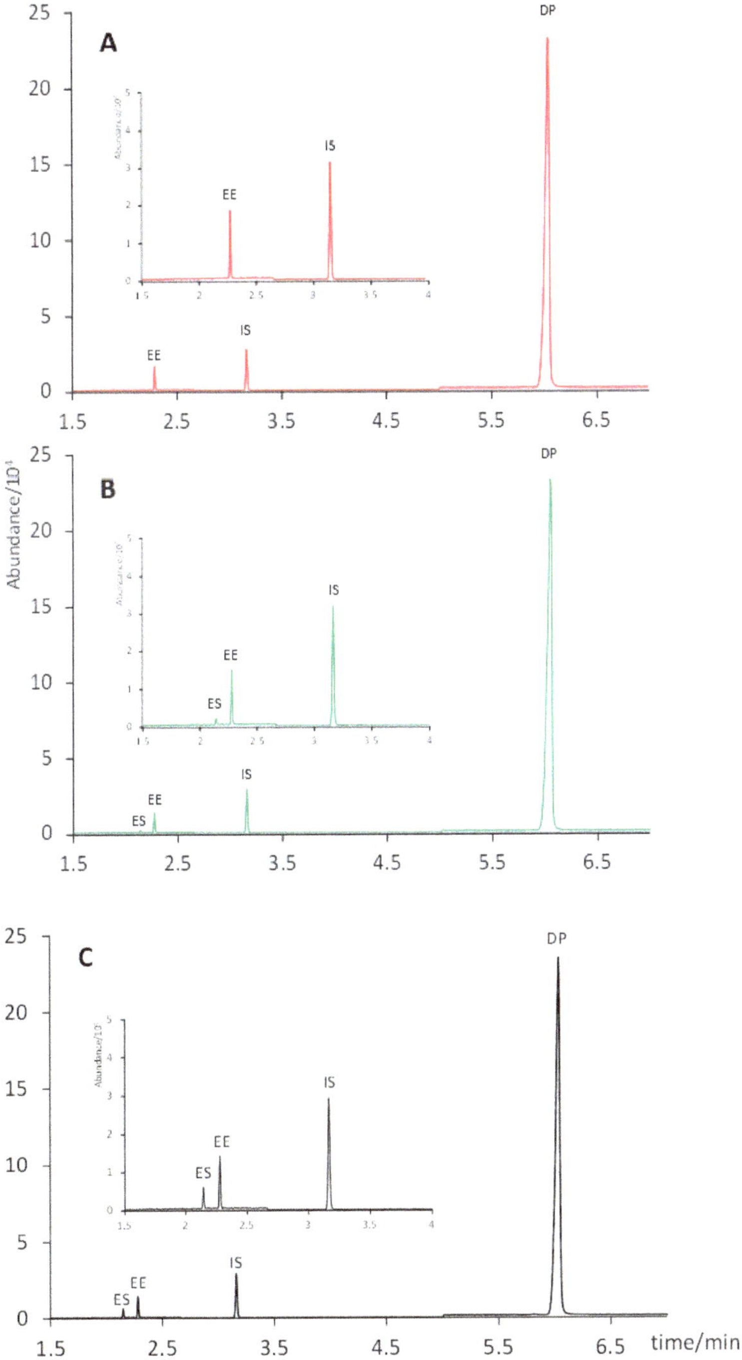

Figure 1. Chromatograms of the standard solutions (6 μg mL^{-1} of EE and 600 μg mL^{-1} of DP) prepared in (**A**) ethyl acetate; (**B**) methanol:ethyl acetate 1:5; (**C**) methanol.

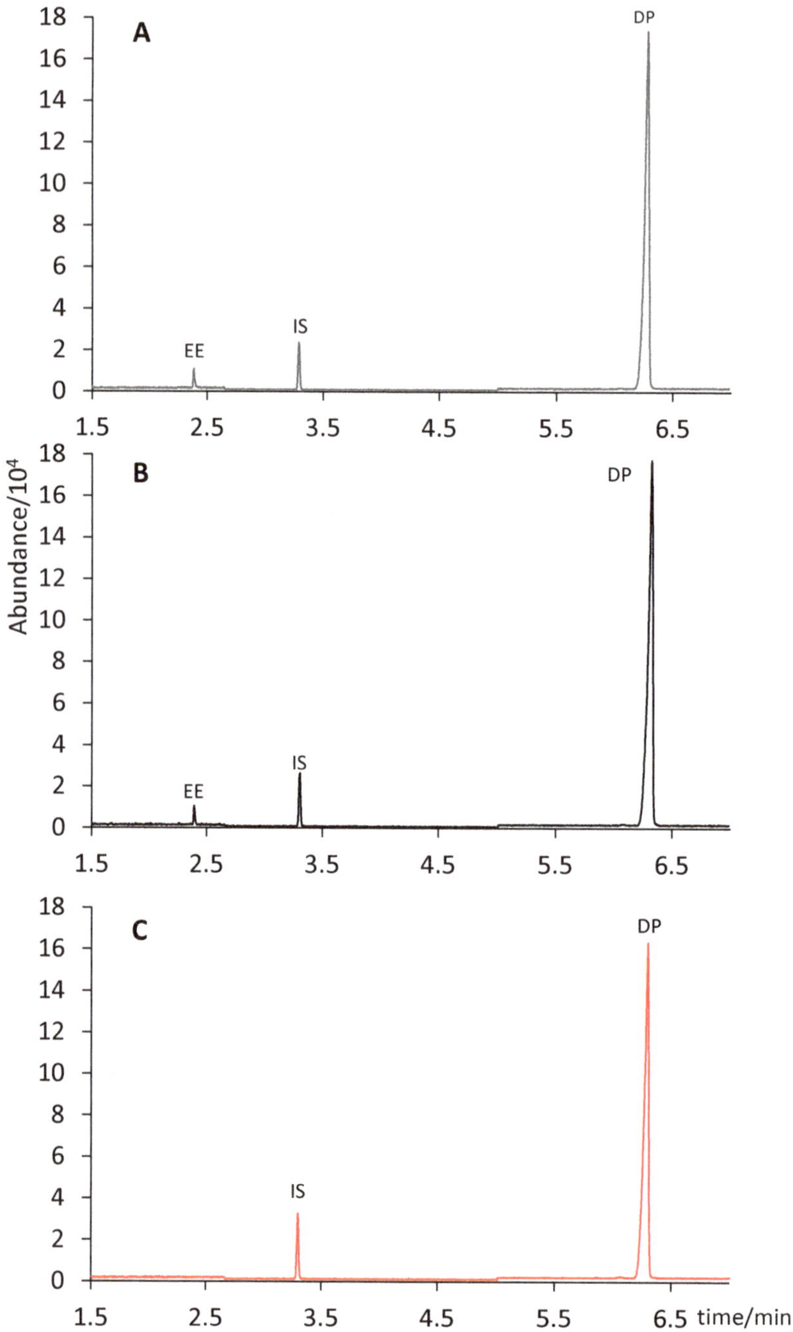

Figure 2. Chromatograms of the tablet extracts obtained by different extraction procedures: (**A**) extraction in 1.0 mL of methanol, evaporation to dryness and reconstitution in 5.0 mL of ethyl acetate; (**B**) extraction in 1.0 mL of methanol, made to a final volume of 5.0 mL with ethyl acetate; (**C**) extraction in 1.0 mL of ethyl acetate, made to a final volume of 5.0 mL also with ethyl acetate.

2.3. Method Validation

Table 1 summarizes the analytical performance results of the proposed method. A good linearity was achieved for both analytes, with R^2 values of 0.9919 for EE and 0.9902 for DP. Figure 3 shows the chromatograms of four standard solutions containing 3, 5, 7, and 9 μg mL^{-1} of EE, and 300, 500, 700, and 900 μg mL^{-1} of DP (chromatograms of intermediate concentrations have been excluded for clarity). Trueness, assessed in terms of percentages of recovery, and intra- and inter-day precision, assessed through % RSD values, were satisfactory: percentages of recovery were 106 ± 8% for EE and 93 ± 9% for DP; and % RSD values were below 4% for EE and below 6% for DP. Instrumental LODs (concentration corresponding to S/N = 3) were 0.25 μg mL^{-1} for EE and 6.6 μg mL^{-1} for DP. Instrumental LOQs (concentration corresponding to a S/N = 10) were 0.82 μg mL^{-1} for EE and 22 μg mL^{-1} for DP.

Table 1. Analytical performance figures.

	EE	DP
Linearity range (μg mL^{-1})	3.0–12.0	300–1200
Regression equation	y = (0.121 ± 0.002)x + (0.01 ± 0.02)	y = (0.045 ± 0.001)x + (0 ± 1)
R^2	0.9919	0.9902
Trueness (% recovery ± SD)	106 ± 8	93 ± 9
Intra-day precision (% RSD)	3.9 [a]	5.4 [a]
	2.5 [b]	4.9 [b]
Inter-day precision (% RSD)	2.7 [a]	4.9 [a]
	0.88 [b]	5.9 [b]
LOD (μg mL^{-1})	0.25	6.6
LOQ (μg mL^{-1})	0.82	22

[a] Only-IS-spiked samples (0.25 mg). [b] Samples spiked with 0.03 mg of EE, 3.00 mg of DP, and 0.25 mg of IS.

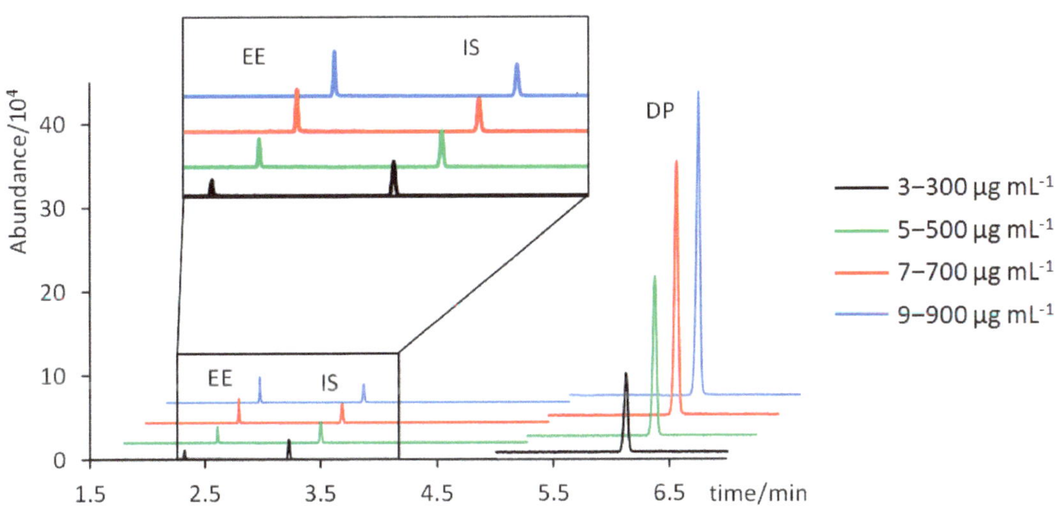

Figure 3. Chromatograms of standard solutions containing 3, 5, 7, and 9 μg mL^{-1} of EE and 300, 500, 700, and 900 μg mL^{-1} of DP.

2.4. Analyte Stability

Analyte stability was assessed in a standard solution and in the sample extracts kept at room temperature (22 °C) for 7 days, as stated in Section 2.5.

Analyte responses at different days (analyte area/IS area, with the IS added every day just before injection) were divided by the corresponding analyte response at the beginning of the study (day 0). In the standard solution, these relative responses varied between 1.03 ± 0.03 and 1.17 ± 0.10 for EE, and between 0.95 ± 0.04 and 1.13 ± 0.16 for DP. In the tablet extracts, they ranged from 0.94 ± 0.05 to 1.00 ± 0.05 for EE, and from 0.97 ± 0.09 to 1.27 ± 0.12 for DP. These results suggest that there was no relevant degradation observed in any case, proving that both the standards and tablet extracts can be stored at room temperature for a minimum of seven days.

2.5. Analysis of Pharmaceutical Preparations

To test the applicability of the developed method, three contraceptive formulations from three different pharmaceutical companies were subjected to the procedure described here. Four tablets from each trademark were processed as detailed in Section 2.2 and analyzed via GC–MS as stated in Section 2.4.

According to the pharmaceutical company, two formulations contained 0.030 mg of EE and 3.0 mg of DP per tablet (Yasmin and Drosure), whereas the third one contained 0.020 mg of EE and 3.0 mg of DP (Stada). As shown in Table 2, no significant differences were observed in any case between the content claimed by the company (considering a 10% interval according to USP [29]) and the amount predicted by our analysis with a confidence interval of 95%: Yasmin (0.027 ± 0.001 mg of EE and 2.93 ± 0.08 mg of DP); Drosure (0.028 ± 0.001 mg of DP and 2.88 ± 0.08 mg of DP); Stada (0.018 ± 0.001 mg of EE and 2.84 ± 0.08 mg of DP). Figure 4 shows a comparison of the extract chromatograms of two different tablet formulations (Yasmin and Stada) with that of a standard solution containing 6 $\mu g\ mL^{-1}$ of EE and 600 $\mu g\ mL^{-1}$ of DP (concentration equivalent to that of an extract from a tablet containing 0.030 mg of EE and 3.0 mg of DP).

Table 2. EE and DP contents (claimed and found) in tablets from different pharmaceutical companies.

Pharmaceutical Formulation	Active Principle Content (mg/tablet)	
	Claimed [a]	Found [b]
Yasmin		
EE	0.030 ± 0.003	0.027 ± 0.001
DP	3.0 ± 0.3	2.93 ± 0.08
Drosure		
EE	0.030 ± 0.003	0.028 ± 0.001
DP	3.0 ± 0.3	2.88 ± 0.08
Stada		
EE	0.020 ± 0.002	0.018 ± 0.001
DP	3.0 ± 0.3	2.84 ± 0.08

[a] Amount stated by the company with a 10% interval, according to the USP (USP 35 NF, 2019, US PHarmacopeial Convention, Inc. Rockville, MD. USA). [b] Amount predicted with a confidence interval of 95%.

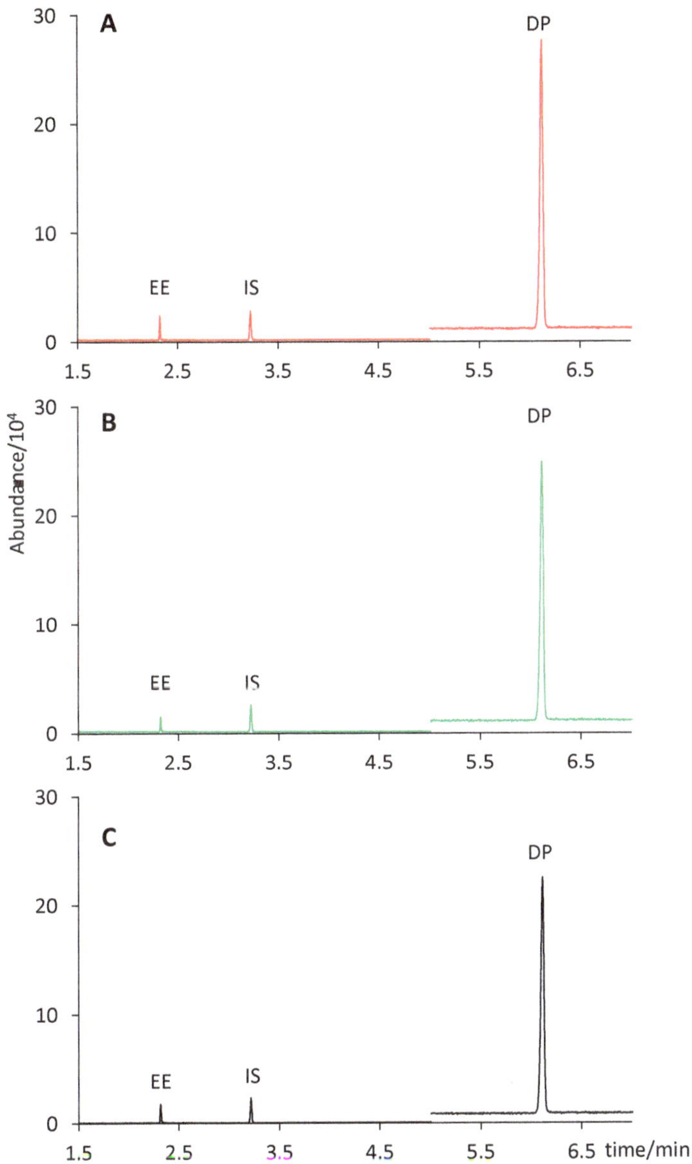

Figure 4. Chromatograms of (**A**) a Yasmin tablet extract; (**B**) a Stada tablet extract; (**C**) a standard solution containing 6 μg mL^{-1} of EE and 600 μg mL^{-1} of DP.

3. Materials and Methods

3.1. Chemicals and Standard Solutions

Ethinyl estradiol (EE, 99.7% purity) and drospirenone (DP, 98.7% purity) were kindly supplied by a local pharmaceutical company. Cholesterol (used as IS, >99% purity), HPLC grade methanol, and ethyl acetate were supplied by Sigma Aldrich (Steinheim, Germany). Individual stock solutions of EE (1000 μg mL^{-1}), DP (5000 μg mL^{-1}), and the IS (1000 μg L^{-1}) were prepared by dissolving the accurately weighed reference standard in either methanol or ethyl acetate. These solutions were stored in the dark at −22 °C and allowed to warm up to room temperature immediately before use. Working standard solutions were prepared daily by diluting the stock solutions in methanol or ethyl acetate to the required working concentrations.

3.2. Samples and Sample Preparation

Contraceptive tablets from three different brands (Yasmin (Bayer España, Spain), Drosure (EFFIK S.A., Madrid, Spain), and Stada (Stada S.L., Barcelona, Spain)), containing 0.03 mg of EE and 3 mg of DP per tablet (Yasmin and Drosure), or 0.02 mg of EE and 3 mg of DP (Stada), were purchased at a local pharmacy. The tablets used for method optimization and validation, containing 0.03 mg of EE and 3.00 mg of DP, were kindly supplied by a local pharmaceutical company. To determine the average tablet weight, twenty tablets were accurately and individually weighed. To prepare a relevant amount of homogenized pill powder (for method optimization), 60 pills were finely powdered and mixed.

Under the final working conditions, one tablet (for commercial tablets) or the amount of powder equivalent to the weight of one tablet (0.0828 g, for method optimization tablets) was accurately weighed into a 5.0 mL volumetric flask. IS (0.25 mg in ethyl acetate) was added to the powder and the mixture was allowed to dry and age overnight in a fume hood, under light-protected conditions. The next morning, the dried mixture was suspended in 1.0 mL of methanol and sonicated for 15 min to extract EE and DP. After cooling to room temperature, the suspension was brought up to a final volume of 5.0 mL with ethyl acetate and centrifuged at 4000 rpm for 10 min. Finally, 1.0 mL of the supernatant was transferred to a glass vial for analysis.

Spiked samples, used to calculate the recovery and relative standard deviation values at a high concentration level, were prepared by mixing an accurate mass of tablet powder with the required amounts of EE, DP, and IS solutions in ethyl acetate. The mixture was homogenized, allowed to age and dry overnight, and processed, as previously described for only-IS-spiked samples.

3.3. Instrumental Configuration

All the chromatographic analyses were conducted using an Agilent 7890A series gas chromatograph (Agilent Technologies, Santa Clara, CA, USA) interfaced to an Agilent 5978C inert XL MSD equipped with an electron ionization source (70 eV). Injection was performed using an MPS2 Multi-Purpose Sampler (Gerstel, Mülheim an der Ruhr, Germany). The inlet was fitted with a liner (71 mm × 2 mm i.d., Gerstel CIS-4) packed with deactivated glass wool. A volume of one microliter of the standard or sample solution was injected in a split mode (split ratio of 1:50), maintaining the injection port at 275 °C. The split flow was set at 20 mL min^{-1} and the septum purge flow at 4 mL min^{-1}. Chromatographic separation was performed using an HP-5MS UI capillary column (30 m × 0.25 mm i.d.; 0.25 µm film thickness) supplied by Agilent Technologies. Helium (99.999% of purity, Air Liquid) was used as carrier gas at a constant flow rate of 2.0 mL min^{-1}. The GC oven temperature was programmed as follows: initial temperature of 250 °C, increase at a rate of 45 °C min^{-1} to 325 °C, and held for 5.7 min (total run time of 7.4 min). The total analysis time per sample was 32.4 min: 15 min for ultrasound-assisted extraction, 10 min for centrifugation, and 7.4 min for the chromatographic analysis.

The MS system consisted of an EI source (70 eV), a single quadrupole analyzer, and an electron multiplier detector. The temperatures of the ionization source, mass analyzer, and transfer line were set at 230 °C, 150 °C, and 300 °C, respectively. The mass analyzer was operated in the selected ion monitoring (SIM) mode, acquiring one quantifier ion and two qualifier ions per compound (1 min of solvent delay). Table 3 shows the three SIM groups and the m/z values acquired for both active principles (EE and DP), the IS, and estrone (ES), an observed by-product of EE during method optimization. The selected dwell times were 20 ms (ES, EE, and IS) and 100 ms (DP). MSD ChemStation, Ver. E.02.00.493 software from Agilent Technologies was used for data acquisition, and NIST_98 database (NIST/EPA/NIH Mass Spectral Library, version 2.0) was used for identification.

Table 3. Retention times, m/z values selected for quantification (in bold), and confirmation of the active principles (EE and DP), the IS and an EE by-product (ES).

Compound	t_R/min	Quantifier and Qualifier/*m/z* Values	SIM Group
ES	2.18	**270**, 146, 185	1
EE	2.32	**213**, 296, 160	
IS	3.19	**275**, 386, 301	2
DP	6.15	**255**, 366, 117	3

3.4. Quantification and Method Validation

To correct for potential instrument variations, the analytes were quantified using the internal standard method. The calibration curves were prepared by diluting the stock solutions (Section 2.1) in ethyl acetate. They consisted of ten concentration levels: 3.0, 4.0, 5.0, 6.0, 7.0, 8.0, 9.0, 10.0, 11.0, and 12.0 μg mL^{-1} for EE, and 300, 400, 500, 600, 700, 800, 900, 1000, and 1200 μg mL^{-1} for DP. The IS was maintained at 50 μg mL^{-1}. Instrumental LOD and LOQ values were calculated as the analyte concentration providing a signal-to-noise ratio (S/N) of 3 (LOD) and 10 (LOQ). These values were estimated from standards solutions in ethyl acetate containing 0.3 μg mL^{-1} of EE and 3 μg mL^{-1} of DP.

Trueness was assessed from recoveries (expressed as percentages) calculated as the difference between the concentrations measured in the extracts from spiked samples and in the extracts from samples containing only the IS, divided by the theoretical spiked level. To this end, six spiked and six only-IS-spiked samples were prepared: the amount of powder equivalent to the weight of one tablet (0.0828 g) was spiked with both the analytes (total mass added: 0.03 mg of EE and 3.00 mg of DP) and IS (0.25 mg), or only with IS (0.25 mg). These were allowed to age overnight, and were then extracted as detailed in Section 2.2.

Precision was assessed by calculating the relative standard deviation (% RSD) obtained from the analysis of spiked samples (0.03 mg of EE, 3.00 mg of DP, and 0.25 mg IS) and only-IS-spiked samples (0.25 mg). Intra-day precision was estimated from the analysis of six samples of each type within the same day, while inter-day precision was assessed from the analysis of four samples of each type performed in three different days over a two-week period.

Linearity was evaluated by calculating the determination coefficient (R^2) value after plotting the peak response (analyte area/IS area) for EE and DP, versus their corresponding concentrations (range: 50–200% of the tablet theoretical concentration).

3.5. Stability Study

To assess the analyte stability, 50.0 mL of a standard solution containing 6 μg mL^{-1} of EE and 600 μg mL^{-1} of DP in ethyl acetate was prepared. With the same aim, 12 tablets extracts were prepared following the procedure described in Section 2.2. Both the standard solution and the sample extracts were stored in closed glass vials, protected from light and kept at a controlled room temperature (22 °C). Three aliquots from the standard solution and three tablet extracts were analyzed via GC–MS, as described in Section 2.3. at the beginning of the study (day 0) and after 1, 3, and 7 days of storage. Prior to analysis, the IS solution was added to each aliquot resulting in a final concentration of 5.7 μg mL^{-1}, 570 μg mL^{-1}, and 47.6 μg mL^{-1} of EE, DP, and IS, respectively. A fresh standard solution in ethyl acetate (also containing 5.7 μg mL^{-1} of EE, 570 μg mL^{-1} of DP, and 47.6 μg L^{-1} of IS) was prepared daily and analyzed together with every set of samples.

4. Conclusions

As stated in the introduction, the analysis of contraceptive formulations containing both EE and DP has been usually performed via LC with spectrophotometric detection, either with a single UV detector or with a UV detector for DP and a fluorescence detector (FD) coupled in series for EE detection. In terms of the speed of analysis, the typical times using LC were approximately 7.0 min [4–6], considering only the chromatographic run time. As an alternative, herein we report the first GC-based method for the simultaneous and accurate quantification of these two synthetic hormones in contraceptive tablets. A total chromatographic run time of 7.4 min was achieved, in the same time range than most LC methods reported in the literature. To the best of our knowledge, this is also the first time that GC coupled to MS has been applied for the determination of DP.

The sample preparation in this method is straightforward, involving an ultrasound-assisted solid–liquid extraction (15 min), followed by dilution, centrifugation (10 min), and a direct injection into the GC–MS system. Due to the high concentrations of the active principles in contraceptive tablets, no derivatization needed to be applied in this case. This eliminated the need for toxic derivatizing reagents and reduced sample manipulation and the uncertainty of the results obtained. The entire method was validated in terms of linearity, trueness, precision, and LODs and LOQs, proving to be consistent and reliable. In addition, the analysis of contraceptive formulations from different pharmaceutical companies (and containing different amounts of active principles per tablet) showed no difference between the measured and the claimed amounts of EE and DP, demonstrating its accuracy and applicability.

Finally, a stability assessment has proven that both the standards and sample extracts, prepared as stated here, can be stored at room temperature for a minimum of seven days, further establishing the practicality and convenience of our procedure.

Hence, this work has proven that GC–MS can be an alternative technique for the determination of EE and DP in contraceptive tablets, giving quality control laboratories an extra tool for content analysis within this type of preparations.

Author Contributions: Conceptualization, J.P., I.G.-M. and J.L.P.P.; Formal analysis, J.P. and I.G.-M.; Funding acquisition, J.L.P.P.; Investigation, J.P. and I.G.-M.; Methodology, J.P. and I.G.-M.; Project administration, J.L.P.P.; Resources, J.L.P.P.; Supervision, I.G.-M.; Validation, J.P. and I.G.-M.; Visualization, J.P.; Writing—original draft, J.P.; Writing—review and editing, I.G.-M. and J.L.P.P. All authors have read and agreed to the published version of the manuscript.

Funding: This work was financially supported by Junta de Castilla y León (project no. SA111P20). The authors would like to thank Chemo S.A. for kindly providing the reference standards of the active principles and the tablets to perform the optimization of the method.

Data Availability Statement: Not applicable.

Conflicts of Interest: The authors declare that they have no known competing financial interests or personal relationships that could have appeared to influence the work reported in this paper.

Sample Availability: Samples of the compounds are not available from the authors.

Abbreviations

CIS-4	Cooled Injection System 4
DP	Drospirenone
DP/20EE	A formulation of 0.02 mg of EE and 3 mg of DP per tablet
DP/30EE	A formulation of 0.03 mg of EE and 3 mg of DP per tablet
EPA	Environmental Protection Agency
ES	Estrone
EE	17α-ethinyl estradiol
FD	Fluorescence detector
GC–MS	Gas Chromatography–Mass Spectrometry
i.d.	Inner Diameter
IS	Internal standard
LC–MS	Liquid Chromatography–Mass Spectrometry
LOD	Limit of Detection
LOQ	Limit of Quantification
MPS2	Multi-Purpose Sampler 2

MS	Mass Spectrometry
NIH	National Institutes of Health
NIST	National Institute of Standards and Technology
RSD	Relative Standard Deviation
SIM	Selected Ion Monitoring
UV	Ultraviolet

References

1. Mathur, R.; Levin, O.; Azziz, R. Use of ethinylestradiol/drospirenone combination in patients with the polycystic ovary syndrome. *Clin. Risk Manag.* **2008**, *4*, 487–492. [CrossRef]
2. Rosenbaum, P.; Schmidt, W.; Helmerhorst, F.M.; Wuttke, W.; Rossmanith, W.; Freundl, F.; Thomas, K.; Grillo, M.; Wolf, A.; Heithecker, R. Inhibition of ovulation by a novel progestogen (drospirenone) alone or in combination with ethinylestradiol. *Eur. J. Contracept. Reprod. Health Care* **2000**, *5*, 16–24. [CrossRef]
3. Huber, J.; Foidart, J.M.; Wuttke, W.; Merki-Feld, G.S.; The, H.S.; Gerlinger, C.; Schellschmidt, I.; Heithecker, R. Efficacy and tolerability of a monophasic oral contraceptive containing ethinylestradiol and drospirenone. *Eur. J. Contracept. Reprod. Health Care* **2000**, *5*, 25–34. [CrossRef]
4. Silva, V.B.; Galdos, A.A.G.; Mothe, C.M.A.; Pallastrelli, M.B.; Prado, M.S.A.; Singh, A.K.; Kedor-Hackmann, E.R.M.; Santoro, M.I.R.M. Simultaneous determination of ethinyl estradiol and drospirenone in oral contraceptive by high performance liquid chromatography. *Braz. J. Pharm. Sci.* **2013**, *49*, 521–528. [CrossRef]
5. Seifeldeen, E.M.; Ethman, M.A.; Aboul-Enein, H.Y. Simultaneous determination of four hormonal compounds in oral contraceptive tablet formulations by high performance liquid chromatography. *J. Liq. Chromatogr. Relat. Technol.* **2016**, *39*, 134–138. [CrossRef]
6. Praveen, C.; Ranganath, M.K.; Divakar, P. Method Development and Validation for Simultaneous Estimation of Ethinyl Estradiol and Drospirenone and Forced Degradation Behavior by HPLC in Combined Dosage Form. *Pharm. Anal. Acta* **2013**, *4*, 231.
7. Talath, S.; Dhaneshwar, S. A Simple and Rapid Validated Stability Indicating HPLC Method for the Determination of Drospirenone in a Pharmaceutical Product. *Indo Am. J. Pharm. Res.* **2017**, *7*, 7401–7410.
8. Sarat, M.; Rambabu, C. A validated simultaneous RP-HPLC method for determination of Desogestrel and Ethinyl Estradiol Tablets. *Int. J. Pharm. Pharm. Sci.* **2012**, *4*, 115–119.
9. Ramakrishna, K. Analytical method development and validation for simultaneous determination and quantification of Ethinyl Estradiol and Gestodene in combined tablet dosage form by RP-HPLC. *J. Chem. Pharm. Res.* **2014**, *6*, 74–80.
10. Ravindra, A.C.; Swamy, K.C.N.; Hima, P.; Kumar, K.V. Validated RP-HPLC Method for Simultaneous Estimation of Levonorgestrel and Ethinylestradiol in Combined Dosage Form. *J. Sci. Innov. Res.* **2013**, *2*, 642–650.
11. Asheesh Singh, P.S. Stability indicating analytical method validation for determination of related substances by rphplc for cyproterone acetate & ethinyl estradiol in cyproterone acetate & ethinyl estradiol tablet. *Int. J. Curr. Adv. Res.* **2018**, *7*, 9815–9820.
12. Antunes da Silva, F.; Lourenço, F.R.; Calixto, L.A. Multiple Optimization for Extraction of Ethinylestradiol, Levonorgestrel, and Their Main Impurities using Mixture Design. *J. Anal. Chem.* **2022**, *77*, 844–852. [CrossRef]
13. Bhaumik, U.; Ghosh, A.; Mandal, U.; Chatterjee, B.; Sarkar, A.K.; Bose, A.; Ray, K.K.; Pal, T.K. Determination of Drospirenone in Human Plasma by LC-Tandem-MS. *Chromatographia* **2008**, *68*, 713. [CrossRef]
14. Idota, N.; Kobayashi, M.; Miyamori, D.; Kakiuchi, Y.; Ikegaya, H. Drospirenone detected in postmortem blood of a young woman with pulmonary thromboembolism: A case report and review of the literature. *Leg. Med.* **2015**, *17*, 109–115. [CrossRef]
15. Mirmont, E.; Boeuf, A.; Charmel, M.; Vaslin-Reimann, S.; Lalere, B.; Laprevote, O.; Lardy-Fontan, S. Development and implementation of an analytical procedure for the quantification of natural and synthetic steroid hormones in whole surface waters. *J. Chromatogr. B Anal. Technol. Biomed. Life Sci.* **2021**, *1175*, 122732. [CrossRef]
16. Xiao, X.-Y.; McCalley, D. Quantitative analysis of estrogens in human urine using gas chromatography/negative chemical ionisation mass spectrometry. *Rapid Commun. Mass Spectrom. RCM* **2000**, *14*, 1991–2001. [CrossRef] [PubMed]
17. Durant, A.A.; Fente, C.A.; Franco, C.M.; Vázquez, B.I.; Cepeda, A. Gas Chromatography-Tandem Mass Spectrometry Determination of 17α-Ethinylestradiol Residue in the Hair of Cattle. Application to Treated Animals. *J. Agric. Food Chem.* **2002**, *50*, 436–440. [CrossRef] [PubMed]
18. Stanford, B.D.; Weinberg, H.S. Isotope dilution for quantitation of steroid estrogens and nonylphenols by gas chromatography with tandem mass spectrometry in septic, soil, and groundwater matrices. *J. Chromatogr. A* **2007**, *1176*, 26–36. [CrossRef]
19. Gonzalez, A.; Avivar, J.; Cerda, V. Estrogens determination in wastewater samples by automatic in-syringe dispersive liquid-liquid microextraction prior silylation and gas chromatography. *J. Chromatogr. A* **2015**, *1413*, 1–8. [CrossRef]
20. Trinh, T.; Harden, N.B.; Coleman, H.M.; Khan, S.J. Simultaneous determination of estrogenic and androgenic hormones in water by isotope dilution gas chromatography-tandem mass spectrometry. *J. Chromatogr. A* **2011**, *1218*, 1668–1676. [CrossRef] [PubMed]
21. Ronderos-Lara, J.G.; Saldarriaga-Noreña, H.; Murillo-Tovar, M.A.; Vergara-Sánchez, J. Optimization and Application of a GC-MS Method for the Determination of Endocrine Disruptor Compounds in Natural Water. *Separations* **2018**, *5*, 33. [CrossRef]
22. Liu, R.; Zhou, J.L.; Wilding, A. Microwave-assisted extraction followed by gas chromatography-mass spectrometry for the determination of endocrine disrupting chemicals in river sediments. *J. Chromatogr. A* **2004**, *1038*, 19–26. [CrossRef]
23. Soliman, M.A.; Pedersen, J.A.; Suffet, I.H. Rapid gas chromatography-mass spectrometry screening method for human pharmaceuticals, hormones, antioxidants and plasticizers in water. *J. Chromatogr. A* **2004**, *1029*, 223–237. [CrossRef]

24. García, I.; Sarabia, L.A.; Ortiz, M.C.; Aldama, J.M. Optimization of the chromatographic conditions for the determination of hormones by gas chromatography with mass spectrometry detection. *Anal. Chim. Acta* **2005**, *544*, 26–35. [CrossRef]
25. García, I.; Sarabia, L.; Cruz Ortiz, M.; Manuel Aldama, J. Building robust calibration models for the analysis of estrogens by gas chromatography with mass spectrometry detection. *Anal. Chim. Acta* **2004**, *526*, 139–146. [CrossRef]
26. Morissette, M.F.; Vo Duy, S.; Arp, H.P.; Sauve, S. Sorption and desorption of diverse contaminants of varying polarity in wastewater sludge with and without alum. *Environ. Sci. Process. Impacts* **2015**, *17*, 674–682. [CrossRef]
27. Yilmaz, B. Simultaneous Determination of Estradiol Valerate and Medroxyprogesterone Acetate in a Tablet Formulation by Gas Chromatography—Mass Spectrometry. *Anal. Sci.* **2010**, *26*, 391–393.
28. Yılmaz, B. GC-MS and HPLC Methods for Determination of Estriol Hormone in Pharmaceutical Preparations. *Austin. J. Anal. Pharm. Chem.* **2015**, *2*, 1054.
29. *USP 35 NF*; US Pharmacopeial Convention. Rockville Inc.: Rockville, MD, USA, 2019.

molecules

MDPI

Article

Multi-Detection Size Exclusion Chromatography as an Advanced Tool for Monitoring Enzyme–Antibody Conjugation Reaction and Quality Control of a Final Product

Adela Štimac [1,2,*], Tihana Kurtović [1,2] and Beata Halassy [1,2]

1 Centre for Research and Knowledge Transfer in Biotechnology, University of Zagreb, Rockefellerova 10, 10000 Zagreb, Croatia
2 Center of Excellence for Virus Immunology and Vaccines, Rockefellerova 10, 10000 Zagreb, Croatia
* Correspondence: adela.stimac@unizg.hr

Abstract: The multi-detection size exclusion chromatography (SEC) has been recognized as an advanced analytical technique for the characterization of macromolecules and process control, as well as the manufacturing and formulation of biotechnology products. It reveals reproducible molecular characterization data, such as molecular weight and its distribution, and the size, shape, and composition of the sample peaks. The aim of this work was to investigate the potential and suitability of the multi-detection SEC as a tool for surveillance over the molecular processes during the conjugation reaction between the antibody (IgG) and horseradish peroxidase (HRP), and demonstrate the plausibility of its application in the quality control of the final product, the IgG-HRP conjugate. Guinea pig anti-Vero IgG-HRP conjugate was prepared using a modified periodate oxidation method, based on periodate oxidation of the carbohydrate side chains of HRP, followed by the formation of Schiff bases between the activated HRP and amino groups of the IgG. The quantitative molecular characterization data of the starting samples, intermediates, and final product were obtained by multi-detection SEC. Titration of the prepared conjugate was performed by the ELISA and its optimal working dilution was determined. This methodology proved to be a promising and powerful technology for the IgG-HRP conjugate process control and development, as well as for the quality control of the final product, as verified by the analysis of several commercially available reagents.

Keywords: size exclusion chromatography; advanced molecular characterization; multi-detection; IgG-HRP conjugate; ELISA; process control; quality control

1. Introduction

Horseradish peroxidase (HRP, EC 1.11.1.7), found in horseradish roots, is the most widely used enzyme for the labelling of antibodies in many different immunoassays, including ELISA, Western blotting, and immunohistochemistry. It is a stable, inexpensive, and easily accessible ~44 kDa glycoprotein which has been used to detect biological compounds by measuring absorption, fluorescence, or luminescence [1–3]. The polysaccharide residues of the HRP molecule may be oxidized with sodium periodate to generate reactive aldehyde groups that can react with the amino groups of an antibody (IgG) through the formation of Schiff base intermediates. These relatively labile intermediates can be stabilized by reductive amination with sodium cyanoborohydride or sodium borohydride [3–6].

In antibody conjugation reactions, it is important to retain the antibody–enzyme complex activity to the greatest extent and prevent the formation of aggregates which reduce the amount of the active ingredient and lower the quality of the final product [1,7]. Therefore, an in-depth characterization of the protein sample and the detection/quantification of its aggregate share are highly recommended. Multi-detection size exclusion chromatography (SEC) has been recognized as a valuable analytical technique for the characterization of macromolecules and process control, as well as the manufacturing and formulation of biotechnology products. It provides a comprehensive insight into the molecular characterization data of the protein sample, including absolute molecular weight (*Mw*), molecular weight distribution or polydispersity, molecular size, intrinsic viscosity (*IV*), and recovery. Moreover, it is possible to obtain additional information on the macromolecular structure, conformation, aggregation, branching, and copolymer/conjugate composition [7–10].

Four detectors most often associated with an advanced multi-detection SEC system are the differential refractive index (RI) detector, the static light scattering (SLS) detector, the differential viscometer detector, and the UV detector. Multi-detection SEC removes the need for column calibration, but it is necessary to calibrate the detector responses by analyzing the stable protein standard with a defined concentration, *Mw*, refractive index increment (d*n*/d*c*), and molar extinction coefficient (d*A*/d*c*) in order to obtain the required detector calibration constants (K_{RI}, K_{UV}, K_{LS}, and K_{Visc}). The responses of the RI and UV detectors are directly proportional to the sample's concentration (Equations (1) and (2)). The exact concentration can be calculated from the RI detector with the knowledge of the sample's d*n*/d*c* and from the UV detector with the knowledge of the sample's d*A*/d*c* [7,9–12].

$$RI\ output\ (mV) = K_{RI} \cdot \mathrm{d}n/\mathrm{d}c \cdot c \cdot injection\ volume \tag{1}$$

$$UVoutput\ (mV) = K_{UV} \cdot \mathrm{d}A/\mathrm{d}c \cdot c \cdot injection\ volume \tag{2}$$

The SLS detector responds to a sample's molecular weight, and, through a combination of the RI and SLS detectors, the absolute *Mw* of the sample's components can be calculated independent of the retention volume or any calibration standards (Equation (3)), as opposed to a column calibration system that can only provide a relative measurement (obtained *Mw* values are relative to the used standards) [7,9–12].

$$LS\ output\ (mV) = K_{LS} \cdot Mw \cdot (\mathrm{d}n/\mathrm{d}c)^2 \cdot c \cdot injection\ volume \tag{3}$$

According to the size exclusion theory, the principle of SEC is the separation of molecules based on their hydrodynamic size (function of both mass and structure), not molecular weight. Therefore, the *Mw* values determined by column calibration SEC and those determined by multi-detection SEC could be different because column calibration SEC introduces into the analysis shape-dependent inaccuracies. In practice, if the calibration curves were constructed with a series of two differently shaped molecules, they would yield different slopes [7].

Four SLS instruments are nowadays available, which differ in the angle at which the intensity of the scattered light is measured: right-angle light scattering (RALS), low-angle light scattering (LALS), hybrid RALS/LALS, and multi-angle light scattering (MALS). Each of them is slightly different and each has advantages and disadvantages. The hybrid RALS/LALS detector offers the advantages of RALS and LALS without any of their disadvantages, making it excellent for measuring the molecular weight of any sample type. The SLS detector response factors are used to calculate the distribution of the weight-averaged (*Mw*), number-averaged (*Mn*), and z-averaged (*Mz*) molecular weights. Using these three values, it is possible to obtain an insight into the entire molecular weight distribution. The *Mw*/*Mn* value corresponds to the polydispersity (*Pd*) of the sample and is related to the size distribution. The *Mw*/*Mn* ratio of 1 indicates that there is a symmetrical and homogeneous size distribution [13]. Furthermore, the SLS measurements show the power in the detection of trace amounts of high-molecular-weight soluble aggregates, because their LS signals have a higher intensity compared to UV and RI signals [9].

The differential viscometer detector measures the changing viscosity of the solution as the sample eluates and provides information about the structure. The *IV* of sample can be directly calculated with the parameters obtained from the RI and viscometer detectors (Equation (4)). Triple detection includes the RI or UV detector, SLS, and viscometer acting together, with each detector providing complementary but different information. Therefore, the hydrodynamic radius (*Rh*) can be indirectly calculated from the data of the RI, SLS, and viscometer detectors, and represents the radius of the theoretical sphere occupied by a sample of the known calculated *Mw* and *IV* [9–12].

$$Viscometer\ output\ (mV) = K_{Visc.} \cdot IV \cdot c \cdot injection\ volume \quad (4)$$

In this study, we investigated the potential and suitability of the multi-detection SEC as a tool for monitoring the conjugation reaction between the antibody (guinea pig anti-Vero IgG) and HRP using the periodate oxidation method and employed it for the advanced molecular characterization of the prepared IgG-HRP conjugate. Starting samples (IgG and HRP), intermediates, and final product, as well as three different commercially available IgG-HRP conjugates, were analyzed. The absolute molecular weight, polydispersity, and composition of the sample peaks were determined, as well as the aggregate content of each final product. In addition, the enzymatic/immunological functionality of the conjugate was examined by the ELISA.

2. Results

2.1. Preparation of IgG-HRP Conjugate with Monitoring of the Conjugation Reaction by Multi-Detection SEC

The purity of guinea pig anti-Vero IgG sample used for the research was confirmed by multi-detection SEC analysis (Figure S1 and Table S1). Protein A affinity chromatography gave a product with ~92% monomers, ~2% dimers, and ~6% aggregate content, and without any contaminants, which is consistent with our previous results [14]. The obtained absolute *Mw* values for the IgG monomer (~153 kDa) and dimer (~300 kDa) were in line with the expected ones. Moreover, the IgG monomer polydispersity value (*Mw*/*Mn*) of 1.002 confirmed the molecular homogeneity and symmetrical size distribution. The obtained *Rh* value for the IgG monomer (Table 1) was in good agreement with the reported values [7,9].

Table 1. Molecular characterization data for IgG monomer, HRP monomer, and final product (IgG-HRP conjugate) obtained by multi-detection SEC. The results are expressed as mean value ± standard error (SE) from *n* measurements.

Sample	*n*	V_R (mL)	*Mw* (kDa)	*Mw*/*Mn*	*IV* (dL/g)	*Rh* (nm)	Share (%)
IgG	3	9.3	152.5 ± 1.2	1.003 ± 0.000	0.042 ± 0.002	4.67 ± 0.09	92.0 ± 0.2
HRP	4	10.9	43.1 ± 0.5	1.004 ± 0.000	0.032 ± 0.001	2.79 ± 0.01	93.8 ± 0.7
IgG-HRP	5	8.6	234.9 ± 3.2	1.028 ± 0.005	0.047 ± 0.002	5.55 ± 0.09	32.1 ± 2.2

Before conjugation, it was a necessity to transfer IgGs into a high-pH matrix since alkaline conditions enhance the formation of Schiff bases [5,15]. Multi-detection SEC results (Table 1, Figure S2, and Table S1) confirmed that IgG molecular parameters were not affected by the alkaline pH. Moreover, the aggregate formation was not detected.

The HRP activation was performed in the dark to prevent sodium periodate breakdown and only for a limited period of time to avoid the loss of enzymatic activity [4,5]. Self-coupling of the activated HRP was minimized through the removal of the periodate excess by ultrafiltration against 1 mM sodium acetate buffer, pH 4.4 [4,16]. Aliquots of the HRP solution before and after activation were examined by multi-detection SEC (Figure S3 and Table S2). According to the obtained results, the oxidation of polysaccharide residues of the HRP did not affect its *Mw*, nor the polydispersity or composition of the sample peaks. The HRP purity was about 94% and the obtained absolute *Mw* value was in accordance with the theoretical one (Table 1).

After the periodate oxidation of the polysaccharide residues, Schiff bases formed between the amino groups of the IgG and the active aldehyde groups of the HRP in an alkaline pH were further converted into stable alkylamine linkages by a reduction with sodium borohydride [4,17]. Finally, ultrafiltration of the resulting IgG-HRP conjugate was performed, giving the final product (UF IgG-HRP).

The representative multi-detection SEC chromatogram of the final product, UF IgG-HRP (Figure 1), shows a peak with a retention volume (V_R) at 8.6 mL, corresponding to the IgG-HRP conjugate, but also a presence of a large fraction of soluble aggregates with the V_R <8 mL. The absolute Mw value for the IgG-HRP conjugate was ~235 kDa (Tables 1 and 2), indicating that, on average, two activated HRP molecules (~43 kDa) were bound to one molecule of IgG (~153 kDa).

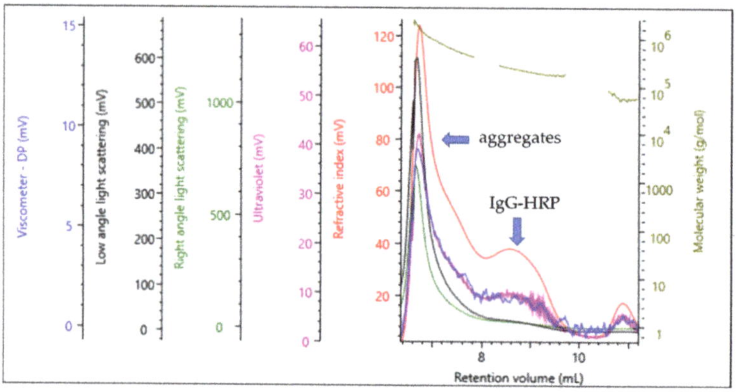

Figure 1. Representative multi-detection SEC chromatogram of the UF IgG-HRP sample—refractive index (red), right-angle light scattering (green), low-angle light scattering (black), viscometer (blue), and ultraviolet (purple). The molecular weight of each species is shown in olive.

Table 2. Molecular characterization data for the IgG-HRP samples before/after reduction and the final product (UF IgG-HRP) obtained by multi-detection SEC. The results are expressed as mean value ± standard error (SE) from n measurements.

	Identity	V_R (mL)	Mw (kDa)	Mw/Mn	Share (%)
IgG-HRP before reduction (n = 2)	aggregate	6.8	902.2 ± 34.0	1.200 ± 0.016	52.76 ± 1.34
	IgG-HRP	8.6	232.2 ± 5.1	1.021 ± 0.002	39.59 ± 1.62
	n/d *	10.9	71.8 ± 0.9	1.019 ± 0.004	7.66 ± 0.28
IgG-HRP after reduction (n = 2)	aggregate	6.8	983.1 ± 51.5	1.239 ± 0.050	55.47 ± 1.46
	IgG-HRP	8.6	246.4 ± 0.9	1.022 ± 0.003	38.64 ± 1.88
	n/d	10.9	83.3 ± 5.6	1.011 ± 0.004	5.90 ± 0.42
UF IgG-HRP (n = 5)	aggregate	6.8	1273.4 ± 49.4	1.285 ± 0.045	61.82 ± 2.52
	IgG-HRP	8.6	234.9 ± 3.2	1.028 ± 0.050	32.06 ± 2.25
	n/d	10.9	64.0 ± 0.6	1.015 ± 0.002	6.12 ± 0.41

* n/d—not defined.

The quantitative properties of different molecular forms present in the conjugation reaction before and after the reduction step, as well in the final UF IgG-HRP sample, provided information on changes triggered by particular process steps (Table 2). The peaks with the V_R at 6.8 mL in all samples were related to the heterogeneous population of higher-order aggregates, as confirmed by the relatively high polydispersity (>1.2). The peaks with the V_R at 8.6 mL corresponded to the IgG-HRP conjugate with the average stoichiometry of 1:2, while those with the V_R at 10.9 mL remained unidentified, but could possibly stem from impurities detected in the HRP sample, according to the obtained absolute Mw value (Figure S3 and Table S2). The reduction step did not affect the Mw or composition of the sample peaks. However, an increase of the aggregate share (>60%) in the IgG-HRP sample after ultrafiltration occurred. In addition, the qualitative comparison of all tested samples

(Figure 2) and quantitative properties of the IgG-HRP conjugate (Table 2) showed that neither free IgG nor free activated HRP was present in the final product.

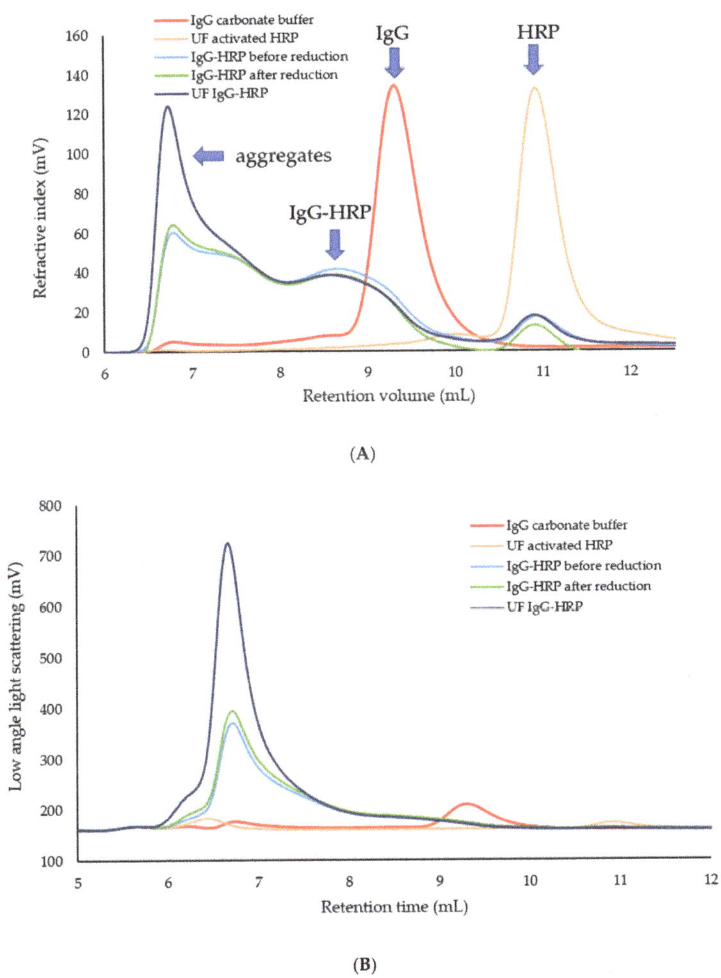

Figure 2. Overlay of multi-detection SEC chromatograms of starting samples (IgG and HRP), intermediates, and final IgG-HRP product (~1.5 mg/mL) obtained by (**A**) refractive index (RI) or (**B**) low-angle light scattering (LALS) detector.

2.2. Evaluation of IgG-HRP Conjugate Functionality

The functionality of the prepared anti-Vero IgG-HRP conjugate (so far termed the UF IgG-HRP) was examined by the ELISA for the quantification of the total Vero cell lysate proteins. In the preliminary experiment, an anti-Vero IgG concentration, used for coating the plate, and an anti-Vero IgG-HRP conjugate dilution were varied to determine the optimal working dilution (Table S3). The prepared conjugate was fully functional in the ELISA, as demonstrated by the excellent regression lines of the Vero lysate standard curves (in the range of 1–30 ng/µL) for the most suitable dilutions of the anti-Vero IgG-HRP conjugate (Figure 3). The optimal concentration of the anti-Vero IgG for plate coating was 10 µg/mL and the most appropriate dilution of the anti-Vero IgG-HRP conjugate (4.4 mg/mL) was 1:2000.

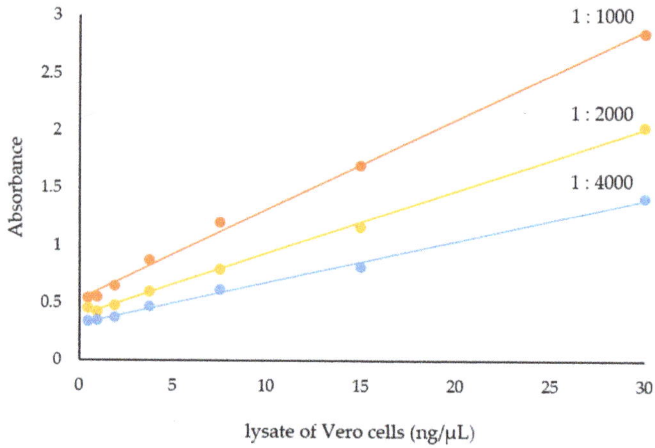

Figure 3. Standard curves for several dilutions of anti-Vero IgG-HRP conjugate in the ELISA for quantification of total Vero cell lysate proteins. The plate was coated with 10 μg/mL of anti-Vero IgG.

2.3. Multi-Detection SEC Analysis of Commercially Available IgG-HRP Conjugates

In order to compare the absolute *Mw*, polydispersity, and composition of sample peaks between various reagents (Table 3), we lastly examined three different commercially available IgG-HRP conjugates from two manufacturers: anti-goat IgG-HRP produced in rabbit (IgG-HRP 1), anti-rabbit IgG-HRP produced in goat (IgG-HRP 2), and anti-guinea pig IgG-HRP produced in rabbit (IgG-HRP 3). Their representative multi-detection SEC chromatograms are presented in Supplementary Materials Figure S4, while the overlay of RI data, including RI data obtained for the prepared in-house UF IgG-HRP conjugate, is shown in Figure 4. The qualitative comparison (Figure 4) and quantitative properties of commercially available IgG-HRP conjugates (Table 3) showed the prevailing peak at 9.4 mL, which could correspond to the free IgG with an absolute *Mw* value of ~150 kDa and a share ranging between 42 and 62%.

Table 3. Molecular characterization data for commercially available IgG-HRP conjugates from multi-detection SEC analysis. The results are expressed as mean value ± standard error (SE) from 3 measurements.

	V_R (mL)	*Mw* (kDa)	*Mw*/*Mn*	*IV* (dL/g)	*Rh* (nm)	Share (%)
IgG-HRP 1 (rabbit anti-goat IgG-HRP)	7.1	965.0 ± 9.2	1.279 ± 0.016	0.076 ± 0.007	10.20 ± 0.35	16.18 ± 1.01
	8.2	321.2 ± 0.7	1.003 ± 0.001	0.059 ± 0.005	6.69 ± 0.21	5.20 ± 0.54
	8.7	225.2 ± 0.6	1.005 ± 0.001	0.054 ± 0.005	5.75 ± 0.18	9.13 ± 1.21
	9.4	145.5 ± 0.4	1.007 ± 0.001	0.046 ± 0.003	4.74 ± 0.10	61.94 ± 0.89
	10.8	68.5 ± 2.1	1.009 ± 0.004	0.040 ± 0.006	3.50 ± 0.20	7.52 ± 0.95
IgG-HRP 2 (goat anti-rabbit IgG-HRP)	6.8	1175.7 ± 5.6	1.205 ± 0.019	0.080 ± 0.000	11.20 ± 0.02	21.27 ± 0.47
	8.0	371.4 ± 0.9	1.006 ± 0.001	0.061 ± 0.001	7.09 ± 0.03	9.71 ± 0.22
	8.7	236.1 ± 0.6	1.005 ± 0.000	0.053 ± 0.000	5.82 ± 0.02	8.99 ± 0.38
	9.4	156.6 ± 0.4	1.005 ± 0.001	0.046 ± 0.001	4.85 ± 0.03	42.36 ± 0.33
	10.9	55.6 ± 0.6	1.032 ± 0.004	0.034 ± 0.002	3.11 ± 0.04	17.67 ± 0.19
IgG-HRP 3 (rabbit anti-guinea pig IgG-HRP)	6.7	1754.5 ± 7.0	1.348 ± 0.045	0.078 ± 0.002	12.50 ± 0.16	15.75 ± 0.99
	8.2	353.2 ± 0.4	1.013 ± 0.003	0.064 ± 0.003	7.08 ± 0.11	8.58 ± 0.73
	8.7	223.7 ± 0.5	1.003 ± 0.001	0.055 ± 0.002	5.78 ± 0.09	6.97 ± 0.67
	9.4	144.8 ± 0.1	1.008 ± 0.001	0.046 ± 0.001	4.74 ± 0.03	56.16 ± 0.77
	10.8	62.7 ± 0.6	1.017 ± 0.004	0.041 ± 0.002	3.43 ± 0.07	12.55 ± 0.57

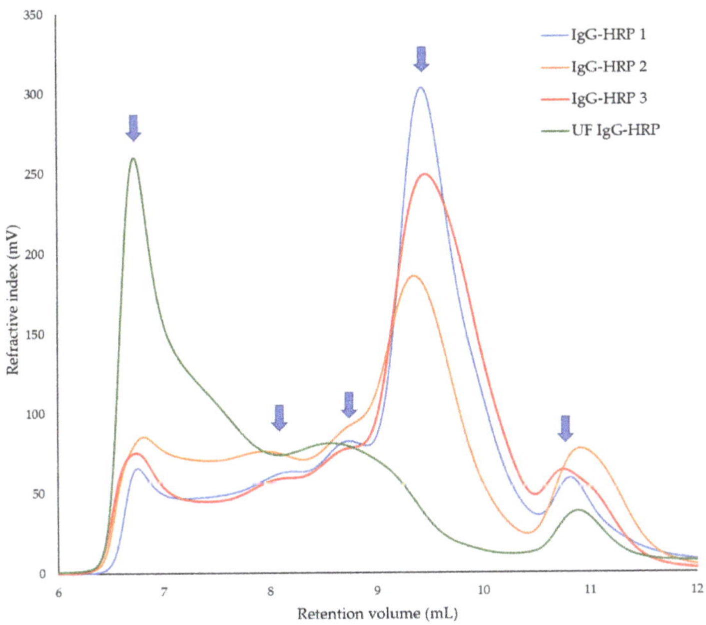

Figure 4. Overlay of RI chromatograms of commercially available IgG-HRP conjugates and prepared UF IgG-HRP conjugate (~3 mg/mL).

From the obtained *Mw* for the IgG and HRP, and peaks with V_R at 8.7 mL and 8.2 mL, we could estimate that the peaks with V_R at 8.7 mL and 8.2 mL could correspond to the IgG-HRP conjugates with the average stoichiometry of 1:2 and 1:4 or 1:5, respectively. The obtained *Mw*/*Mn* values indicated a high degree of molecular homogeneity and symmetrical size distribution of those peaks. The total share for both types of IgG-HRP conjugates (with the average stoichiometry of 1:2 and 1:4 or 1:5) in each of the three tested samples was below 19%. All samples also contained a population of higher-order aggregates with the V_R < 6.7 mL (share from 16–21%). The peaks with the V_R at 10.8 mL (share from 8–18%) remained unidentified. The obtained *Rh* values for IgG-HRP conjugates (Table 3) increased compared to the IgG monomer, as expected, in accordance with the increase in the *Mw* of the conjugates and the estimated average stoichiometry of their components.

3. Discussion

An ideal coupling reaction procedure would be the one generating a homogeneous antibody–enzyme conjugate composed of one molecule of enzyme linked with one molecule of antibody in which most of the immunological and enzymatic activity is preserved. However, even when performing the reaction under strictly controlled conditions, the formation of undesirable conjugate types with different stoichiometric ratios could not be avoided. Therefore, it is necessary to pay attention to the retention of antibody and enzyme activity and the prevention of extensive aggregate formation, which reduces the amount of the active ingredient in the sample and may cause precipitation [1,18]. In this study, we aimed to demonstrate the potential of the multi-detection SEC in the advanced molecular characterization of the coupling reaction between IgG and HRP in order to gain insight into the process yield and the conjugate quality with respect to the preservation of its immunological and enzymatic activity. The IgG-HRP conjugate was prepared by the modified periodate oxidation method, which was first described by Nakane and Kawaoi [6]. The protocols for the periodate oxidation method described in the literature vary according to the concentration of the IgG, HRP, and required reagents, the $NaIO_4$/HRP and HRP/IgG molar ratio, the type of reducing agent, the duration of the reaction steps, the pH at which the reaction steps are performed, and the methods used to purify the IgG-HRP conjugate and remove the unconjugated enzyme [4,5,16,19].

In this work, individual reaction steps during the preparation of the IgG-HRP conjugate were monitored by analyzing the starting samples, intermediates, and final product with multi-detection SEC (Figure 2). The absolute molecular weight of IgG and HRP as input compounds was in accordance with the expected values (Table 1). Their purity was exceptionally high (>92%) and more than suitable for the coupling reaction.

The results showed that the oxidation of the polysaccharide residues of HRP with sodium periodate did not affect the *Mw* or composition of the sample peaks. In addition, the reaction did not induce any self-coupling of the activated HRP (Figure S3 and Table S2). Although HRP contains only two available primary ε-amine groups (two lysine residues) and its self-coupling is not a common phenomenon [1], the risk was minimized by keeping the activated enzyme in a slightly acidic medium [16].

A multi-detection SEC analysis (Figures 1 and 2) showed that the conjugation reaction produced a heterogeneous sample containing IgG-HRP conjugates with the average stoichiometry of 1:2, but also unwanted populations, composed either of aggregates or higher *Mw* conjugation products. The possible reason for their occurrence might be the formation of a large number of highly active aldehyde groups on the HRP molecules [6,18] which, when activated, bridge the IgGs, causing their cross-linking [19,20]. Nakane and Kawaoi demonstrated that one IgG molecule can bind a maximum of five to six activated HRP molecules [6]. The excessive cross-linking could be decreased by reducing the concentration of IgG and/or HRP, adjusting the IgG/HRP molar ratio, or shortening the reaction time. In order to optimize the reaction conditions and improve the yield, the influence of these factors on the conjugation could be easily monitored by multi-detection SEC, which, in only one analysis, generates multiple quantitative data of the sample, including absolute *Mw*, polydispersity, and composition.

In addition, the *Mw* or composition of the sample peaks were not affected by the reduction step. However, the ultrafiltration step led to the increase of aggregate content (Table 2, Figure 2), possibly as a consequence of a combination of several factors, such as the high protein concentration, enhanced solution viscosity, and physical stress, which implies the filtration rate, fluid flow-induced shear stress, interfacial interactions, and extensive contact with the membrane surface [21–24]. Such a finding suggests that ultrafiltration might not be the best method for the matrix exchange and goes in favor of dialysis, as described in the majority of published protocols [4,6,16,17]. Namely, the ultrafiltration process involves a transient increase in the concentration of the sample in contrast to dialysis, which might be a trigger for aggregation.

In addition to the characterization of our in-house anti-Vero IgG-HRP conjugate, another aim was to gain insight into its compliance with those that are commercially available. For that purpose, we have randomly selected three conjugates from two manufacturers. The multi-detection SEC chromatograms of the purchased conjugates showed very similar profiles which indicates consistency in their preparation (Figure 4, Table 3). All three conjugates contained a low share of two types of IgG-HRP conjugates (<19%) and a significant amount of free IgGs (42–62%). In contrast, our in-house preparation had a greater share of the IgG-HRP conjugates (32%) than the purchased ones and fully coupled IgGs. Additionally, the aggregates' share appeared to be roughly three-fold lower in the purchased conjugates than in ours. The observed differences are probably the consequence of variations in the protocols that were used for their preparation. However, the applicability of multi-detection SEC, rather than a comparison of different conjugation reaction processes, was the main focus of this paper. It is interesting to note that the estimated *Mw* of the putative IgG monomer in two commercial conjugates, both prepared from rabbit IgGs, was ~144 kDa which is in complete agreement with the *Mw* of the rabbit IgG previously determined by Rayner et al. [25].

The optimal relative size of the antibody–enzyme complex preferably should be adapted to the assay application. Nakane and Kawaoi demonstrated that conjugates containing an average of two activated HRP molecules per one IgG molecule were more suitable than highly labelled conjugates (with more than four activated HRP molecules per one IgG molecule) for immunohistochemical staining where penetration of the complex through membrane barriers is an important consideration [6]. In contrast, high-molecular weight IgG-HRP conjugates could be an appropriate choice for ELISA procedures, where high sensitivity is important, and washing off excess conjugate is not a problem. However, two critical points still remain. Firstly, the aggregates reduce the active ingredient share [5]. Secondly, the presence of free IgGs affects the sensitivity of immunoassays. Therefore, the composition of the IgG-HRP conjugate sample directly influences the degree of its quality as a reagent.

The functionality of the prepared in-house anti-Vero IgG-HRP conjugate, as well as its optimal working dilution for ELISA, was determined. It is a well-known fact that the optimal dilutions of commercially available IgG-HRP conjugates vary, irrespective of the declared concentration. According to our results, the conjugates' working range might be influenced by the differences in their content as well, primarily the share of the (enzymatic/immunological) active substance and aggregates, affecting the sensitivity and specificity of the final product.

In this work, it was demonstrated that advanced molecular characterization with the multi-detection SEC could offer an in-depth insight into the conjugation reaction. Many other standard manufacturing procedures could also benefit from its implementation. Multi-detection SEC might serve as a routine technique in quality control or provide guidelines that could contribute to the process improvement, providing full molecular characterization due to its ability to simultaneously measure and characterize the sample's composition on multiple levels.

4. Materials and Methods

4.1. Reagents and Chemicals

Horseradish peroxidase and $NaBH_4$ were from Sigma-Aldrich (St. Louis, MO, USA). $NaIO_4$ was from Merck (Darmstadt, Germany). A daily supply of ultrapure water was obtained from a PureLab Classic purification system (Elga, High Wycombe, UK). Chemicals for buffers and solutions were from Kemika (Zagreb, Croatia), unless stated otherwise.

Three different commercially available IgG-HRP conjugates from two manufacturers were analyzed with multi-detection SEC: anti-goat IgG-HRP produced in rabbit (IgG-HRP 1), anti-rabbit IgG-HRP produced in goat (IgG-HRP 2), and anti-guinea pig IgG-HRP produced in rabbit (IgG-HRP 3).

4.2. Purification of Antibodies

Antibodies were purified from a guinea pig (immunized with lysate of Vero cells) serum using protein A HiTrap MabSelect Xtra column (1 mL; GE Healthcare, Chicago, IL, USA). Serum was 2-fold diluted with the binding buffer (20 mM phosphate buffer, 0.15 M NaCl, pH 7.2) and loaded (2 mL/run) to the pre-equilibrated protein A column at a flow rate of 2 mL/min on an ÄKTApurifier 100 system equipped with P-900, UV-900, and pH/C-900 (GE Healthcare, USA) at room temperature (RT). The absorbance was monitored at 280 nm. The bound antibodies were eluted from the column with 20 mM glycine, 0.15 M NaCl, and pH 2.3. Eluted fractions from all runs were pooled, concentrated, and transferred to the binding buffer using a Vivacell device (Sartorius, Göttingen, Germany) with a 50 kDa molecular weight cut-off (MWCO) polyethersulfone membrane, resulting in polyclonal anti-Vero IgG sample that was used as a reagent for preparation of the IgG-HRP conjugate and for plate coating in the ELISA.

4.3. Protein Concentration Estimation

All protein (antibodies and total Vero cell lysate proteins) concentrations were estimated spectrophotometrically using the following equation:

γ [mg/mL] = ($A_{228.5\ nm} - A_{234.5\ nm}$) $\times f \times$ dilution, where Ehresmann's factor "f" of 0.3175 was used [26]. Appropriate dilution of each sample was independently prepared a minimum of three times to obtain the mean value for further calculation of yield and purity. Absorbance measurements were performed on a Multiskan Spectrum (Thermo Fisher Scientific, Waltham, MA, USA).

4.4. Preparation of IgG-HRP Conjugate

Conjugate of HRP and guinea pig polyclonal antibodies was prepared by a modified periodate oxidation method [3,4,16,17]. Briefly, freshly prepared 0.1 M sodium (meta) periodate solution (220 equivalents) was added to 4 mg/mL HRP solution (1 equivalent) in water and stirred gently on a shaker (300 rpm) for 20 min at room temperature (RT) in dark. Excess of periodate was removed by ultrafiltration of the activated HRP solution against 1 mM sodium acetate buffer, pH 4.4, in centrifugal filter unit with 10 kDa MWCO membrane (Sartorius, Göttingen, Germany). Before conjugation reaction, pH of ultrafiltrated and activated HPR solution was adjusted to alkaline pH by adding 0.2 M sodium carbonate buffer, pH 9.5 (40 µL to 1 mL HRP solution). Moreover, anti-Vero IgG sample was transferred into 10 mM sodium carbonate buffer, pH 9.5, using a 50 kDa MWCO ultrafiltration device (Sartorius, Göttingen, Germany). Then, 4 mg/mL IgG solution (20 mg, 1 equivalent) was added to the activated HPR solution (2 equivalents) and stirred on a shaker (300 rpm) for 2 h at RT in dark. After the addition of 4 mg/mL freshly prepared sodium borohydride solution (210 equivalents), the mixture was stirred (300 rpm) for 90 min at RT in dark. Following filtration through a 0.45 µm PVDF syringe filter (Merck Millipore, Darmstadt, Germany), solution of IgG-HRP conjugate was ultrafiltrated and transferred to PBS using a 100 kDa MWCO ultrafiltration device (Sartorius, Göttingen, Germany). Aliquots of starting samples, all intermediates, and final product were collected for multi-detection SEC and protein concentration estimation.

4.5. Multi-Detection SEC

Size exclusion chromatography was carried out on TSKGel G3000SWXL column (7.8 $\times$ 300 mm; Tosoh Bioscience, Japan) with a TSK gel SWXL guard column (6.0 $\times$ 40 mm; Tosoh Bioscience, Japan). HPLC analysis was performed with 0.1 M phosphate–sulphate running buffer, pH 6.6, at a flow rate of 1 mL/min at RT on a HPLC Prominence system (Shimadzu, Kyoto, Japan). The samples were centrifuged (Eppendorf, Hamburg, Germany) at 3000$\times$ g for 10 min and loaded to the column in a volume of 50 µL/run. Total run time was 17 min. Advanced detection was obtained on Omnisec Reveal (Malvern Panalytical Ltd., Malvern, United Kingdom) multi-detector module which consists of refractive index (RI) detector, UV/Vis absorbance detector, hybrid right-angle light scattering, low-angle light scattering (RALS/LALS) detector, and intrinsic viscosity detector. Calibration was performed using a bovine serum albumin (BSA) standard (Thermo Scientific, Waltham, MA, USA). The resulting chromatograms were analyzed using RI, RALS, LALS, and viscometer detectors. The dn/dc values, calculated from the sample concentrations, were used to determine the molecular weight of the peaks using Omnisec software (version 11.32). Each sample was analyzed at least twice.

4.6. ELISA for Evaluation of IgG-HRP Conjugate Functionality

Indirect ELISA was performed according to the previously published procedure [27]. Briefly, flat-bottomed high-binding ELISA plate (Corning, New York, SAD) was coated with 100 µL/well of guinea pig anti-Vero IgG sample (1 µg/mL or 10 µg/mL) in 0.05 M carbonate buffer, pH 9.0, and incubated overnight at RT. The plate was washed and blocked with 250 µL/well of 2% (*w/v*) BSA (Sigma-Aldrich, Saint Louis, MO, USA) in PBS with 0.05% (*v/v*) Tween 20 for 2 h at 37 °C. After blocking, a lysate of Vero cells of known concentration was applied as a standard in two-fold serial dilutions (100 μL/well) in duplicates, starting from 30 ng/μL and incubated overnight at RT. Plates were washed and 100 μL/well of guinea pig anti-Vero IgG-HRP conjugate in 1:1000, 1:2000, 1:4000, 1:16,000, or 1:32,000 dilution was added and incubated for 2 h at 37 °C. Following washing, 100 μL/well of 0.6 mg/mL *o*-phenylenediamine dihydrochloride (OPD) solution in citrate–phosphate buffer, pH 5.0, containing 0.5 μL 30% H_2O_2/mL of OPD solution, was added and incubated for 30 min at RT in the dark. The enzymatic reaction was stopped by addition of 50 μL/well of 12.5% (*w*/*v*) H_2SO_4 and the absorbance was measured on a Multiskan Spectrum (Thermo Fisher Scientific, Waltham, MA, USA) at 492 nm.

5. Conclusions

In this study, we investigated the potential of multi-detection SEC as a tool for monitoring the conjugation reaction between the antibody and HRP, as well as the advanced characterization of prepared or commercially available IgG-HRP conjugates. The absolute molecular weight, polydispersity, size, composition of sample peaks, and aggregate content were determined in all reaction stages. Based on the absolute molecular weight, the average stoichiometric relationship between the IgG and HRP in the conjugate was estimated. This study showed that multi-detection SEC could be an appropriate method for monitoring the conjugation reaction in order to gain insight into the process quality with the aim of the enhancement of the active component's yield, and, consequently, the improvement of the reagent's sensitivity and specificity. Furthermore, it seems to be a promising and powerful technology for the quality control of many other protein–protein conjugates, providing a detailed, precise, and full molecular characterization.

Supplementary Materials: The following supporting information can be downloaded at: https://www.mdpi.com/article/10.3390/molecules28114567/s1, Figure S1: Representative multi-detection SEC chromatograms of IgG sample in phosphate buffer—refractive index (red), right-angle light scattering (green), low-angle light scattering (black), and viscometer (blue). The molecular weight of each species is shown in olive; Figure S2: Overlay of RI data obtained for the representative IgG sample in phosphate and carbonate buffer by multi-detection SEC; Figure S3: Overlay of RI data obtained for the representative HRP sample before and after activation by multi-detection SEC; Figure S4: Representative multi-detection SEC chromatograms of commercially available IgG-HRP conjugates: (A) anti-goat IgG-HRP produced in rabbit (IgG-HRP 1), (B) anti-rabbit IgG-HRP produced in goat (IgG-HRP 2), and (C) anti-guinea pig IgG-HRP produced in rabbit (IgG-HRP 3); Table S1: Summary of the results from multi-detection SEC analysis of the representative IgG sample (presented in Figure S2) in phosphate and carbonate buffer; Table S2: Summary of the results from multi-detection SEC analysis of the representative HRP sample before and after activation (presented in Figure S3); Table S3: Absorbance readings from the ELISA assay obtained by two different concentrations of anti-Vero IgG (1 μg/mL and 10 μg/mL) and various dilutions of anti-Vero IgG-HRP conjugate. A lysate of Vero cells of known concentration was applied as a standard in two-fold serial dilutions, starting from 30 ng/μL.

Author Contributions: Conceptualization, A.Š., T.K. and B.H.; methodology, A.Š.; formal analysis, A.Š.; investigation, A.Š.; resources, B.H.; writing—original draft preparation, A.Š.; writing—review and editing, T.K. and B.H.; supervision, B.H.; project administration, B.H.; funding acquisition, B.H. All authors have read and agreed to the published version of the manuscript.

Funding: This research was funded by the European Regional Development Fund, grant number KK.01.1.1.01.0006, "Strengthening the capacity of CerVirVac for research in virus immunology and vaccinology".

Institutional Review Board Statement: Not applicable.

Informed Consent Statement: Not applicable.

Data Availability Statement: The data presented in this study are available upon request from the corresponding author.

Conflicts of Interest: The authors declare no conflict of interest.

Sample Availability: Not available.

References

1. Hermanson, G.T. Enzyme Modification and Conjugation. *Bioconjug. Tech.* **2013**, *22*, 951–957. [CrossRef]
2. Veitch, N.C. Horseradish Peroxidase: A Modern View of a Classic Enzyme. *Phytochemistry* **2004**, *65*, 249–259. [CrossRef] [PubMed]
3. Wisdom, G.B. Conjugation of Antibodies to Horseradish Peroxidase. *Methods Mol. Biol.* **2005**, *295*, 127–130. [CrossRef] [PubMed]
4. Pavliuchenko, N.; Hazarnian, V.; Bassil, M. Modification of Periodate Oxidation Method to Produce HRP-IgG Conjugate and Test Its Stability Overtime. *Am. J. Mol. Biol.* **2019**, *9*, 52–63. [CrossRef]
5. Hermanson, G.T. Antibody Modification and Conjugation. *Bioconjug. Tech.* **2013**, *20*, 867–920. [CrossRef]
6. Nakane, P.K.; Kawaoi, A. Peroxidase-Labeled a New Antibody Method Conjugation. *J. Histochem. Cytochem.* **1974**, *22*, 1084–1091. [CrossRef]
7. Malvern. *Application Note: Characterisation of IgG Monomers & Their Aggregates*; Malvern Instruments Limited: Malvern, UK, 2015.
8. Dhankher, A.; Hernandez, M.E.; Howard, H.C.; Champion, J.A. Characterization and Control of Dynamic Rearrangement in a Self-Assembled Antibody Carrier. *Biomacromolecules* **2020**, *21*, 1407–1416. [CrossRef]
9. Hartmann, W.K.; Saptharishi, N.; Yang, X.Y.; Mitra, G.; Soman, G. Characterization and Analysis of Thermal Denaturation of Antibodies by Size Exclusion High-Performance Liquid Chromatography with Quadruple Detection. *Anal. Biochem.* **2004**, *325*, 227–239. [CrossRef]
10. Miercke, L.J.W.; Robbins, R.A.; Stroud, R.M. Tetra Detector Analysis of Membrane Proteins. *Curr. Protoc. Protein Sci.* **2014**, *2014*, 29.10.1–29.10.30. [CrossRef]
11. Wen, J.; Arakawa, T.; Philo, J.S. Size-Exclusion Chromatography with on-Line Light-Scattering, Absorbance, and Refractive Index Detectors for Studying Proteins and Their Interactions. *Anal. Biochem.* **1996**, *240*, 155–166. [CrossRef]
12. Williams, K. How Does Multi-Detector GPC/SEC Work? Materials Talks. Available online: https://www.materials-talks.com/how-does-multi-detector-gpc-sec-work/ (accessed on 24 May 2023).
13. Malvern. *White Paper: Static Light Scattering Technologies for GPC/SEC Explained*; Malvern Instruments Limited: Malvern, UK, 2013; pp. 1–28.
14. Lukačević, S.M.; Kurtović, T.; Balija, M.L.; Brgles, M.; Steinberger, S.; Marchetti-Deschmann, M.; Halassy, B. Quality-Related Properties of Equine Immunoglobulins Purified by Different Approaches. *Toxins* **2020**, *12*, 789. [CrossRef]
15. Hermanson, G.T. Zero-Length Crosslinkers. *Bioconjug. Tech.* **2013**, *4*, 259–273. [CrossRef]
16. Beyzavi, K.; Hampton, S.; Kwasowski, P.; Fickling, S.; Marks, V.; Clift, R. Comparison of Horseradish Peroxidase and Alkaline Phosphatase-Labelled Antibodies in Enzyme Immunoassays. *Ann. Clin. Biochem.* **1987**, *24*, 145–152. [CrossRef]
17. Eivazi, S.; Majidi, J.; Maleki, L.A.; Abdolalizadeh, J.; Yousefi, M.; Ahmadi, M.; Dadashi, S.; Moradi, Z.; Zolali, E. Production and Purification of a Polyclonal Antibody against Purified Mouse Igg2b in Rabbits towards Designing Mouse Monoclonal Isotyping Kits. *Adv. Pharm. Bull.* **2015**, *5*, 109–113. [CrossRef]
18. Avrameas, S.; Ternynck, T.; Guesdon, J.-L. Coupling of Enzymes to Antibodies and Antigens. *Scand. J. Immunol.* **1978**, *8*, 7–23. [CrossRef]
19. Tijssen, P.; Kurstak, E. Highly Efficient and Simple Methods for the Preparation of Peroxidase and Active Peroxidase-Antibody Conjugates for Enzyme Immunoassays. *Anal. Biochem.* **1984**, *136*, 451–457. [CrossRef]
20. Kumar, K.R. Anti-Human IgG-Horseradish Peroxidase Conjugate Preparation and Its Use in ELISA and Western Blotting Experiments. *J. Chromatogr. Sep. Tech.* **2014**, *5*, 211. [CrossRef]
21. Rosenberg, E.; Hepbildikler, S.; Kuhne, W.; Winter, G. Ultrafiltration Concentration of Monoclonal Antibody Solutions: Development of an Optimized Method Minimizing Aggregation. *J. Memb. Sci.* **2009**, *342*, 50–59. [CrossRef]
22. Callahan, D.J.; Stanley, B.; Li, Y. Control of Protein Particle Formation during Ultrafiltration/Diafiltration through Interfacial Protection. *J. Pharm. Sci.* **2014**, *103*, 862–869. [CrossRef]
23. Allmendinger, A.; Mueller, R.; Huwyler, J.; Mahler, H.C.; Fischer, S. Sterile Filtration of Highly Concentrated Protein Formulations: Impact of Protein Concentration, Formulation Composition, and Filter Material. *J. Pharm. Sci.* **2015**, *104*, 3319–3329. [CrossRef]
24. Kim, K.J.; Chen, V.; Fane, A.G. Aggregation during Ultra Filtra Tion. *Biotechnology* **1993**, *42*, 260–265.
25. Rayner, L.E.; Kadkhodayi-Kholghi, N.; Heenan, R.K.; Gor, J.; Dalby, P.A.; Perkins, S.J. The Solution Structure of Rabbit Igg Accounts for Its Interactions with the Fc Receptor and Complement C1q and Its Conformational Stability. *J. Mol. Biol.* **2013**, *425*, 506–523. [CrossRef] [PubMed]
26. Ehresmann, B.; Imbault, P.; Well, J.H. Spectrophotometric Determination of Protein Concentration in Cell Extracts Containing TRNA's and RRNA's. *Anal. Biochem.* **1973**, *54*, 454–463. [CrossRef] [PubMed]
27. Sviben, D.; Forčić, D.; Kurtović, T.; Halassy, B.; Brgles, M. Stability, Biophysical Properties and Effect of Ultracentrifugation and Diafiltration on Measles Virus and Mumps Virus. *Arch. Virol.* **2016**, *161*, 1455–1467. [CrossRef]

molecules

MDPI

Article

Development and Validation of Green and High-Throughput Microwell Spectrophotometric Assay for the Determination of Selective Serotonin Reuptake Inhibitors in Their Pharmaceutical Dosage Forms

Ibrahim A. Darwish * and Nourah Z. Alzoman

Department of Pharmaceutical Chemistry, College of Pharmacy, King Saud University, P.O. Box 2457, Riyadh 11451, Saudi Arabia; nalzoman@ksu.edu.sa

* Correspondence: idarwish@ksu.edu.sa

Abstract: This study describes the development and validation of a new green and high-throughput microwell spectrophotometric assay (MW-SPA) for the determination of three selective serotonin reuptake inhibitors (SSRIs) in their pharmaceutical dosage forms. These SSRIs are fluoxetine, fluvoxamine, and paroxetine, the most prescribed drugs for the treatment of depression. The proposed assay was based on the formation of orange-colored *N*-substituted naphthoquinone derivatives upon the reaction of SSRIs with 1,2-naphthoquinone-4-sulphonate (NQS) in alkaline media. The assay was conducted in 96-microwell assay plates, and the absorbances of the reaction products were measured by a microplate reader at their maximum absorbance wavelengths. The optimum conditions of the reaction were refined and established. Under these conditions, calibration curves were generated, and linear regression equations were computed. The linear relations between the absorbances and drug concentrations were linear with good correlation coefficients (0.9992–0.9997) in the range of 2–80 μg/mL. The assay limits of detection were in the range of 1.5–4.2 μg/mL. The precision was satisfactory as the values of relative standard deviation did not exceed 1.70%. The accuracy of the assay was ≥98.2%. The proposed MW-SPA was successfully applied to the analysis of the SSRIs in their pharmaceutical dosage forms with acceptable accuracy and precision; the label claims were 99.2–100.5% (±0.96–1.35%). The results of the proposed MW-SPA were compared with those of the official/pre-validated assays by statistical analysis with respect to the accuracy (by *t*-test) and precision (by F-test). No significant differences were found between the calculated and theoretical values of the t- and F-tests at the 95% confidence level, proving similar accuracy and precision in the determination of SSRIs by both assays. The greenness of the proposed assay was confirmed by two metric tools. In addition, the assay is characterized with a high-throughput property which enables the simultaneous analysis of many samples in a short time. Therefore, the assay is a valuable tool for rapid routine application in pharmaceutical quality control units for the determination of the investigated SSRIs.

Keywords: green and high-throughput analysis; microwell assay; pharmaceutical quality control; spectrophotometry; SSRIs

1. Introduction

Depression is a prevalent psychiatric or mental disorder that affects approximately 5% of adults worldwide. It is mediated by changes in the levels of some central neurotransmitters and/or their biochemical functions. The change in the balance of these neurotransmitters (norepinephrine, serotonin, or both) occurs due to an impaired synthesis of neurotransmitters, increased breakdown or metabolism, and/or increased pump uptake [1]. The use of antidepressant drugs is the most commonly used approach for the management of depression. These drugs reduce the reuptake of neurotransmitters and also the generation of cyclic adenosine monophosphate, which is involved in the pathogenesis of depression. There are four main classes of antidepressant drugs, namely tricyclic antidepressants, monoamine oxidase inhibitors, selective serotonin reuptake inhibitors (SSRIs), and other miscellaneous antidepressants [2].

SSRIs are mostly prescribed as the first-line treatment of depression and other psychiatric disorders. The wide use of SSRIs is attributed to their safety, potency, and tolerability. SSRIs exert their antidepressant efficacy by selective inhibition of the reuptake of serotonin by the presynaptic receptors. This inhibitory effect leads to an increase in the level of serotonin at the synapse and thus stimulates the central nervous system. The Food and Drug Administration (FDA) has approved nine SSRI drugs for the treatment of depression. These drugs are fluoxetine, fluvoxamine, paroxetine, sertraline, citalopram, escitalopram, vilazodone, dapoxetine, and vortioxetine. These SSRIs have different chemical structures, leading to differences in their physicochemical and pharmacokinetic characteristics [3]. SSRIs have a similar antidepressant potency to that of tricyclic antidepressants; however, SSRIs have better safety and acceptability in patients. In addition, SSRIs have no sedative side effects, are significantly safe in overdose cases, and have a long half-life, which enables their use at once-daily and/or once-weekly dosing. These combined benefits are the supporting factors standing behind the wide use of SSRIs as antidepressant drugs [4].

The remarkably potent and improved safe therapy for depression with SSRIs is principally dependent on the active drug content and content uniformity of their pharmaceutical dosage forms. To ensure these requirements are met, a proper assay is required for quality control of their dosage forms. Different assays relying on varying techniques have been described for the quality control of dosage forms of SSRIs. These assays/techniques were reviewed and published in some review articles [5,6]. Among these assays, spectrophotometric assays are the most widely used, as indicated from the number of published articles describing spectrophotometric assays for SSRIs [7–19]. The great interest in spectrophotometric assays for SSRIs is due to their simplicity, convenience, low cost, ready automation with spectrophotometric analyzers, and availability of the technique in almost all quality control laboratories in pharmaceutical manufacturing companies [17–19]. However, the existing spectrophotometric assays for SSRIs employ the conventional analytical practice, which uses volumetric flasks/cuvettes in the analysis and is conducted in a manual fashion. Accordingly, the throughputs of these assays are limited and do not fulfill the important needs of quality control laboratories for processing batches containing a large number of pharmaceutical formulations [20–22]. In addition, these assays consume large volumes of organic solvents, which are costly and have a harmful impact on the environment and, more importantly, on the health of analysts [23–26]. Therefore, these assays do not meet the needs of pharmaceutical industries for assays with high throughput [20–22] and do not apply the principles of the green analytical chemistry (GAC) approach [27].

This study describes a new green and high-throughput microwell spectrophotometric assay (MW-SPA) for the quality control of SSRIs in pharmaceutical laboratories. The assay was developed and validated for three SSRIs, which are fluoxetine (FLX), fluvoxamine (FXM), and paroxetine (PXT). These drugs were selected as representative examples of SSRIs because they are the most important and widely prescribed SSRIs in the treatment of depression. These drugs are official in the British Pharmacopoeia [28] and United States Pharmacopeia [29]. The chemical structures of FLX, FXM, and PXT are given in Figure 1, and their chemical names, molecular formulae, and molecular weights are given in Table 1. The assay was based on the reaction of the investigated SSRIs, via their primary or secondary amino groups, with 1,2-naphthoquinone-4-sulphonate (NQS) in an alkaline medium, forming orange-colored *N*-substituted naphthoquinone products. The reaction was conducted in 96-well transparent plates, and the absorbances of the colored reaction products were measured by a microplate reader in absorbance mode. The proposed MW-SPA was proven to meet the principles of the GAC approach and fulfills the demands of high-throughput analysis for the pharmaceutical industry.

2.2.6. Stability of the Chromogen

Using the optimal conditions identified earlier, the reactions between SSRIs and NQS were completed within 10 min, and the absorbances did not change after 25 min of standing. The stability of the chromogen over time was also examined by measuring the absorption intensity of the reaction solution at various time intervals. The results showed that the absorbance of the chromogen remained stable for at least 2 h, which improved the convenience and applicability of the assay for a large number of samples.

Under the aforementioned optimum conditions, the reactions between SSRIs and NQS were completed within 10 min, and the absorbances no longer changed after standing for up to 25 min. The effect of time on the stability of the chromogen was studied by following the absorption intensity of the reaction solution (after dilution) at different time intervals. It was found that the absorbance of the chromogen remained stable for at least 2 h. This increased the convenience of the method as well as making it applicable for a large number of samples.

A summary of the optimization of the variables affecting the reaction of NQS with the investigated SSRIs is given in Table 2.

Table 2. Summary of the optimization of variables affecting the reaction of NQS with the investigated SSRIs.

Variable	Studied Range	Optimum Condition		
		FXM	**FLX**	**PXT**
Measuring wavelength (nm)	400–600	470	490	490
NQS (%, *w*/*v*)	0.05–1.25	0.5	0.5	0.5
pH	4–13	9	11	9
Temperature (°C)	25–50	40 [a]	25	25
Time (min)	2–25	10	10	10
Solvent	Different [b]	Water	Water	Water
Stability of chromogen (hour)	0.5–2	2	2	2

[a] The subsequent experiments involved in assay procedures were carried out at 25 °C. [b] Solvents tested: water, methanol, ethanol, isopropanol, acetone, acetonitrile, dimethyl sulfoxide, and 1,4-dioxane.

2.3. Validation of the Proposed MW-SPA

2.3.1. Calibration and Sensitivity

Using the optimal conditions outlined in Table 2, calibration curves for determining SSRIs through their reaction with NQS were constructed by plotting absorbances against their concentrations (Figure 7). The regression equation was A = a + bC, where A represents the absorbance at λ_{max} and C represents the concentration of the SSRI in µg/mL. Linear relations with small intercepts (0.0015–0.0502) and high correlation coefficients (0.9992–0.9997) were obtained for the three SSRIs in the concentration range of 5–80 µg/mL for FXM and 2–40 µg/mL for FLX and PXT. The limits of detection (LODs) and limits of quantification (LOQs) were determined according to the International Conference of Harmonization (ICH) guidelines for validating analytical procedures [30]. The following formula was used: LOD or LOQ = κSDa/b, where κ = 3.3 for LOD and 10 for LOQ, SDa is the standard deviation of the intercept, and b is the slope. The LOD values were 4.2, 1.5, and 1.8 µg/mL for FXM, FLX, and PXT, respectively. The LOQ values were 12.7, 4.5, and 5.5 µg/mL for FXM, FLX, and PXT, respectively.

The parameters for the analytical performance of the proposed method are summarized in Table 3.

2.3.2. Reproducibility and Accuracy

To assess the reproducibility of the proposed MW-SPA, five separate solutions of the working standard SSRI solutions were analyzed. The assay showed satisfactory reproducibility, as the relative standard deviation (RSD) did not exceed 1.70% (Table 4). This precision level is sufficient for the routine analysis of the investigated drugs in quality control laboratories.

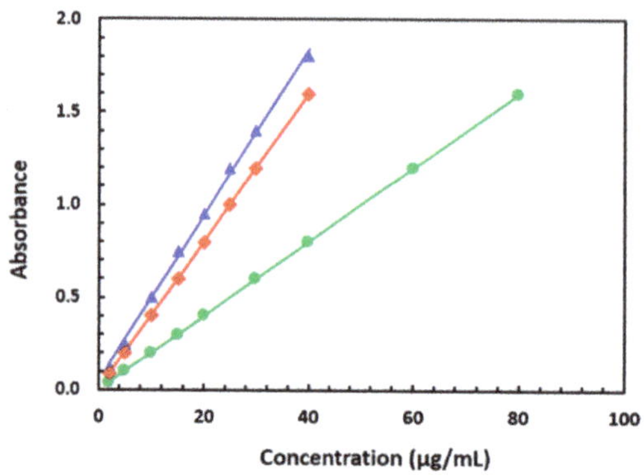

Figure 7. The calibration curves for the determination of FXM (•), FLX (▲), and PXT (♦) by the proposed MW-SPA via their reaction with NQS reagent.

Table 3. Statistical parameters for the determination of SSRIs by the proposed MW-SPA based on their reaction with NQS.

Parameter	FXM	FLX	PXT
λ_{max} (nm)	470	490	490
Linear range (µg/mL)	5–80	2–40	2–40
Intercept	0.0061	0.0502	0.0015
SD of intercept	0.0253	0.0201	0.0225
Slope	0.0199	0.0442	0.0412
SD of slope	0.0054	0.0091	0.0211
Correlation coefficient (r)	0.9996	0.9997	0.9992
LOD (µg/mL)	4.2	1.5	1.8
LOQ (µg/mL)	12.7	4.5	5.5

Table 4. Replicate analysis of SSRIs by the proposed MW-SPA for the determination of SSRIs by the proposed MW-SPA.

Sample Number	Absorbance		
	FXM (20 µg/mL)	FLX (20 µg/mL)	PXT (20 µg/mL)
1	0.453	0.953	0.804
2	0.458	0.958	0.815
3	0.451	0.951	0.799
4	0.471	0.971	0.815
5	0.459	0.959	0.799
Mean	0.458	0.958	0.806
SD	0.008	0.008	0.008
RSD (%)	1.70	0.81	1.01

To assess the accuracy of the proposed assay, recovery studies were conducted by adding known concentrations of the SSRIs to the sample and analyzing them using the proposed assay. The recovery values were found to be within the range of 97.8–102.2 ± 0.56–1.94% (Table 5), indicating the accuracy of the assay.

Table 5. Recovery studies for determination of SSRIs by the proposed MW-SPA.

SSRI Concentration (μg/mL)	Recovery (% ± SD) [a]		
	FXM	FLX	PXT
5	101.0 ± 1.94	98.8 ± 0.56	99.5 ± 1.85
10	99.3 ± 1.54	99.5 ± 1.04	98.9 ± 1.20
15	98.5 ± 0.87	101.5 ± 1.15	101.3 ± 0.92
20	101.6 ± 1.54	102.2 ± 0.95	97.8 ± 0.84
25	98.2 ± 1.86	99.0 ± 0.76	99.7 ± 1.50

[a] Values are the mean of three determinations.

2.3.3. Robustness and Ruggedness

To assess the robustness of the assay, the impact of small variations in the assay variables on its analytical performance was evaluated. In these experiments, one parameter was changed while keeping the others constant, and the recovery percentage was calculated each time. The results showed that slight variations in the method variables did not have a significant impact on the procedure, as the recovery values ranged between 97.5–101.5 ± 0.78–1.87% (Table 6). This indicates the reliability of the proposed assay during routine analysis of SSRIs.

Table 6. Robustness of the proposed MW-SPA for determination of SSRIs by their reaction with NQS reagent.

Parameters	Recovery (% ± SD) [a]		
	FXM	FLX	PXT
Recommended conditions [b]	99.2 ± 1.45	98.5 ± 1.45	101.2 ± 1.23
NQS concentration (%, *w*/*v*)			
0.25	100.1 ± 1.36	99.5 ± 1.74	98.6 ± 0.93
0.75	99.4 ± 1.57	97.9 ± 1.45	101.4 ± 1.05
Buffer solution (pH)			
8.8	98.2 ± 1.26	ND [c]	98.4 ± 1.65
9.2	99.4 ± 0.89	ND	101.2 ± 1.49
10.8	ND	98.9 ± 1.87	ND
11.2	ND	101.5 ± 1.67	ND
Reaction time (min)			
5	97.5 ± 1.58	99.4 ± 1.62	98.9 ± 1.57
15	99.8 ± 1.60	101.1 ± 1.42	100.5 ± 0.78

[a] Values are the mean of three determinations. [b] The recommended conditions are given in the Experimental section. [c] ND: not determined.

Ruggedness was also tested by applying the proposed assay to the analysis of SSRIs under the same operational conditions by two independent analysts on three different days. The results obtained from analyst-to-analyst and day-to-day variations were reproducible, with the RSD not exceeding 1.74% (Table 7).

Table 7. Ruggedness of the proposed MW-SPA for determination of SSRIs.

Parameters	Recovery (% ± RSD) [a]		
	FXM	FLX	PXT
Analyst-to-analyst			
Analyst-1	99.4 ± 0.91	99.7 ± 1.62	99.4 ± 0.96
Analyst-2	96.6 ± 0.82	101.1 ± 1.74	98.5 ± 1.24
Day-to-day			
Day-1	100.4 ± 1.33	100.2 ± 0.62	101.1 ± 1.36
Day-2	98.7 ± 0.92	99.7 ± 1.23	100.5 ± 1.54
Day-3	101.6 ± 1.45	99.8 ± 1.07	99.4 ± 1.25

[a] Values are the mean of three determinations ± SD.

2.4. Application of the Proposed MW-SPA

The results mentioned above demonstrate that the proposed assay produced satisfactory results when used to analyze bulk SSRIs. Therefore, the SSRI contents in their pharmaceutical dosage forms were analyzed using both the proposed assay and the official methods [28,29]. The label claim percentages obtained using the proposed method were 99.2–101.3 ± 0.96–1.61% (Table 8). To compare the results with those obtained from the official assay, statistical analysis was performed to assess accuracy (using a *t*-test) and precision (using an F-test). The results showed no significant differences between the calculated and theoretical values of the *t*- and F-tests at a 95% confidence level, indicating similar accuracy and precision in the determination of SSRIs by both assays.

Table 8. Determination of SSRIs in their pharmaceutical dosage forms using the proposed MW-SPA and the official assays.

Product	Label Claim (%) ± SD [a]		F-Value [b]	t-Value [b]
	Proposed Assay	Official Assay [c]		
Prozac capsules	99.8 ± 1.28	99.6 ± 1.22	1.00	1.10
Fluzac capsules	100.1 ± 1.61	99.7 ± 1.43	1.64	1.27
Salipax capsules	99.8 ± 1.24	100.2 ± 0.85	2.35	2.13
Flutin capsules	100.5 ± 0.96	100.4 ± 0.94	1.02	3.47
Octozac capsules	99.2 ± 1.06	101.3 ± 0.48	2.20	4.88
Faverin tablets	99.9 ± 1.35	100.3 ± 1.25	1.68	1.17
Seroxate tablets	100.1 ± 1.26	99.6 ± 1.12	2.62	1.27

[a] Values are the mean of five determinations ± SD. [b] Theoretical values for t- and F-tests at 95% confidence limit (n = 5) were 2.78 and 6.39, respectively. [c] Reference [28] for FXM and reference [29] for FLX and PXT.

2.5. Greenness of the Proposed MW-SPA

Generally, assays involving microwell plates and microplate readers mostly follow the principles of GAC. This is attributed to the fact that these assays consume small volumes of solvents/reagents and produce low volumes of waste. To accurately assess the greenness of the proposed MW-SPA for SSRIs, two different metric tools were used. These tools were GAPI [31] and AGREE [32].

The GAPI metric tool [31] is capable of providing a reliable and comprehensive assessment of the ecological impact of an entire analytical procedure based on 15 parameters classified into five categories. These categories cover all aspects of the analytical process, including sample handling, sample preparation, the use of solvents/reagents, instrumentation, and the general type of assay. The results are displayed as a pictogram consisting of 15 sections, each assigned a color (green, yellow, or red). The green color indicates a safe procedure, while the red color indicates a non-green procedure.

The AGREE metric tool [32] is a new method for assessing the greenness of analytical procedures. This tool employs user-friendly software that yields comprehensive, flexible, informative, and easily interpretable results. The AGREE software evaluates 12 parameters based on the principles of GAC. The results are automatically generated and presented as a circular pictogram consisting of 12 sections, each assigned a specific color ranging from deep green (score = 1) to deep red (score = 0). The overall score, which is a fraction of unity, is automatically calculated and displayed in the center of the pictogram.

In the GAPI pictogram (Figure 8), three parameters (1, 7, and 15) related to sample collection/preparation (1), the use of solvents/reagents (7), and waste treatment (15) are shown in red. This is due to the use of an off-line sample collection/preparation, the use of buffer and NQS reagent, and the lack of intrinsic treatment for the assay's waste, respectively. It is worth noting that the university has an integrated system for the safe disposal of all organic/hazardous substances. However, the waste treatment parameter of the proposed assay was shown in red to provide a fair assessment of the assay, regardless of the integrated university system.

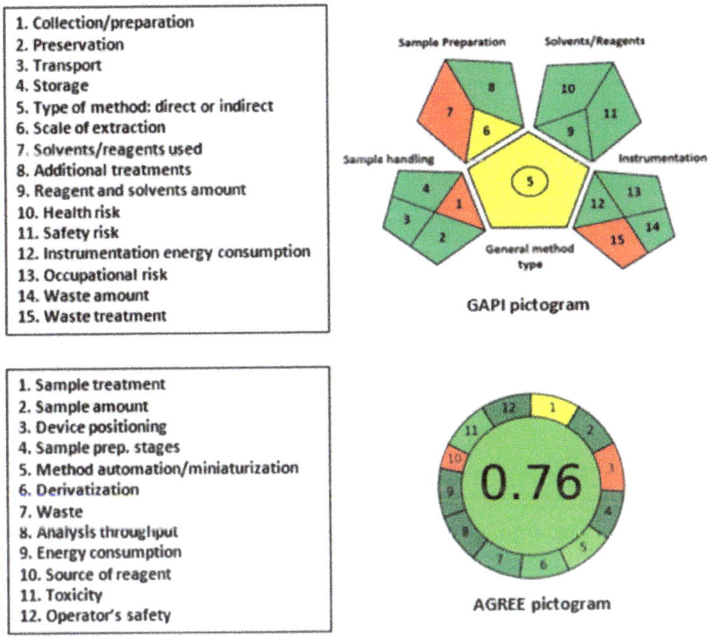

Figure 8. Results of GAPI and AGREE analyses for evaluation of the greenness of the proposed MW-SPA for SSRIs

In the AGREE pictogram (Figure 8), parameters number 3 (device positioning) and 10 (source of reagents) took red colors because the sample treatment was carried out off-line and the microplate reader was not operated in an automated way. The other parameters took yellow or varying degrees of green colors. The automatically generated number in the center of the AGREE pictogram for overall assessment was 0.76 out of 1, which confirmed the overall greenness of the proposed assay.

In conclusion, the overall evaluation of greenness of the proposed MW-SPA fulfills the requirements for GAC for routine application in pharmaceutical quality control laboratories for analysis of the investigated SSRIs.

3. Experimental

3.1. Instruments

The SpectraMax® M5 microplate reader was manufactured by Molecular Devices, LLC (San Jose, CA, USA). The reader is controlled by SoftMax® Pro Enterprise software (version 7.1), which was provided with the reader. Additionally, the JB1603-C/FACT digital balance is a product of Mettler-Toledo International Inc., located in Zürich, Switzerland. All spectrophotometric spectra were recorded using a V-530 ultraviolet-visible spectrophotometer manufactured by JASCO Co., Ltd. (Kyoto, Japan). The pH meter used in the experiments was the Model Jenway 350 m, manufactured by Bibby Scientific Ltd. (Essex, UK).

3.2. Standard Materials and Dosage Forms

The standards used in the study, namely fluoxetine HCl (FLX), fluvoxamine maleate (FXM), and paroxetine HCl (PXT), were obtained from their respective manufacturers, namely Hetero Drugs Ltd. in Hyderabad, India, for FLX; Solvay Pharma in Suresnes, France, for FXM; and SmithKline Beecham Pharmaceuticals in Brentford, England, for PXT. The appearance and solubility of the drugs were described before [28,29]. The dosage forms of FLX, FXM, and PXT used in this investigation with their manufacturers and strengths (mg/unit; tablet or capsule) are summarized in Table 9.

Table 9. The pharmaceutical dosage forms of the investigated SSRIs.

Brand Name (Dosage Form)	Manufacturer (Address)	Active Ingredient (SSRI Drug)	Label Claim (mg/unit)
Prozac capsules	Eli Lilly & Co., Ltd. (Hampshire, UK)	FLX	20
Fluzac capsules	Riyadh Pharma (Riyadh, Saudi Arabia)	FLX	20
Salipax capsules	Mepha Ltd. (Aesch-Basilea, Switzerland)	FLX	20
Flutin capsules	Egyptian International Pharmaceutical Industries Co. (Cairo, Egypt)	FLX	20
Octozac capsules	October Pharma, S.A.E. (Cairo, Egypt)	FLX	20
Faverin tablets	Solvay Pharma (Suresnes, France)	FXM	50
Seroxate tablets	GSK plc, SmithKline Beecham Pharmaceuticals (London, UK)	PAR	20

3.3. *Reagents, Buffer Solutions, and Tools*

1,2-Naphthoquinone-4-sulphonate (NQS) was purchased from Sigma-Aldrich Chemicals Co. (St. Louis, MI, USA). The NQS solution was 0.5% (*w*/*v*, in water). Clark and Lubs buffer solutions of different pH values were used. Transparent microwell plates with 96 wells were acquired from Corning/Costar Inc. (Cambridge, MA, USA). Variable volumes and adjustable multi-channel pipettes were purchased from Sigma-Aldrich Chemicals Co. (St. Louis, MI, USA). The solvents used in the study were spectroscopic-grade and were obtained from Fisher Scientific (California, CA, USA).

3.4. *Preparation of Standard and Sample Solutions*

3.4.1. Standard Solutions

An accurate weight (20 mg) of each drug was dissolved in 10 mL distilled water to obtain a stock solution of 2 mg/mL. An aliquot (1 mL) of the stock solution was diluted with water to obtain working solutions in the range of 2–40 µg/mL for FLX and PXT and in the range of 5–80 µg/mL for FXM.

3.4.2. Dosage form Sample Solutions

To prepare the sample, either 10 tablets or the contents of 10 capsules were weighed and ground into a fine powder. An accurately weighed amount of the powder, equivalent to 50 mg of the active ingredient, was dissolved in 50 mL of distilled water. The resulting solution was filtered. An aliquot of the filtrate (1 mL) was then diluted with distilled water to obtain appropriate concentrations for analysis within the linear range of each SSRI.

3.5. *General Procedure of MW-SPA*

Aliquots (100 µL) of SSRI solution containing concentrations in the range of 2–40 µg/mL for FLX and PXT and 5–80 µg/mL for FXM were transferred into separate wells of the 96-well assay plates. Then, 50 µL of buffer solution of pH 9 for FXM and PXT and of pH 11 for FLX was added, followed by 50 µL of NQS solution (0.5%, *w*/*v*). The reaction solution was allowed to proceed for 10 min at room temperature (25 ± 5 °C). The absorbance of the solution in each well was measured by the microplate reader at 470 nm for FXM and at 490 nm for FLX and PXT. The blank wells were treated in a similar way, except 100 µL of chloroform was dispensed in each well instead of the standard drug solution.

4. Conclusions

The current study presents the development and validation of an MW-SPA for three SSRIs, namely FLX, FXM, and PXT. The proposed assay employs the formation of colored products upon the condensation reactions of the SSRIs, via their amino groups, with NQS reagent. The proposed assay offers several advantages over previously reported spectrophotometric assays for SSRIs. These advantages include the use of low volumes of samples/solvents (which is cost-effective and environmentally friendly), ease of use, and high throughput (making it suitable for the pharmaceutical industry). The assay is useful for the routine analysis of SSRIs in dosage forms with satisfactory accuracy and precision, and it also expands the utilization of microwell assays assisted with microplate readers for quantifying various pharmaceuticals in the pharmaceutical industry. The limitation of the proposed assay is its application to the analysis of the cited SSRIs in their biological fluids. Current work is going to develop microwell assays for the determination of the cited SSRIs in biological samples.

Author Contributions: Conceptualization, I.A.D.; Methodology, N.Z.A.; Validation, N.Z.A.; Investigation, I.A.D. and N.Z.A.; Resources, N.Z.A.; Data curation, I.A.D.; Writing—original draft, N.Z.A.; Writing—review & editing, I.A.D. All authors have read and agreed to the published version of the manuscript.

Funding: This research was funded by Researchers Supporting Project Number (RSP2023R215), King Saud University, Riyadh, Saudi Arabia.

Data Availability Statement: All data are available in the article.

Acknowledgments: The authors extend their appreciation to the Researchers Supporting Project (Number RSP2023R215), King Saud University, Riyadh, Saudi Arabia, for funding this work.

Conflicts of Interest: The authors declare no conflict of interest.

Sample Availability: Not applicable.

References

1. World Health Organization. Depression. Available online: https://www.who.int/health-topics/depression#tab=tab_1 (accessed on 5 May 2023).
2. Ford, S.M. *Introduction to Clinical Pharmacology*, 12th ed.; Lippincott Williams & Wilkins: Philadelphia, PA, USA, 2021.
3. Preskorn, S.H. *Clinical Pharmacology of Selective Serotonin Reuptake Inhibitors*, 1st ed.; Professional Communications, Inc.: New York, NY, USA, 1996.
4. Sabri, R. *The Addictive Potential of Selective Serotonin Reuptake Inhibitors: Addictive Potential of Selective Serotonin Reuptake Inhibitors Relative to Benzodiazepines in UK Psychiatric Patients*; LAP Lambert Academic Publishing: London, UK, 2011.
5. Şentürk, Z.; Saka, C.; Teğin, I. Analytical methods for determination of selective serotonin reuptake inhibitor antidepressants. *Rev. Anal. Chem.* **2011**, *30*, 87–122. [CrossRef]
6. Jia, E.; Bartlett, M.G. Recent advances in liquid chromatographic methods for the determination of selective serotonin reuptake inhibitors and serotonin norepinephrine reuptake inhibitors. *Biomed. Chromatogr.* **2020**, *34*, e4760. [CrossRef] [PubMed]
7. Darwish, I.A.; Refaat, I.H. Spectrophotometric analysis of selective serotonin reuptake inhibitors based on formation of charge-transfer complexes with tetracyanoquinodimethane and chloranilic acid. *J. AOAC Int.* **2006**, *89*, 326–333. [CrossRef]
8. Soliman, S.M.; Mohamed, T.A. Spectrophotometric and spectrofluorimetric determination of some selective serotonin reuptake inhibitors [SSRIs] through ternary complex formation with eosin and copper [II] in pharmaceuticals. *J. Drug Res. Egypt* **2011**, *32*, 111–121.
9. Emin, C.; Rodica, S.; Ionescu, A.M.; Safta, A.; Tomescu, C.L. Validation of a spectrophotometric method of some antidepressant drugs. *Eur. J. Natural Sci. Med.* **2023**, *6*, 1–9.
10. Annapurna, V.; Jyothi, G.; Kumari, T.R.; Sailaja, B.B.V. Spectrophotometric determination of various drugs using chloranilic acid as chromogenic reagent—II. *J. Chem.* **2010**, *7*, 306858. [CrossRef]
11. Rao, A.L.; Rajeswari, K.R.; Sankar, G.G. Spectrophotometric methods for the determination of selected drugs in pharmaceutical formulations. *J. Chem. Pharm. Res.* **2010**, *2*, 280–282.
12. Annapurna, M.M.; Pradhan, D.B. New derivative spectrophotometric methods for the determination of fluoxetine—An antidepressant drug. *Chem. Sci. Trans.* **2012**, *1*, 697–701. [CrossRef]
13. Constantinescu, I.C.; Florea, M.; Neagu, A.F. Development of a spectrophotometric method for determination of fluoxetine hydrochloride in bulk and pharmaceutical dosage forms. *Farmacia* **2015**, *63*, 816–820.
14. Mukhopadhyay, A.; Pradhan, K.K.; Samanta, R. Development and validation of an UV-spectrophotometric method for the estimation of fluoxetine in pure and tablet dosage forms. *Int. J. Pharm. Sci. Res.* **2014**, *5*, 3418–3424.
15. Darwish, I.A. Development and validation of spectrophotometric methods for determination of fluoxetine, sertraline, and paroxetine in pharmaceutical dosage forms. *J. AOAC Int.* **2005**, *88*, 38–45. [CrossRef] [PubMed]
16. Shaza, A.; Alhaj, S.A. Spectrophotometric method for simultaneous estimation of dapoxetine and some phosphodiesterase-5 inhibitors in new combinations. *Res. J. Pharm. Technol.* **2019**, *12*, 5193–5198.
17. Görög, S. *Ultraviolet-Visible Spectrophotometry in Pharmaceutical Analysis*; CRC Press: New York, NY, USA, 2018.
18. Ahmed, S.; Rasul, A.; Masood, Z. *Spectrophotometry in Pharmaceutical Analysis*; LAP Lambert Academic Publishing: London, UK, 2011.
19. Gore, M.G. (Ed.) *Spectrophotometry and Spectrofluorimetry: A Practical Approach*, 2nd ed.; Oxford University Press: Oxford, UK, 2000.
20. Wang, P.G. (Ed.) *High-Throughput Analysis in the Pharmaceutical Industry*, 1st ed.; CRC Press: Boca Raton, FL, USA, 2019.

21. Mennen, S.M.; Alhambra, C.; Allen, C.L.; Barberis, M.; Berritt, S.; Brandt, T.A.; Campbell, A.D.; Castañón, J.; Cherney, A.H.; Christensen, M.; et al. The evolution of high-throughput experimentation in pharmaceutical development and perspectives on the future. *Org. Process Res. Dev.* **2019**, *23*, 1213–1242. [CrossRef]
22. Welch, C.J. High throughput analysis enables high throughput experimentation in pharmaceutical process research. *React. Chem. Eng.* **2019**, *4*, 1895–1911. [CrossRef]
23. Wennborg, H.; Bonde, J.P.; Stenbeck, M.; Olsen, J. Adverse reproduction outcomes among employees working in biomedical research laboratories. *Scand. J. Work Environ. Health* **2005**, *28*, 5–11. [CrossRef]
24. Lindbohm, M.L.; Taskinen, H.; Sallman, M.; Hemminki, K. Spontaneous abortions among women exposed to organic solvents. *Am. J. Indust. Med.* **1990**, *17*, 449–463. [CrossRef]
25. Wennborg, H.; Bodin, L.; Vainio, H.; Axelsson, G. Pregnancy outcome of personnel in Swedish biomedical research laboratories. *J. Occup. Environ. Med.* **2000**, *42*, 438–446. [CrossRef]
26. Kristensen, P.; Hilt, B.; Svendsen, K.; Grimsrud, T.K. Incidence of lymphohaematopoietic cancer at a university laboratory: A cluster investigation. *Eur. J. Epidemiol.* **2008**, *23*, 11–15. [CrossRef]
27. Keith, L.H.; Gron, L.U.; Young, J.L. Green analytical methodologies. *Chem. Rev.* **2017**, *107*, 2695–2708. [CrossRef] [PubMed]
28. *British Pharmacopoeia*; The Stationary Office Ltd.: London, UK, 2005; pp. 860, 878–879, 1515–1516.
29. *The United States Pharmacopeia 31, The National Formulary 26*; US Pharmacopeial Convention Inc.: Rockville, MD, USA, 2008; pp. 2225–2227, 2920–2923.
30. ICH (International Council for Harmonisation of Technical Requirements for Pharmaceuticals for Human Use). *ICH Harmonised Guideline, Validation of Analytical Procedure: Q2(R2)*; ICH: Geneva, Switzerland, 2022.
31. Płotka-Wasylka, J. A new tool for the evaluation of the analytical procedure: Green Analytical Procedure Index. *Talanta* **2018**, *181*, 204–209. [CrossRef]
32. Pena-Pereira, F.; Wojnowski, W.; Tobiszewski, M. AGREE-analytical greenness metric approach and software. *Anal. Chem.* **2020**, *92*, 10076–10082. [CrossRef] [PubMed]

molecules

MDPI

Review

Pyrrolizidine Alkaloids as Hazardous Toxins in Natural Products: Current Analytical Methods and Latest Legal Regulations

Agnieszka Lis-Cieplak [1], Katarzyna Trześniowska [1], Krzysztof Stolarczyk [2] and Elżbieta U. Stolarczyk [1,*]

1 Spectrometric Methods Department, National Medicines Institute, Chełmska 30/34, 00-725 Warsaw, Poland; a.lis@nil.gov.pl (A.L.-C.); k.trzesniowska@nil.gov.pl (K.T.)
2 Faculty of Chemistry, University of Warsaw, Pasteura 1, 02-093 Warsaw, Poland; kstolar@chem.uw.edu.pl
* Correspondence: e.stolarczyk@nil.gov.pl

Abstract: Pyrrolizidine alkaloids (PAs) are toxic compounds that occur naturally in certain plants, however, there are many secondary pathways causing PA contamination of other plants, including medicinal herbs and plant-based food products, which pose a risk of human intoxication. It is proven that chronic exposure to PAs causes serious adverse health consequences resulting from their cytotoxicity and genotoxicity. This review briefly presents PA occurrence, structures, chemistry, and toxicity, as well as a set of analytical methods. Recently developed sensitive electrochemical and chromatographic methods for the determination of PAs in honey, teas, herbs, and spices were summarized. The main strategies for improving the analytical efficiency of PA determination are related to the use of mass spectrometric (MS) detection; therefore, this review focuses on advances in MS-based methods. Raising awareness of the potential health risks associated with the presence of PAs in food and herbal medicines requires ongoing research in this area, including the development of sensitive methods for PA determination and rigorous legal regulations of PA intake from herbal products. The maximum levels of PAs in certain products are regulated by the European Commission; however, the precise knowledge about which products contain trace but significant amounts of these alkaloids is still insufficient.

Keywords: pyrrolizidine alkaloids; chromatographic methods; electrochemical sensors; herbal products; carcinogenesis; genotoxicity; public health

1. Introduction

Natural products have been used for centuries in traditional medicine to treat and prevent diseases. Nowadays, many drugs are obtained from plant raw material, especially from medicinal herbs, and natural substances isolated from medicinal plants are considered candidates for new drugs [1]. Moreover, other natural sources, including fungi, lichens, bacteria, and marine organisms, also provide valuable material for the pharmaceutical industry. On the other hand, due to the growing interest in healthy lifestyles, the use of supplements and medicinal products of plant origin is becoming popular among consumers. The market for foods classified as dietary supplements is developing dynamically. There is a misconception that these products are always safe and side-effect free. Importantly, these products are consumed without proper medical supervision. To make dietary supplements safe for consumers, relevant legal regulations and market control are needed.

Contamination of food products and plant-based medicines with pyrrolizidine alkaloids (PAs) emerged recently as a significant issue at the international level [2–18] as PAs have been detected in honey and plant material such as tea, herbs, vegetables, cereals, salads, and spices. Therefore, it becomes the subject of numerous discussions, research projects, reports, and legal regulations. Numerous aspects of PA chemistry, metabolism, and toxicity were discussed in high-quality review articles [8,15,19–30]. The occurrence of PAs in food and herbal products is a very important problem, and due to the health hazards, effective methods of their identification and determination should be constantly developed. Techniques such as high-performance liquid chromatography (HPLC), gas chromatography (GC), and thin-layer chromatography (TLC) are used for the detection and identification of PAs [31]. Spectroscopic techniques are utilized as well [31]. However, these methods do not provide sufficient identification accuracy or sensitivity to determine PA contamination on trace levels. For that purpose, more advanced methods were recently developed. These protocols mostly rely on the application of liquid chromatography combined with mass spectrometry detection (LC-MS) [31,32]. LC-MS methods combine reliability and high sensitivity with ease of sample preparation. This is because LC-MS ensures very selective PA identification and determination, even in complicated matrixes. Therefore, recently, it has become the method of choice in PA analysis [33–41].

This review provides information on the occurrence of PAs in the environment along with insight into secondary pathways of PA contamination of plant products. It focuses on analytical techniques applied for PA determination in medicinal and food products of plant origin. Special emphasis is put on liquid chromatography (LC) methods combined with mass spectrometry (MS) detection. However, other methods are discussed as well. Recently published reviews [25,27–29,41–43] summarize methods for determining PA, mainly methods with MS detection using various types of analyzers, which are characterized by different levels of sensitivity. The most modern and sensitive type of detector is currently Orbitrap, and there are relatively few reports on methods using this type of analyzer in other reviews. This review presents PA testing methods using both basic types of analyzers and the most modern ones, based on the latest literature reports. This review also provides much more detailed information on the mechanisms of PA-induced cytotoxicity and genotoxicity, as well as their LD_{50} values, which are not included in existing reviews. It also provides a broad summary of European regulations regarding these dangerous pollutants.

2. Natural Occurrence of PAs, Health Risks, and Possible Routes of Human Intoxication

In general, alkaloids are a diverse class of naturally occurring organic compounds. They are characterized primarily by the presence of at least one nitrogen atom in the structure. The inherence of nitrogen in the form of a heteroatom in the rings, in the form of an amino group or, less often, an amide group, causes the basicity of these compounds. This broad group also includes related ones that have neutral or even slightly acidic properties. Alkaloids may also contain other heteroatoms such as sulfur and, less commonly, phosphorus, chlorine, and bromine [44]. These compounds are synthesized by a wide range of organisms; they can be found in bacteria, fungi, plants, and animals [45,46]. Alkaloids have a wide range of pharmacological activities, such as antimalarial, antiasthmatic, anticancer, cholinomimetic, vasodilatory, antiarrhythmic, analgesic, antibacterial, and antihyperglycemic effects. Due to their strong biological activity, many alkaloids are used in traditional or modern medicine or serve as starting points for drug discovery [47].

Alkaloid synthesis is a secondary metabolic process. The necessary compounds for alkaloid synthesis are pyruvic acid and acetyl-CoA. Figure 1 shows the synthesis of alkaloids in the metabolic system of plants. The synthesis of alkaloids in the plant is a multi-step process. The first stage is photosynthesis, which produces the monosaccharide D-glucose, which is used as a substrate in the next stage of the Krebs cycle. PAs are compounds containing nitrogen in fused heterocyclic rings, produced in the ornithine metabolism pathway (see Figure 1), while L-ornithine is derived from L-glutamate [21]. There were hundreds of structurally different PAs discovered. The occurrence of PAs, their propagation, toxicity, and chemistry are described in detail in the next paragraphs.

PAs are synthesized by a wide variety of plant species. They were identified in over 6000 plants [48], in families: *Boraginaceae* (all genera), *Asteraceae* (*Senecioneae*, *Eupatorieae*), and *Fabaceae* (*Crotalaria*), as natural toxins providing protection against animals feeding on plants [23,49]. PA-containing plants are common weeds and are considered invasive and harmful to the environment because they may contaminate the raw plant material. On the crop fields, PA-containing plants and their parts or seeds can contaminate soil and get into harvested cereals, herbs, or vegetables. Accidental mixing of PA-containing plants with plants intended for fodder may lead to contamination of prepared feeds and grains that are subsequently eaten by animals. That makes food of animal origin, such as milk and eggs, a health hazard [23,48–51]. Bees can ingest PAs containing pollen, and then they produce contaminated honey [52]. Therefore, it leads to PA contamination of the whole food chain.

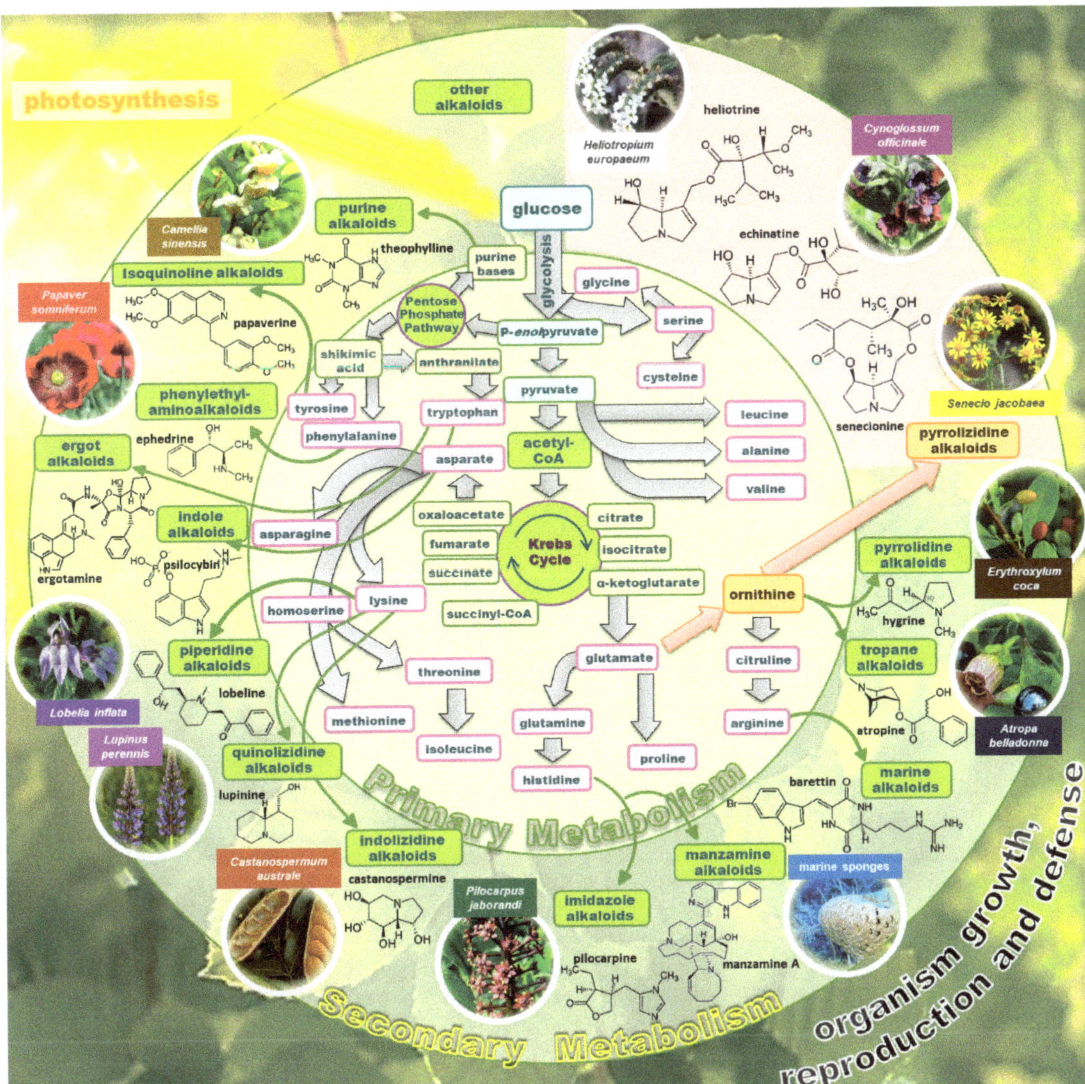

Figure 1. A block scheme of the alkaloid biosynthesis as a secondary metabolism of plants. The pathway of PAs biosynthesis is marked with orange arrows and boxes. Based on [21]. The figure was prepared using GIMP 2.8.14 (GNU General Public License) software.

The European Food Safety Authority (EFSA) recognized PAs as potential toxic components of feed and food, which can become a significant public health problem due to the high risk of contamination of food of plant or animal origin [9]. Possible routes of intoxication with PAs are shown in Figure 2. Generally, PA intoxication is possible through ingestion of PA-containing herbal products or PA-contaminated foods, such as tea, herbs, vegetables, spices, and salads. PA-containing plants are numerous and widespread. Human intoxication can occur through the consumption of contaminated basic food products and some herbal remedies [15].

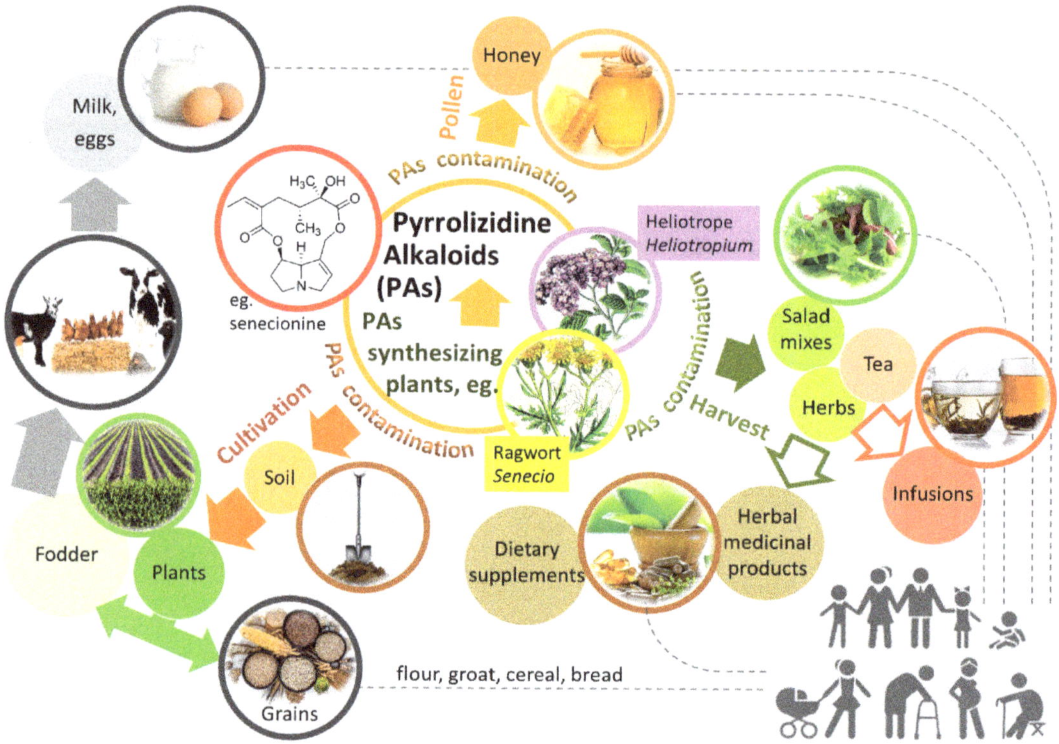

Figure 2. The possible pathways of human exposure to PAs. The figure was prepared using GIMP 2.8.14 (GNU General Public License) software.

3. Chemistry of PAs

So far, more than 500 PAs have been found and their structures determined [53]. Taking into account the form of *N*-oxides, over 900 structures are known. PAs are heterocyclic compounds, they include a group of basic ester compounds (mono- and diesters), which structurally include a combination of amino alcohols with mono- or dicarboxylic acids [22]. They are pyrrolizidine or necine derivatives, esters, and diesters [21]. PAs undergo acidic or basic hydrolysis, giving basic necine-type amino alcohols that can be assigned as PA groups according to the necine base: otonecine, retronecine, heliotridine, and platynecine. The biosynthesis of PAs begins with the amino acid ornithine, which leads to the generation of putrescine and then spermidine. One molecule of putrescine and spermidine is transformed into homospermidine [54]. Homospermidine is deaminated and creates the necine base skeleton, esterified with a necic acid [55]. The necine base forms esters with small organic acids and generates cyclic PAs (retronecine and otonecine type) and open-ringed PAs (heliotridine type) (Figure 3). In detail, a molecule of PAs consists of the core structure—pyrrolizidine, a bicyclic aliphatic hydrocarbon consisting of two fused five-membered rings with a nitrogen atom between them and in many structures with a double bond in the 1,2 positions (sub-group of PAs, 1,2-unsaturated). The main toxic effects of PAs are on the liver and lungs. 1,2-unsaturated PAs are genotoxic and cause liver cancer in experimental animals. The Scientific Panel on Contaminants in the Food Chain

(CONTAM Panel) of the European Food Safety Authority published a scientific opinion on the risks to public health related to the presence of pyrrolizidine alkaloids in food and feed. The CONTAM Panel concluded that only 1,2-unsaturated pyrrolizidine alkaloids are toxic and may act as genotoxic carcinogens in humans [9,56]. The necine base is often retronecine, heliotridine, or otonecine. Necic acids are varied organic acids, when they are dicarboxylic, they form macrocyclic PAs. PAs demonstrate great structural diversity. With the large number of necic acids, which can be combined with a set of necine bases, a huge structural diversity of PAs is possible [22]. Moreover, modifications including *N*-oxidation of the tertiary nitrogen of the necine base, hydroxylation of the necine base and the necic acid, and acetylation of hydroxy groups further enhance these possibilities. Chemical structures of PAs listed in Annex 1 to the Commission Regulation (EU) 2023/915 [57] are presented in Table 1. The chemical structures of PAs, which should also be monitored in food and feed, according to the EFSA opinion [8,9], are presented in Table 2. *N*-oxidation is a special type of modification because it is reversible. *N*-oxidation of the tertiary amine nitrogen significantly changes the properties of a native PA. The main point is that PANOs become more polar and highly water soluble. In plants, what is most important is that the major fraction of PAs is present as PANOs [22].

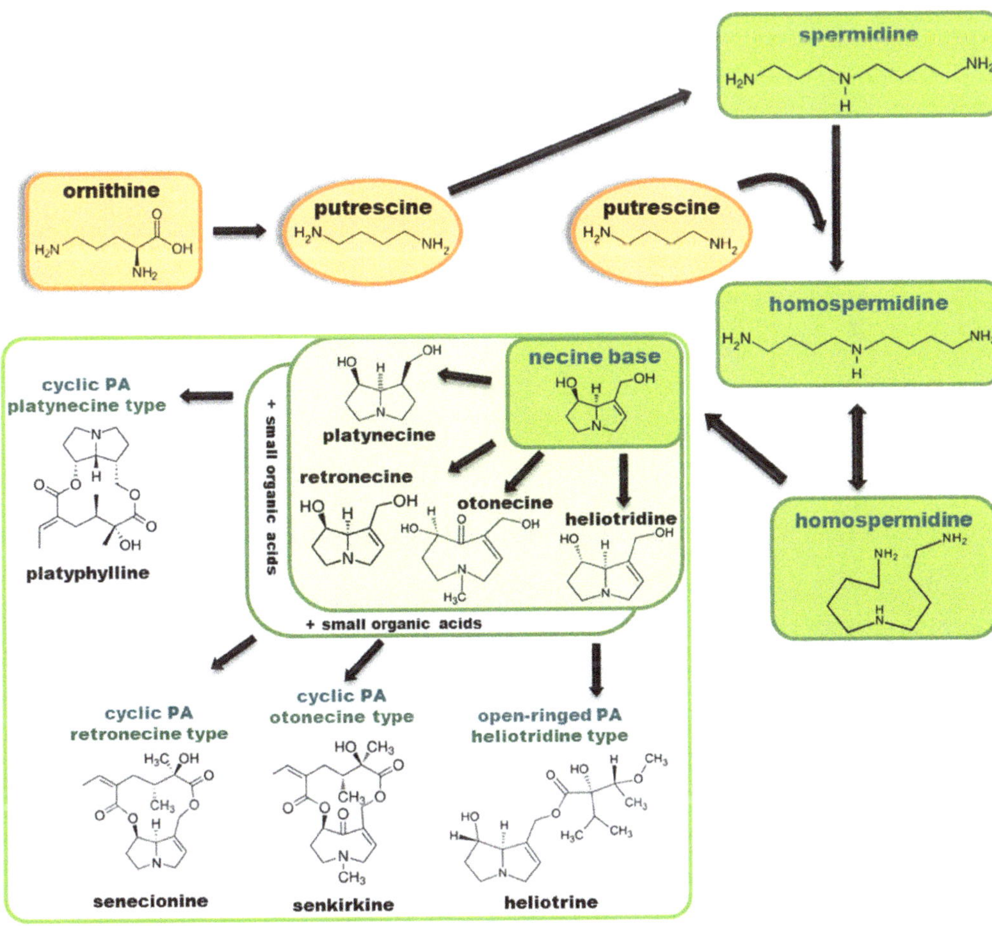

Figure 3. Biosynthesis of the most representative examples of PAs. Based on [19,55]. The figure was prepared using GIMP 2.8.14 (GNU General Public License) software.

Table 1. The chemical structures of PAs listed in Annex 1 to the Commission Regulation (EU) 2023/915. Twenty-one PAs whose concentration should be monitored (**n**-numbers) and an additional 14 PAs (**na** numbers) who are known to co-elute with one or more of the requiring investigation 21 PAs.

No.	Name	Alkaloid Chemical Structure	Chemical Structure of Corresponding *N*-Oxide
1, 2	Echimidine, echimidine-*N*-oxide	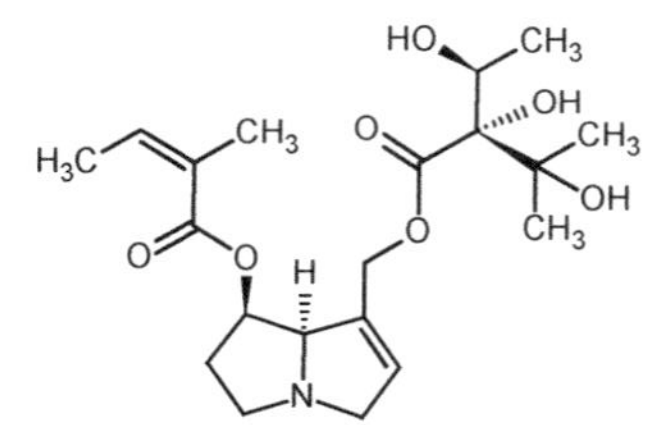	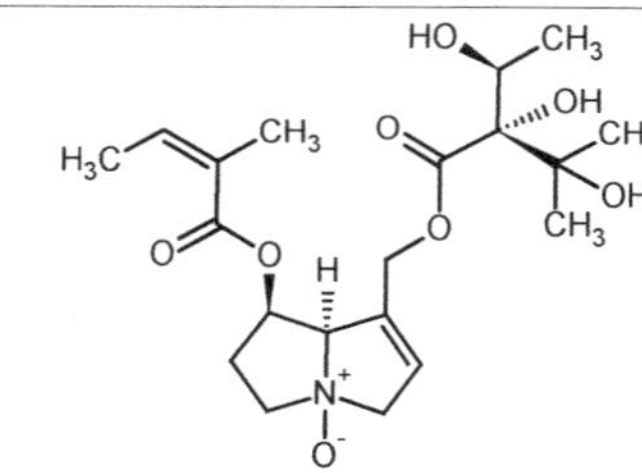
Possible co-elution of **1** and **2** with, respectively:			
1a, 2a	Heliosupine, heliosupine-*N*-oxide	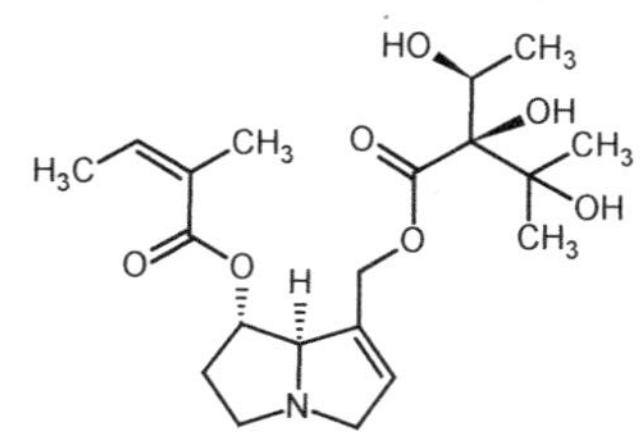	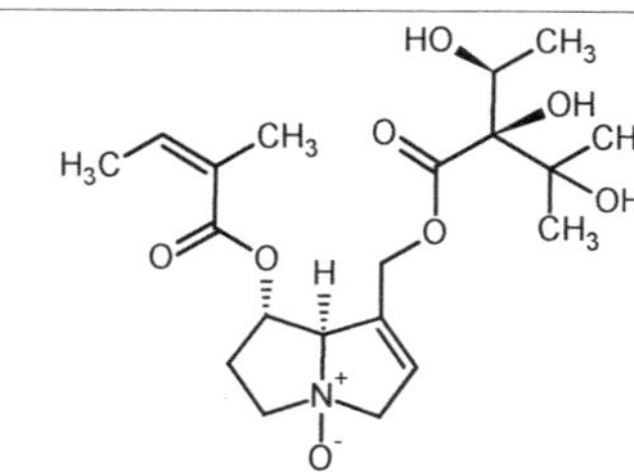
3, 4	Heliotrine, heliotrine-*N*-oxide	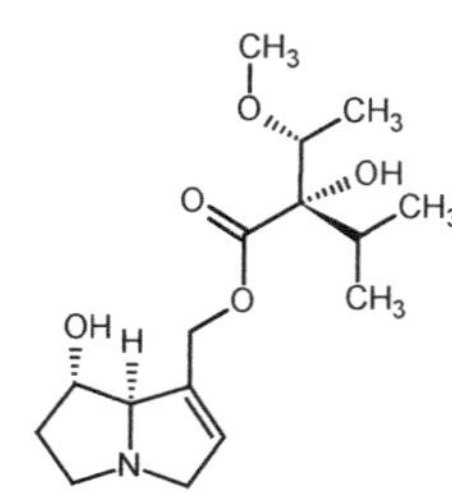	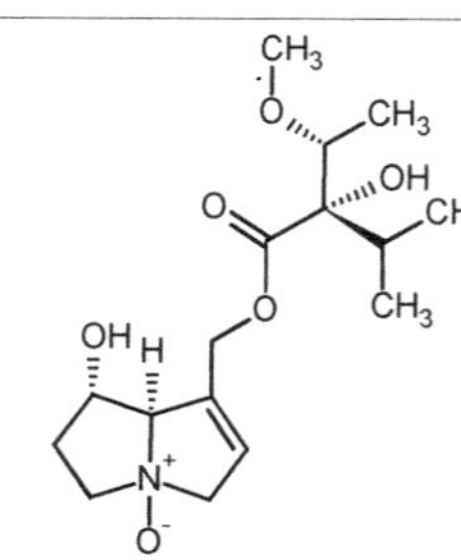

Table 1. *Cont.*

No.	Name	Alkaloid Chemical Structure	Chemical Structure of Corresponding N-Oxide
5, **6**	Intermedine, intermedine-*N*-oxide	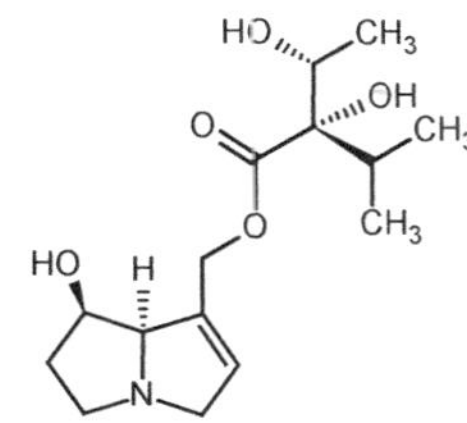	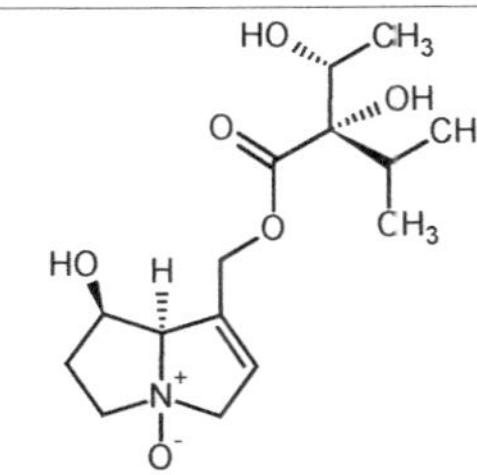
7, **8**	Lycopsamine, lycopsamine-*N*-oxide	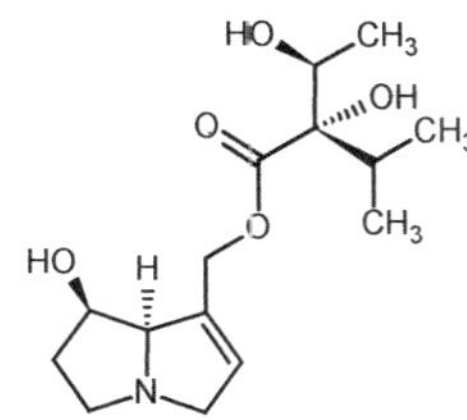	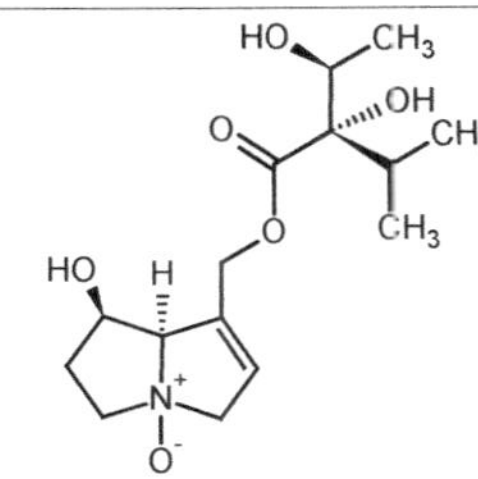
Possible co-elution of **5**, **6**, **7**, and **8** with, respectively:			
3a, **4a**	Indicine, indicine-*N*-oxide	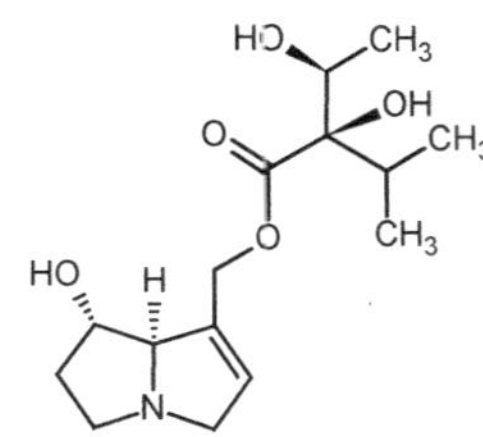	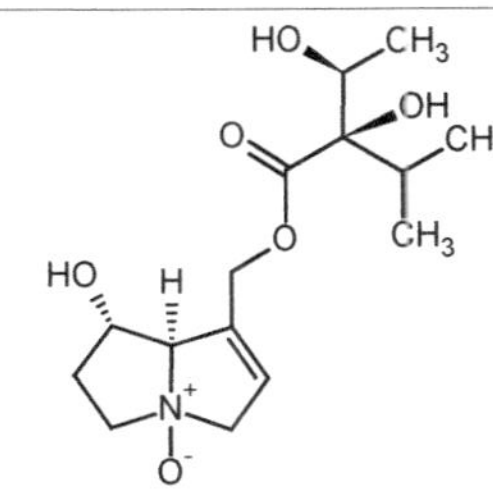

Table 1. *Cont.*

No.	Name	Alkaloid Chemical Structure	Chemical Structure of Corresponding *N*-Oxide
5a, 6a	Echinatine, echinatine-*N*-oxide	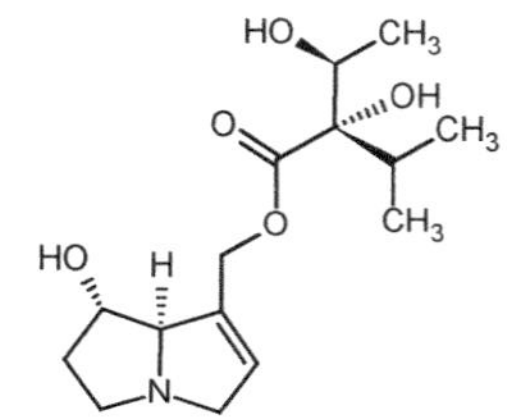	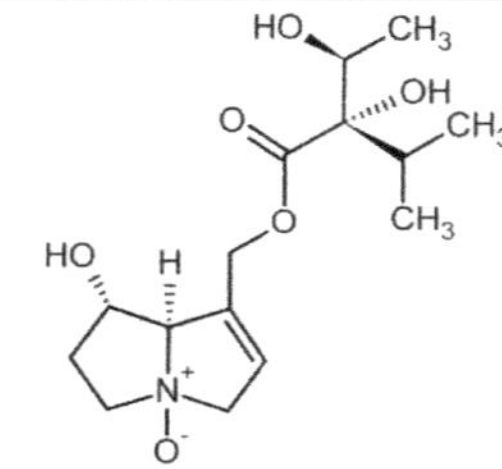
7a, 8a	Rinderine, rinderine-*N*-oxide	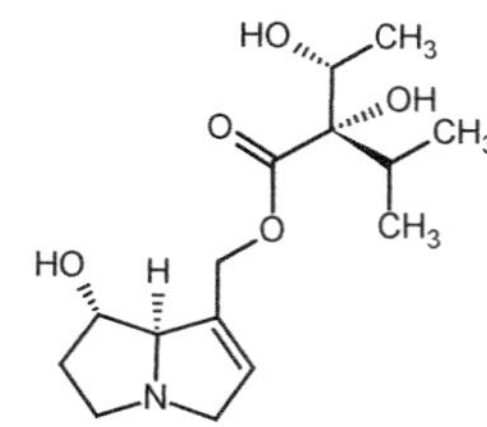	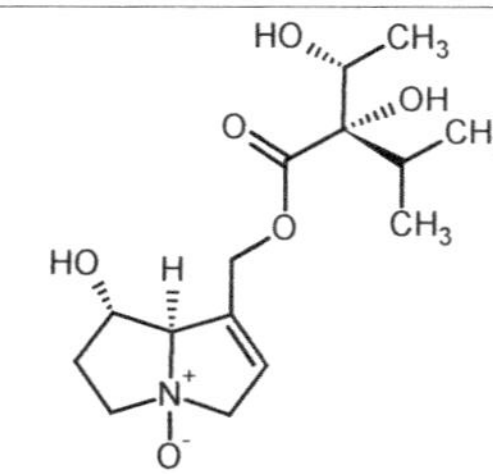
9, 10	Retrorsine, retrorsine-*N*-oxide	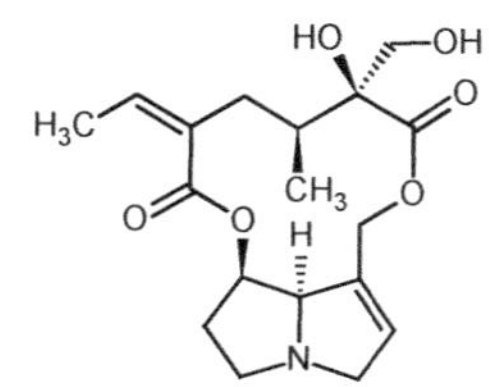	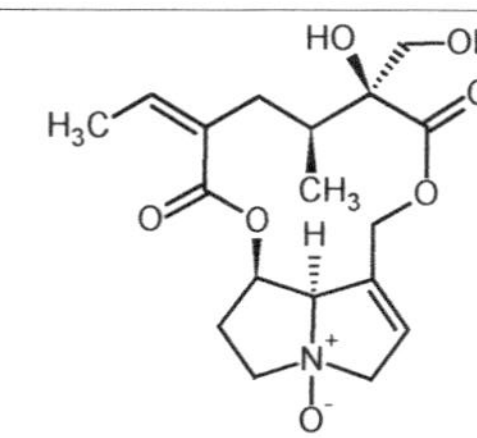

Table 1. *Cont.*

No.	Name	Alkaloid Chemical Structure	Chemical Structure of Corresponding N-Oxide
	Possible co-elution of **9** and **10** with, respectively:		
9a, 10a	Usaramine, usaramine-N-oxide	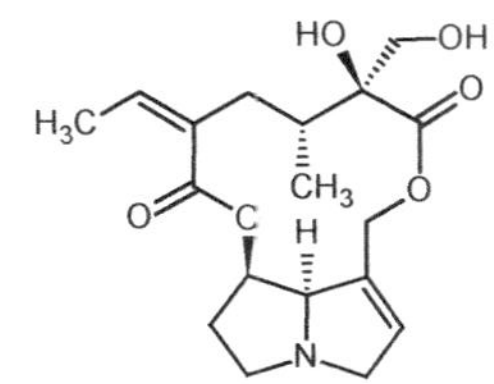	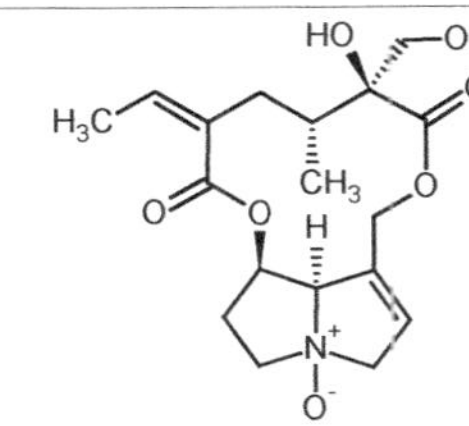
11, 12	Senecionine, sencionine-N-oxide	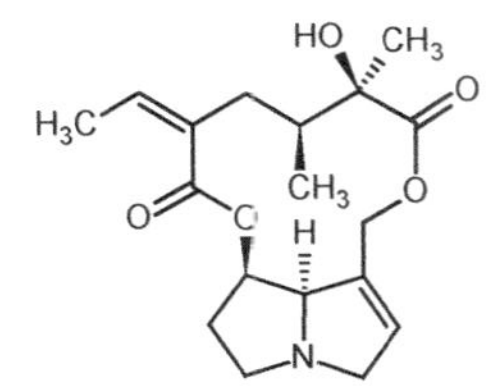	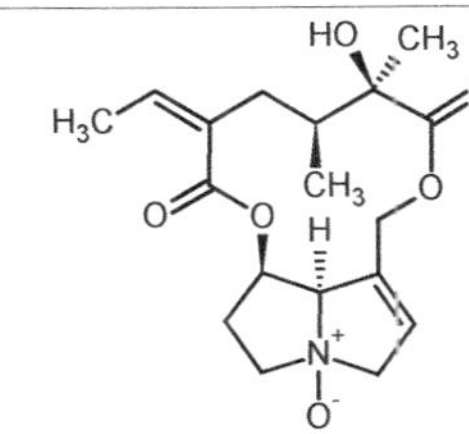
13, 14	Senecivernine, senecivernine-N-oxide	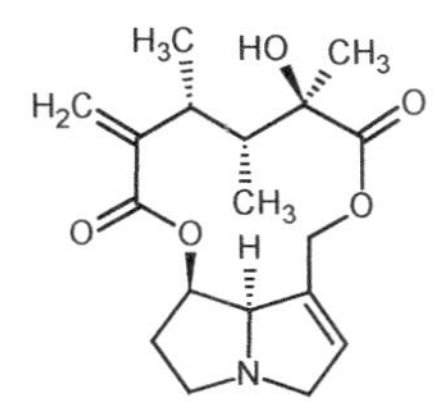	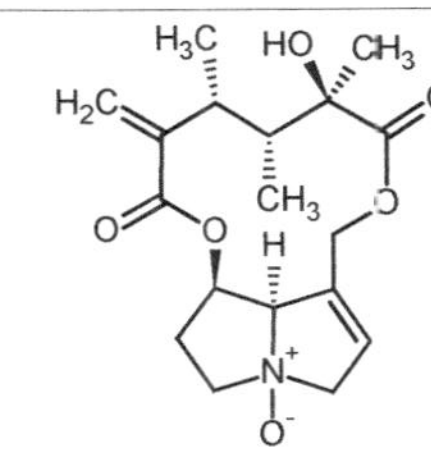

Table 1. *Cont.*

No.	Name	Alkaloid Chemical Structure	Chemical Structure of Corresponding *N*-Oxide
	Possible co-elution of **11**, **12**, **13** and **14** with:		
11a, 12a	Integerrimine, integerrimine-*N*-oxide	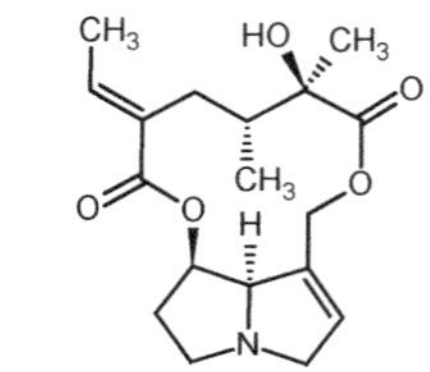	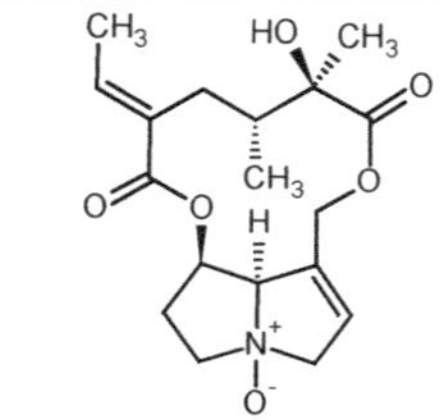
15, 16	Seneciphylline, seneciphylline-*N*-oxide	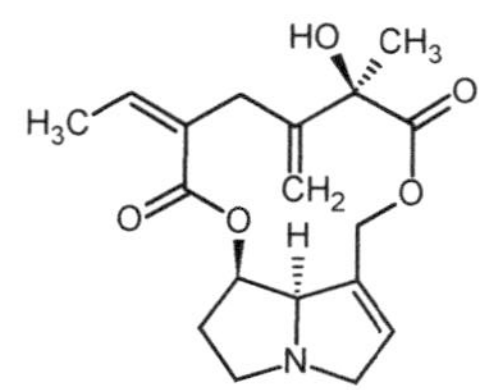	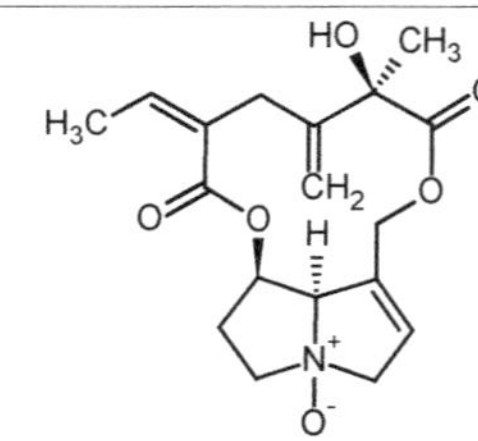
	Possible co-elution of **15** and **16** with, respectively:		
13a, 14a	Spartioidine, spartioidine N-oxide	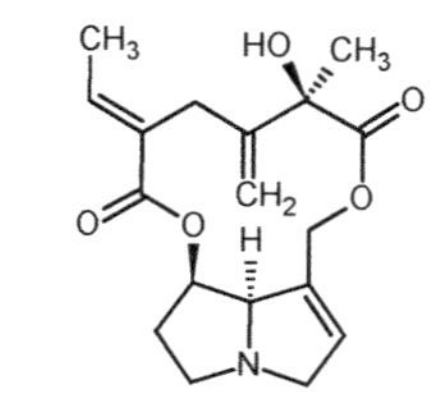	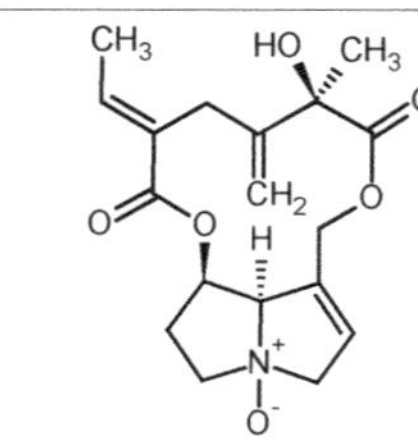

Table 1. *Cont.*

No.	Name	Alkaloid Chemical Structure	Chemical Structure of Corresponding N-Oxide
17, 18	Europine, europine-*N*-oxide	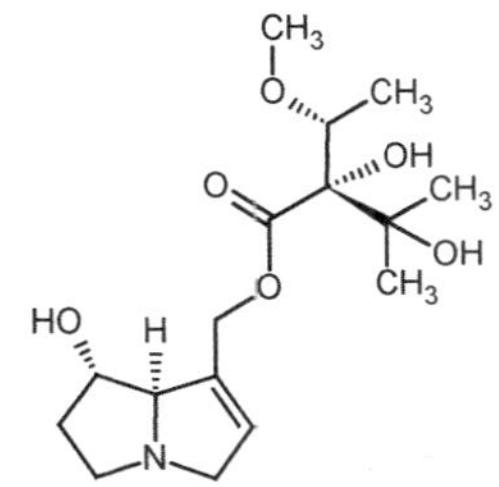	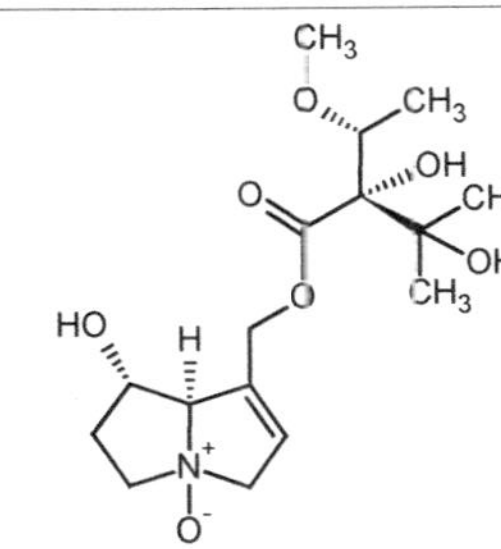
19, 20	Lasiocarpine, lasiocarpine-*N*-oxide	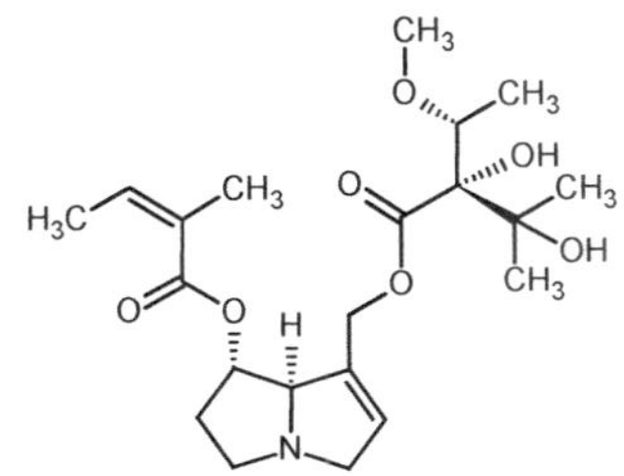	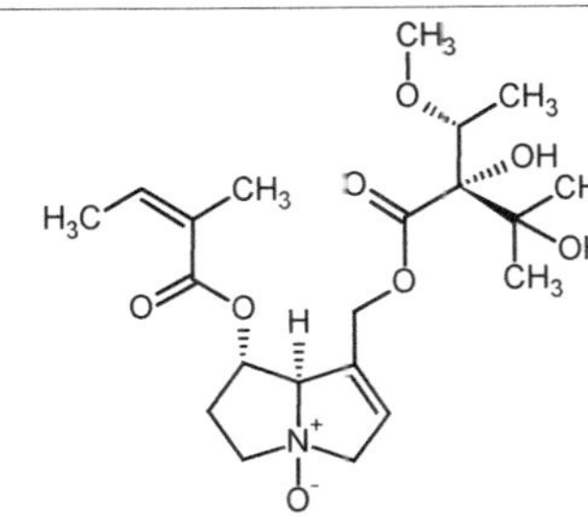
21	Senkirkine	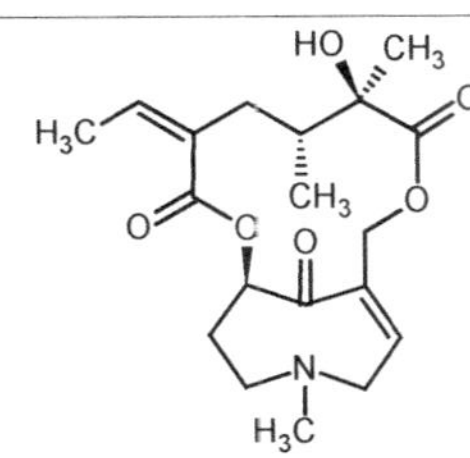	

Table 2. Chemical structures of PAs not listed in Table 1 (nb numbers), which, according to the EFSA opinion, should also be monitored in food and feed [8].

No.	Name	Alkaloid Chemical Structure	Chemical Structure of Corresponding N-Oxide
1b, 2b	Erucifoline, erucifoline-*N*-oxide	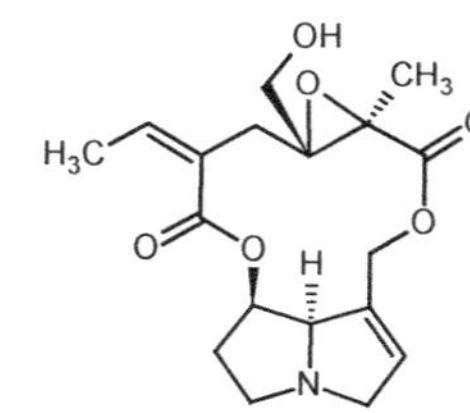	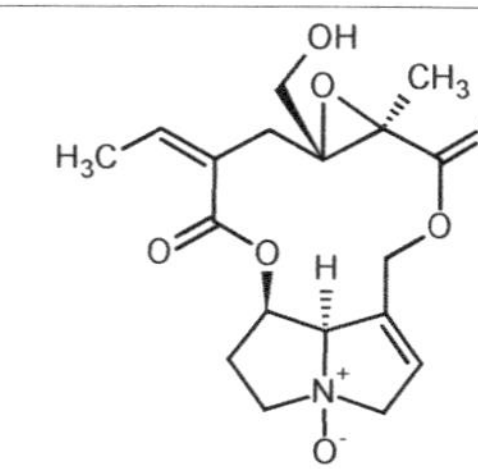
3b, 4b	Monocrotaline, monocrotaline-*N*-oxide	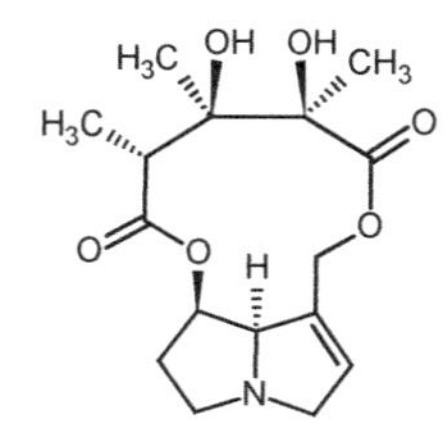	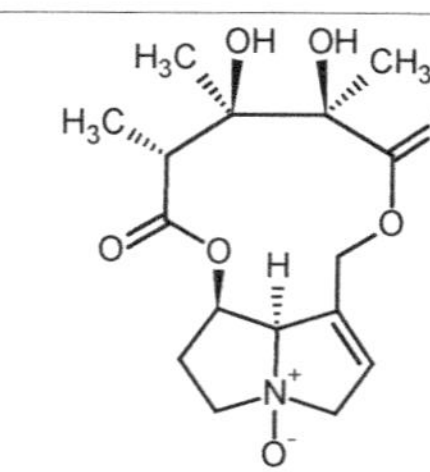
5b, 6b	Trichodesmine, trichodesmine-*N*-oxide	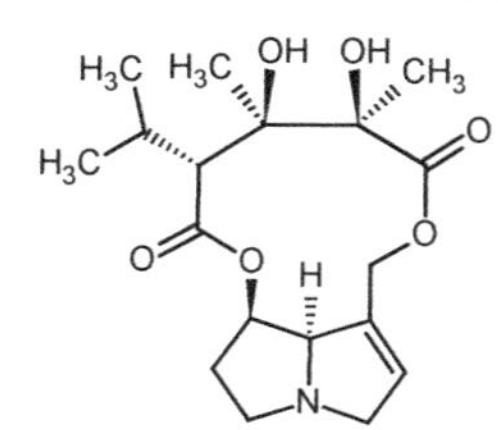	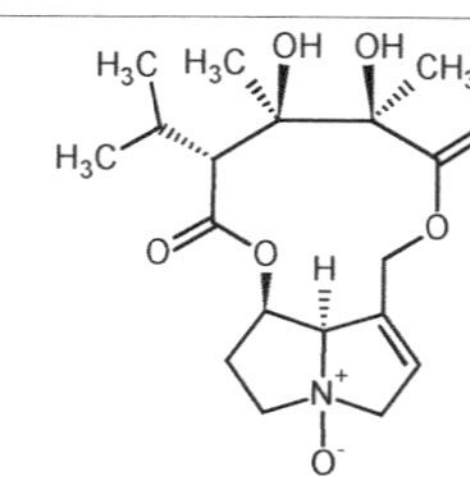

Table 2. *Cont.*

No.	Name	Alkaloid Chemical Structure	Chemical Structure of Corresponding N-Oxide
7b, 8b	Jacobine, jacobine-*N*-oxide	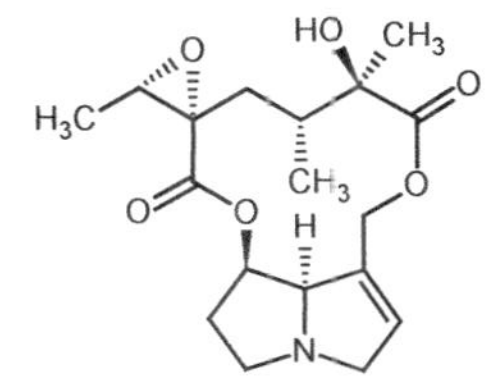	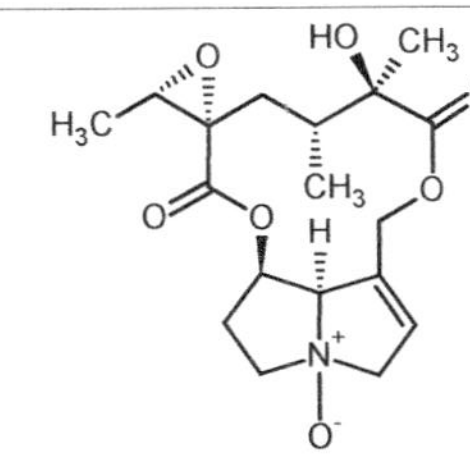
9b, 10b	Jaconine, jaconine-*N*-oxide	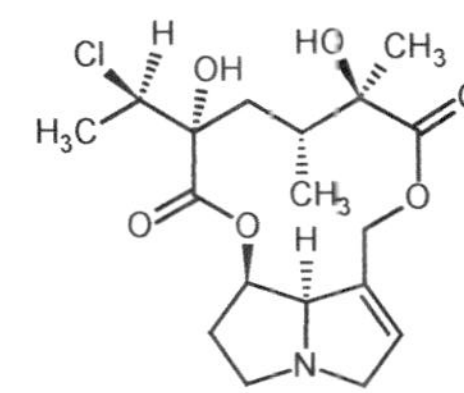	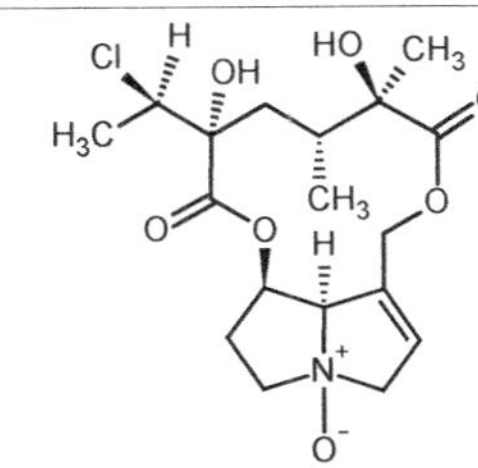

The determination of PANO concentration is necessary because, in vivo, these *N*-oxides can be biotransformed into the corresponding PA-free bases. This process takes place after ingestion in the gastrointestinal tract and in the liver, mediated by, respectively, the intestinal microbiota and hepatic cytochrome P450 monooxygenases [58]. When reduced, *N*-oxide-derived PAs are subsequently metabolized to the hepatotoxic pyrrole. Therefore, the total PA content of the tested material may be underestimated unless both PA and PANO contents are determined.

4. Toxicity of PAs and Regulation Limits

PAs are potentially hepatotoxic compounds that lead to human poisoning through the food chain (such as tea, herbs, botanical preparations, spices, and vegetables). The toxicity of PAs depends on their physical properties and their metabolism in the liver [59]. PAs very often produce pyrrolizidine alkaloid *N*-oxides (PANOs), of lower toxicity, that cannot be directly converted to hydroxypyrrolidines. PANOs generally exhibit low toxicity but undergo toxic processes in vivo and cause toxification through biotransformation to the corresponding PAs [60]. PANOs are reduced to free bases in the intestines after their ingestion, which was described further. The hepato- and cytotoxicity mechanisms of PAs are shown in Figure 4.

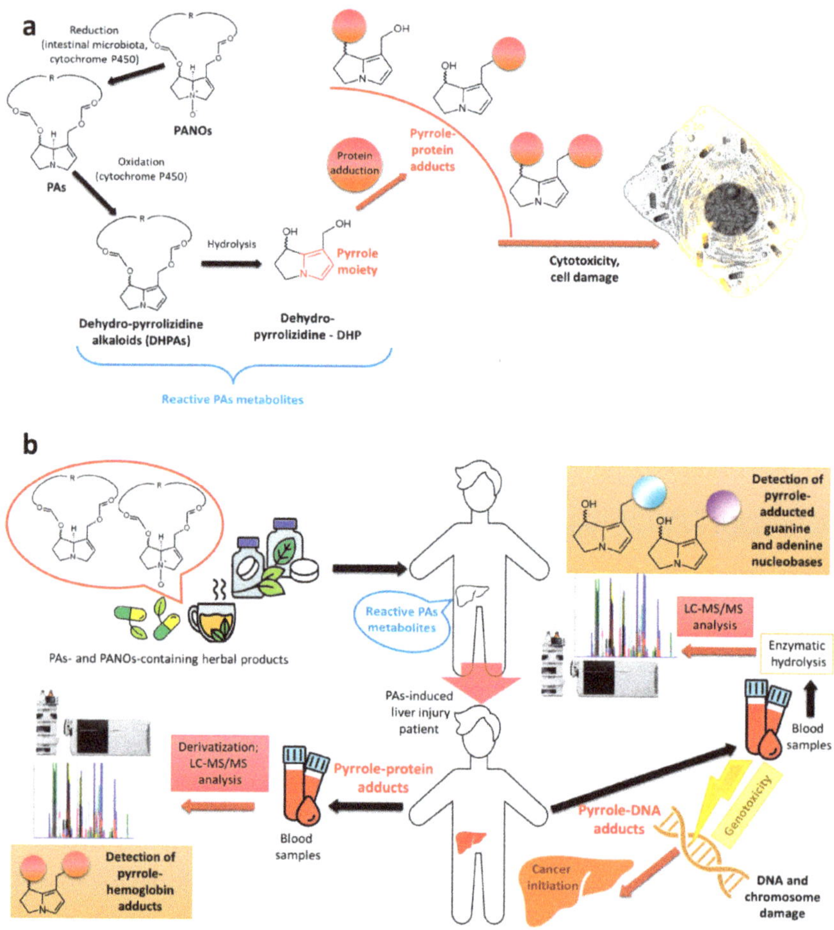

Figure 4. Mechanism of metabolic processes of PAs, which leads to hepato- and cytotoxicity (**a**) and hepato- and genotoxicity (**b**). Adapted from [61,62]. The figure was prepared using GIMP 2.8.14 (GNU General Public License) software.

The oxidation of PAs is a metabolic process that occurs in the liver. This process is crucial as it leads to the formation of reactive PA metabolites [62] (see Figure 4a). During oxidation processes in the liver, a hydroxyl group is attached to the alkaloid molecule and the carbon adjacent to the nitrogen atom. The alkaloid thus formed is unstable and is immediately dehydrated into the dehydro-pyrrolizidine alkaloids (DHPAs). As a result, a second double bond is formed in the necine molecule. At a later stage, after hydrolysis, an aromatic pyrrole moiety is generated (dehydropyrrolizidine—DHP), with the ester groups removed. In the final process, there are formed pyrrole–protein adducts and pyrrole-DNA adducts, which are responsible for cytotoxicity and genotoxicity (see Figure 4b) [61,63]. PA-induced liver injury is suspected to be associated with the consumption of PA-containing herbal products, and the pyrrole–DNA adducts were detectable in patients' blood samples [61].

PAs may cause acute toxicity, mutagenicity, chromosomal aberrations, the formation of abnormal cross-links between DNA strands and DNA–protein bonds, and megalocytosis [23,64,65]. They are responsible for: the formation of cancer cells [56,66,67], disturbances of the liver metabolism [23,66], liver necrosis [19], fibrosis [19], cirrhosis [19], photosensitization [19,23], diarrhea [19,23], incoordination [8,28], aggressive behavior [8], body weight loss [8,23], loss of appetite [8,23]. Poisoning with PAs is usually asymptomatic. By the time symptoms of liver damage appear, the disease process has already developed and caused death in a short time [23,68]. The LD_{50} values of the most important PAs are known [2]. However, the increase or inhibition of cytochrome P450 activity by drugs may also change the toxicity of PAs. PA metabolites react with SH groups located in glutathione or cysteine. Therefore, a diet rich in glutathione, taurine, cysteine, and methionine may reduce the toxicity of ingested PAs [69]. PA toxicity also depends on exposure time, dose, and the organism's susceptibility. PAs cause acute toxicity within 1–6 days, while doses of 0.1 mg/kg body weight per day cause chronic toxicity. In humans, the toxic dose ranges from 0.1–10 mg/kg body weight per day [10]. Table 3 shows the LD_{50} of the more important PAs studied in vivo in animal models [2] and predicted using the computer software TOPKAT (Discovery Studio 2019 (Accelrys, Inc., San Diego, CA, USA) [70].

Table 3. LD_{50} values of the most important PAs.

	In Silico Oral/Rat LD_{50} (mg/kg Body Weight)	**In Vivo Intraperitoneal LD_{50} (mg/kg Body Weight)**	**Species**
Compound/Ref.:	[70]	[2]	
Echimidine	616	200	rat, male
Echinatine	250	350	rat, male
Europine	-	>1000	rat, male
Heleurine	616	140	rat, male
Heliosupine	708	60	rat, male
Heliotridine	-	1500	rat, male
Heliotrine	56	296	rat, male
Heliotrine	-	478	rat, female
Indicine	264	>1000	rat, male
Intermedine	264	1500	rat, male
Jacobine	461	138	rat, female
Jaconine	-	168	rat, female
Lasiocarpine	555	77	rat, male
Lasiocarpine	-	79	rat, female
Lycopsamine	239	1500	rat, male
Monocrotaline	-	154	mouse
Monocrotaline	731	109	rat, male
Monocrotaline	-	230	rat, female
Otonecine	467	-	-
Platyphylline	443	252	rat, male
Retronecine	242	-	-
Retrorsine	320	34–38	rat, male
Retrorsine	-	153	rat, female
Riddelliine	616	80	rat, male
Rinderine	486	550	rat, male

Table 3. Cont.

	In Silico Oral/Rat LD_{50} (mg/kg Body Weight)	In Vivo Intraperitoneal LD_{50} (mg/kg Body Weight)	Species
Senecionine	127	50	rat, male
Seneciphylline	264	77	rat, male
Seneciphylline	-	83	rat, female
Senecivernine	592	-	-
Senkirkine	275	220	rat, male
Spectabiline		50	rat, male
Supinine	215	450	rat, male
Symphytine	-	130	rat, male
Usaramine	264	-	-
Trichodesmine	324	-	-

On 31 May 2016, the regulatory authorities of the EU (European Medicines Agency—EMA and Herbal Medicinal Products Committee—HMPC) issued a public statement regarding the contamination of herbal medicinal products/traditional products by PAs [10]. Following a review of the available data, the EMA (HMPC) considered a harmonized approach to implementing appropriate controls for the markets in EU countries. A contamination level of herbal medicinal products leading to a daily intake of a maximum of 1.0 μg PAs per day during a transitional period of 3 years was considered acceptable from a public health point of view. After this period, producers of herbal medicinal products should be required to take the necessary measures to reduce the contamination to a level resulting in a daily intake not exceeding 0.35 μg PAs per day [10,14,71]. The report concluded that contamination of herbal products (food or medicines) with PAs is not a new matter, but new, sensitive analytical methods can now detect very low levels of PAs. Tea and herbal teas are the largest contributors to human exposure to pyrrolizidine alkaloids, as well as pollen-based supplements. It was found that the exposure to pyrrolizidine alkaloids associated with the consumption of honey is lower. It has also been found that herbal dietary supplements may contribute significantly to human exposure to PAs, but incidence data are insufficient [10]. Generally, the human intake of PAs through food and herbal medicinal products has probably remained constant over the last few years, and statistically, the incidence of liver hemangiosarcoma in humans is very low. EMA has emphasized that once the problem with PA contamination of herbal medicinal products has been identified, regulatory actions to mitigate the problem must be considered.

The European Food Safety Authority (EFSA) prepared in 2016 a report on chronic and acute dietary exposure to PAs in the European population through the consumption of foods of plant origin [11]. This scientific report focuses on the 28 PAs selected based on the EFSA scientific opinion from 2011 [9], which identified key PAs in tea and herbal infusions. The 28 selected PAs include echimidine, heliotrine, lycopsamine, intermedine, erucifoline, senecionine, seneciphylline, monocrotaline, jacobine, senecivernine, retrorsine, europine, lasiocarpine, senkirkine, and their *N*-oxide forms. A total of 274,632 analytical results on PAs in food samples were available, accounting for a total of 19,332 food samples. The number of PAs analyzed per sample ranged between one and 28. To avoid underestimation of the presence of PAs, only those samples with a minimum number of PAs were selected. Special attention was paid to the presence of two additional PAs, riddelliine and riddelliine-*N*-oxide, due to their toxicity. These two PAs were analyzed in 301 samples of tea and herbal infusions, and in all cases, they were reported below the Limit of Quantification (LOQ). Food samples were mainly tea and herbs for infusion, and honey samples A total of 294 samples of food supplements were also available. In addition to honey samples, 825 food samples of animal origin were also part of this data set, with 97% of them having all analyzed PAs as left-censored data. Previous studies demonstrated that the levels of PAs in animal-derived food are much lower than those that can be found in food commodities such as tea and herbal infusions. Among 746 samples of animal origin, only occasional low levels of PAs in milk samples were found, mostly with single PAs in their free base form.

Except for two egg samples, PAs were absent in the milk products, eggs, meat, and liver samples analyzed.

On 27 July 2017, the EFSA Panel on Contaminants in the Food Chain (CONTAM) published a statement on the risks to human health related to the presence of pyrrolizidine alkaloids in honey, tea, herbal teas, and dietary supplements [12]. The CONTAM Panel established a new benchmark of 237 μg/kg body weight per day to assess the carcinogenic risks associated with PAs and concluded that exposure to PAs poses a potential risk to human health, especially for people who frequently consume large amounts of tea, herbal teas, and herbal based medicines. The conclusion was also that the younger segment of the population is particularly vulnerable.

In the document of the Standing Committee on Plants, Animals, Food and Feed (PAFF Committee) of 17 April 2018, there is also a reference to PAs [13]. According to the Committee, consideration should be given to setting maximum PA levels for the following foods: tea and herbal infusions, tea for babies and small children, herbal dietary supplements derived from plants containing PAs, and dietary supplements accidentally contaminated with plants containing PAs, honey, and pollen-based dietary supplements.

The toxicity studies on PAs, based on animal experiments [69,72] as well as on human cell lines [70], demonstrated their hepatotoxicity [58,59,61,69,70,72], genotoxicity [62,67,69] and also carcinogenic potential [66]. For these reasons, the EFSA has consistently identified PAs as a serious health risk and has been placing them among the substances requiring careful monitoring in food products. Commission Regulation (EU) 2020/2040 of 11 December 2020 [14] sets maximum levels for PAs in food products, namely for teas and herbal infusions (with lower limits for infants and young children), certain food supplements, pollen and pollen products, dried herbs, borage leaves, and cumin seeds. They refer to the lower limit of the sum of 35 pyrrolizidine alkaloids: intermedine, lycopsamine, intermedine *N*-oxide, lycopsamine *N*-oxide, senecionine, senecivernine, senecionine *N*-oxide, senecivernine *N*-oxide, seneciphylline, *N*-oxide seneciphylline, retrorsine, retrorsine *N*-oxide, echimidine, echimidine *N*-oxide, lasiocarpine, lasio carpine *N*-oxide, senkirkine, europine, europine *N*-oxide, heliotrine, heliotrine *N*-oxide, indicine, echinatine, rinderine, indicine-*N*-oxide, echinatine-*N*-oxide, rinderine-*N*-oxide, integerrimine, integerrimine-*N*-oxide, heliosupine, heliosupine-*N*-oxide, spartioidine, spartioidine-*N*-oxide, usaramine, usaramine *N*-oxide. Specifically for food supplements, the maximum levels of alkaloids are as follows: Food supplements containing herbal ingredients, including extracts: 400 μg/kg; pollen-based food supplements: 500 μg/kg. In the case of food, the maximum levels of PAs are as follows: Dried herbs and cumin seeds: 400 μg/kg; tea (*Camellia sinensis*) and flavored tea: 150 μg/kg for adults and 75 μg/kg for infants; herbal infusions (dried product) (here are differences in details): 200 μg/kg or 400 μg/kg when concerns herbal teas made from rooibos, anise, lemon balm, chamomile, thyme, peppermint, lemon verbena (dried product) and mixtures exclusively composed of these dried herbs. Limits for any herbal teas for babies are invariably the same as for teas, 75 μg/kg.

In the document Commission Regulation (EU) 2023/915 dated 25 April 2023 these maximum levels for PAs were maintained [57]. Legally, the foodstuffs listed in the Annex to the Regulation placed on the market before 1 July 2022 may be marketed until 31 December 2023. After 1 July 2022, each product covered by the regulation should obligatorily meet the legal requirements in this area.

A European Pharmacopoeia Commission at the European Directorate for the Quality of Medicines (EDQM) published on 1 July 2021 in Supplement 10.6 of the European Pharmacopoeia (Ph. Eur.) the new general chapter "Contaminant pyrrolizidine alkaloids (2.8.26)" [73]. This general chapter, which describes 28 target PAs, allows for the use of any procedure consisting of chromatography coupled with MS/MS or high-resolution MS that meets the validation requirements given in the chapter. This approach was adopted because there is considerable variation in the composition and matrices of the herbal drugs, as well as in the applicable limits, making it difficult to describe all the methods suitable for quantitative analysis of the target PAs [73]. Performance criteria for method validation for PAs are given in the document Commission Implementing Regulation (EU) 2023/2783 dated 14 December 2023 [74] and in the Guidance Document on Performance Criteria for Methods of Analysis for Mycotoxins and Plant Toxins in Food and Feed from the European Union Reference Laboratory (EURL) [75].

Also, the United States Pharmacopeia (USP) has adopted a new Chapter on Pyrrolizidine Alkaloids <1567> Pyrrolizidine Alkaloids as Contaminants in USP-NF 2023 Issue 3 [76].

All the regulators indicate that the presence of PAs in food products can be minimized or prevented by the application of good agricultural and harvest practices. The European Tea and Herbal Infusions Industry has developed a Code of Practice to prevent and reduce pyrrolizidine alkaloid contamination in agricultural commodities used in the manufacture of tea and herbal infusions. This is designed to minimize contamination of materials at the primary producer level [77]. To prevent and reduce PA contamination, management practices such as effective weed control and careful monitoring of animal feed are crucial. It is also important to note that total eradication of PA-containing plants is not feasible or ecologically desirable [3].

5. Methods for PA Determination

5.1. Sample Preparation for the PAs Containing Materials

PA determination is closely related to the appropriate sample preparation. Efficient PA extraction from biological material is required for a valid determination. Analysis of trace amounts of PAs requires proper sample preparation to increase the PA concentration and remove compounds that interfere with the analysis. Various techniques for PA extraction from medicinal plants are described in the literature [24,31,32]. The applied sample preparation method should efficiently extract the PAs as well as PANOs at the same time. Therefore, the use of polar organic solvents or aqueous solutions is preferred due to the high polarity of PANOs [15,32]. For the PAs and PANOs extraction from different matrices, the most frequently used procedures are solid-liquid extraction (SLE), e.g., sample/methanol [78], and liquid–liquid extraction (LLE), e.g., chloroform/methanol [79]. SLE methods are variations based on maceration or percolation, with additional use of other factors such as sonication, high pressure, or solvent modification. Maceration is the process when the herbal material is continuously soaked with solvent, while during percolation, the solvent flows through the plant material. Generally used solvents are dilute aqueous acids: 0.05 M sulfuric acid, 0.15 M hydrochloric acid solution, 0.5% formic acid, and polar organic solvents, e.g., acidified methanol or acetonitrile [31]. Organic solvents such as chloroform and dichloromethane may also be used, but this approach is problematic because it requires additional steps. The extract should be dissolved in an aqueous acidic solution and washed with a non-polar solvent, e.g., chloroform, to remove less polar material (such as fats, waxes, and terpenes). The addition of ammonia makes the solution strongly basic, and the PAs are extracted back into an organic solvent. Repetition of this process provides extracts clean enough for GC-MS [32]. PANOs are less soluble in relatively non-polar solvents, and to determine the entire profile of PA free bases and *N*-oxides, in the first step, PA free bases are extracted alone, and in the second step, after reduction of the *N*-oxides, the total content of PAs is obtained. The proportion of the *N*-oxides can then be determined by calculating the difference between these two measurements [22,80].

Sample purification is often a necessary step, and solid-phase extraction (SPE) is then used, e.g., solid-phase extraction cartridges (SPE) [81]. There are two main types of SPE: strong cation exchange (SCX) SPE and reversed-phase SPE. SCX-SPE is a silica-based benzenesulfonic acid-based filler. Its negatively charged sulfonic acid group has a strong cation exchange capacity, and the benzene ring has a certain hydrophobic retention. SCX extracts positively charged basic compounds, such as amines. Reversed-phase SPE is a slightly selective separation technique. Reversed-phase sorbents, mainly based on octadecylsilane ligands (C18), can retain most molecules with hydrophobic character, making them very useful for extracting analytes that are very diverse in structure within the same sample [82]. Mixed-mode sorbents (a combination of reversed-phase and cation-exchange interactions) are also used in SPE cartridges. For this purpose, SPE is commonly used for the purification of PAs/PANOs from food samples [31]. The technique that is very often used for the determination of PAs/PANOs in food samples is QuEChERS (acronym of "Quick, Easy, Cheap, Effective Rugged and Safe"). This procedure assumes the simultaneous extraction and purification of the samples and is suitable for extracting a large number of compounds. This procedure is miniaturized and can be successfully applied to the analysis of 21 PAs/PANOs in oregano, significantly reducing the amount of reagents used by ten times in comparison to the classic methodology [83,84]. A broad range of extraction procedures to determine PAs/PANOs in dried plants and food supplements was presented in the review [42].

5.2. LC-and GC-Methods for PAs Analysis

Following the successful extraction, sample solutions can be analyzed using chromatographic separation techniques HPLC and GC coupled with various types of detection.

The structural diversity of PAs and PANOs is a challenge for analysts. According to EFSA requirements, the total sum and individual amounts of PAs should be determined in plant material [11]. Fast qualitative tests to measure total PA content, instrumental methods to determine the PA profile in samples, and sensitive quantification of these alkaloids are needed. Spectroscopic techniques can be utilized in such studies. However, these methods are not sensitive enough to determine trace amounts of PAs. Ultraviolet–visible spectroscopy (UV–Vis) methods and colorimetry are used for PA detection, rather than for quantitative measurements [31]. Nuclear magnetic resonance (NMR) spectroscopy methods were applied to PA analysis for structural identification and identity confirmation of novel PAs [85]. The use of immunoassays for the determination of PAs is rare, but the classical enzyme-linked immunosorbent assay (ELISA) can be used for the analysis of PAs [86]. The technique is hampered by the presence of some cross-reactivity, preventing easy detection of target PAs. The most popular and useful techniques are chromatographic techniques. Preparative HPLC and TLC can be used for PA isolation [32]. TLC, GC, and HPLC methods for PA analysis are also used; however, these methods are suitable for working with materials containing quite a high PA content. A recent review of the analysis of PAs in medicinal plants refers to plants from genera that naturally produce these alkaloids: *Tussilago* (coltsfoot), *Symphytum* (comfrey), *Senecio*, *Petasites* (butterburs), *Lithospermum* (gromwell), *Heliotropium* (bloodstone), *Cynoglossum*, *Borago*, *Brachyglottis Anchusa*, and *Alkanna* [24]. To determine trace amounts of PAs, methods based on mass spectroscopy should be used, which is recommended by the Ph. Eur. [73].

5.3. Electrochemical Methods for PAs Analysis

Alkaloid detection methods that use detection techniques such as ELISA, and MS/MS are laborious, time-consuming, expensive, and require the use of complex equipment and staff training. Currently, fast, simple, and accurate methods are being developed to monitor alkaloids in real samples. These benefits are related to the widespread use of electrochemical sensors modified with various materials [87–89]. Electrochemical detection was found to be favorable because it is easy to use, affordable, rapid, and highly sensitive, and the analysis can be performed on-site [43,90]. Various electrochemical techniques can be applied for quantitative analysis, mainly cyclic or stripping voltammetry (CV), differential pulse voltammetry (DPV), and more recently, electrochemical impedance spectroscopy (EIS). Using electrochemical biosensors, researchers can identify specific analytes present in biological samples and convert biochemical signals into electrical signals, simplifying quantification [87–89]. Electrochemical sensors are also convenient because they can be easily miniaturized, demonstrate excellent sensitivity, and are simple to construct and use. Electrodes are used in detection and are mainly composed of carbonaceous materials such as glassy carbon (GCE), pyrolytic graphite (HOPG), screen-printed carbon electrodes (SPCE), and even still mercury electrodes [91]. By modifying the surfaces of these electrodes with various materials, good selectivity can be achieved [92]. Improvement of sensitivity and resolution is obtained by using the DPV compared to CV, which is important, especially in detecting traces of alkaloids. EIS is very useful in studying surface processes, kinetics, and mechanisms of alkaloid reactions. Recently, the surfaces of electrodes have been additionally decorated with nanoparticulate materials to increase the sensitivity and selectivity of the sensor. Nanomaterials are selected so that their physical and chemical properties match the detected analyte as much as possible. Particular attention is paid then to their chemical composition, crystal structure, orientation of the crystallographic axis, morphology, and dimensions of the nanoparticles. Inorganic and organic nanomaterials used for electrode modification include carbon and metallic nanoparticles, polymer materials, and others [43,90]. Nanomaterials have different shapes, such as nanoflowers, nanowires, nanorods, or nanofibers. Due to their specific surface, high conductivity, and electrocatalytic properties, they are widely used to improve detection limits and specificity. The modified electrodes are useful in testing alkaloids in biological, pharmacological, and agri-food matrices. The determination of alkaloids in biological samples: serum, urine, blood, etc., is crucial in forensics and clinical applications [43,90]. It is important to determine alkaloids in human body fluids, especially when alkaloid poisoning is suspected. The analysis must be performed immediately, and the analyte has to be identified at low concentrations [93–95]. Nanomaterial-modified electrodes are also used for the determination of various alkaloids in pharmaceutical samples because the analytical procedures are simple, fast, and accurate. Trace amounts of various alkaloids can be determined individually or simultaneously. To our knowledge, among the very large number of sensors for the determination of alkaloids, only a few sensors designed specifically for the determination of PAs can be found in the literature.

Erdem and coworkers [96] developed a simple and inexpensive electrochemical test based on a single-use sensor for the quantification of senecionine (SEN) in food. SEN was immobilized on the surface of a pencil graphite electrode. The SEN oxidation signal was used to evaluate the sensor using differential pulse voltammetry (DPV). The selectivity of the sensor was also checked in the presence of other similar PAs such as intermedine, lycopsamine, and heliotrine. The detection limit was 5.45 μg/mL. Electrochemical detection of SEN had high sensitivity and good selectivity. The sensor was also tested by examining its use in flour and herbal tea products.

Yang and coworkers [97] developed a visual, easy-to-use, and cost-effective mesoporous silica-based electrochemiluminescence (MPS-ECL) sensor for point-of-care (POC) testing of PAs. ECL activity was found to depend on the PA structure. The intensity of ECL also varies for different PAs in order: monocrotaline > senecionine *N*-oxide > retrorsine > senkirkine. The POC sensors had excellent linearity, low detection limits (0.02 μM–0.07 μM), and good recovery, indicating good accuracy and practicality. The portable and low-cost sensor is user-friendly and can be used to test PAs in drugs, food products, and clinical samples, which shows promise in the preliminary assessment of PA-induced health risks. The sensor is repeatable and temperature stable and was used to perform on-site PA screening in milk, tea, herbal medicines, and human serum samples.

5.4. LC-MS and GC-MS Methods for PAs Analysis

PA contamination of herbal products is usually at low levels, so sensitive analytical methods based on mass spectrometry (MS), such as liquid chromatography-mass spectrometry (LC-MS) and gas chromatography-mass spectrometry (GC-MS), are required for their determination. LC-MS is now the preferred method for the determination of PAs [31,32]. LC-MS and LC-MS/MS methods have become the most popular approaches to the identification and quantification of PAs as they combine reliability and high sensitivity with ease of sample preparation. A range of mass spectrometer types can be used, including single quadrupoles (MS), ion traps (IT), triple quadrupoles (QqQ; MS/MS), and time of flight (ToF) instruments. The ionization method is primarily electrospray ionization (ESI). Atmospheric pressure chemical ionization (APCI) can also be applied, but it is less sensitive. High sensitivity is provided by the positive electrospray ionization modes (ESI +) [31]. In LC-MS methods, the experiment can be run in single-ion monitoring (SIM) or scan mode.

The detection and quantification of PAs using SIM were successfully performed [98]. In LC-MS/MS methods, collision-induced dissociation (CID) of the PA molecular ion provides fragment ions that can be used in selected reaction monitoring (SRM) or multiple reaction monitoring (MRM). Applying MRM detection that uses the transition from the molecular ion to the specific fragments of the molecule, the highest sensitivity and specificity in PA analysis could be obtained [33–41]. SIM detection on MS/MS also was applied for PAs analysis [99]. SRM provided greater sensitivity and selectivity than high resolution SIM on a single quadrupole [100]. High-resolution mass spectrometry (HRMS) is the latest approach to the analysis of complex sample matrices, such as plant-based products. The increased resolution of HRMS instrumentation enables the resolution of isotope distributions and the generation of fragmentation paths. It uses mass analyzers such as ToF and Orbitrap, which have high mass resolving power that can be used to generate high-quality results. Nevertheless, HRMS instrumentation does not replace the standard low-resolution mass spectrometers found in many research laboratories [101]. The identification and quantification of 25 PAs and *N*-oxides using LC-Q-ToF/MS was performed [102]. A new type of mass spectrometer, the Orbitrap, has brought a significant change to high-resolution mass spectrometry. PAs in botanical samples and PA environmental degradation products were examined using the Orbitrap MS [103–107]. GC-MS analysis of PAs was performed on a single quadrupole spectrometer and was limited to single PA [108], qualitative analysis [109], and the analysis of volatile PAs [110].

An overview of the methods for PA determination in various food products is given in Table 4.

Table 4. Analytical separation and detection techniques for the determination of PAs in herbal material.

Sample Type	PAs	Separation Technique	Chromatographic Conditions		Detection	LOD/LOQ	Ref.
			Column	Elution			
Senecio brasiliensis beehive pollen honey	senecionine senecionine *N*-oxide retrorsine *N*-oxide	HPLC	C_{18} 100 mm × 3.0 mm, 3.5 µm (manufacturer undefined)	Mobile phase A: water with 0.1% formic acid Mobile phase B: acetonitrile with 0.1% formic acid Gradient: 98% A from 0 to 2.0 min, 85% A from 2.0 to 5.0 min, 50% A from 5.0 to 8.0 min, 10% A from 8.0 to 9.0 min, 98% A from 9.0 to 11.0 min.	Q-TRAP; MS/MS; Mode: ESI + MRM	-	[34]
bee-collected pollen teas herbal infusions	acetyllycopsamine echimidine group europine heliotrine intermedine lasiocarpine lycopsamine group retrorsine group senecionine group seneciphylline group senkirkine trichodesmine	UHPLC	Accu-core™ RP-MS (Thermo Scientific, Waltham, MA, USA) 100 mm × 2.1 mm, 2.6 µm	Mobile phase A: 0.1% formic acid in water Mobile phase B: methanol/acetonitrile 1:1 (*v*/*v*). Gradient: from 3% to 4% B (0–1 min), from 4% to 17% B (1–6 min), 17% B held for 2 min, from 17% to 44% B (8–10.5 min), from 44% to 95% B (in 0.1 min), 95% B held for 1 min, from 95% to 3% B in 0.1 min re-equilibration to 3% B for 4 min.	QqQ; MS/MS; Mode: ESI + MRM	LOD: 2.4–5.3 ng/g LOQ: 4.0–9.0 ng/g LOD: 0.04–0.08 ng/mL LOQ: 0.07–0.14 ng/mL	[111]
oregano	intermedine europine lycopsamine europine *N*-oxide intermedine *N*-oxide lycopsamine *N*-oxide retrorsine retrorsine *N*-oxide seneciphylline heliotrine heliotrine *N*-oxide senecivernine senecionine seneciphylline *N*-oxide senecivernine *N*-oxide senecionine *N*-oxide echimidine echimidine *N*-oxide lasiocarpine lasiocarpine *N*-oxide senkirkine	UHPLC	Luna Omega Polar C_{18} (Phenomenex, Torrance, CA, USA), 100 mm × 2.1 mm, 1.6 µm	Mobile phase A: 0.2% formic acid and 5 mM ammonium acetate in water Mobile phase B: 10 mM ammonium acetate in methanol Gradient: 5% B (0–0.5 min), 5–50% B (0.5–7 min), 50% B (7–7.5 min), 50–100% B (7.5–11 min), 100% B (11–12 min), 100–5% B (12–14 min). re-equilibrated with the initial composition for 1 min.	IT; MS/MS; Mode: ESI + TIC	LOD: 0.1–7.5 ng/g LOQ: 0.5–25.0 ng/g	[83]

Table 4. *Cont.*

Sample Type	PAs	Separation Technique	Chromatographic Conditions		Detection	LOD/LOQ	Ref.
			Column	Elution			
black tea, green tea dark tea *Chrysanthemum* weed	heliotrine heliotrine-*N*-oxide retrorsine retrorsine-*N*-oxide senecionine senecionine-*N*-oxide jacobine jacobine-*N*-oxide intermedine intermedine-*N*-oxide seneciphylline seneciphylline-*N*-oxide europine senkirkine	UHPLC	Waters Acquity UPLC HSS T3 (Waters, Milford, MA, USA) 100 mm × 2.1 mm, 1.8 µm	Mobile phase A: methanol buffered with 0.1% formic acid and 1 mM ammonium formate. Mobile phase B: Water buffered with 0.1% formic acid and 1 mM ammonium formate. Gradient: MPB was applied: 0–1 min at 90%, 1–4 min from 90% to 40%, 4–7 min from 40% to 30%, 7–7.1 min from 30% to 2%, 7.1–11 min at 2%, 11–11.1 from 2% to 90%, held for 2.9 min before the next run.	QqQ; MS/MS; Mode: ESI + MRM	LOD: 0.001–0.4 ng/g LOQ: 1–5 ng/g	[112]
honey	echimidine intermedine lycopsamine retrorsine retrorsine *N*-oxide senecionine senecionine *N*-oxide echimidine *N*-oxide erucifoline erucifoline *N*-oxide europine europine *N*-oxide heliotrine heliotrine *N*-oxide intermedine intermedine *N*-oxide jacobine jacobine *N*-oxide lasiocarpine lasiocarpine *N*-oxide lycopsamine lycopsamine *N*-oxide monocrotaline monocrotaline *N*-oxide seneciphylline seneciphylline *N*-oxide senkirkine trichodesmine	UHPLC	Waters Acquity UPLC BEH C18 (Waters, Milford, MA, USA) 100 mm × 2.1 mm, 1.7 µm	Mobile phase A: 6.5 mM ammonium hydroxide in water Mobile phase B: 6 5 mM ammonium hydroxide in acetonitrile Gradient: 0 to 2 min: 0% B; 2 to 10 min: 0 to 50% B, maintained to 2 min; 12 to 14 min: 50 to 100% B, maintained to 16 min; 16 to 19 min: 100 to 0% B, maintained to 23 min.	QTOF-MS/MS; Mode: ESI +	LOD: 1–7 ng/g LOQ: 10–20 ng/g	[35]

Table 4. *Cont.*

Sample Type	PAs	Separation Technique	Chromatographic Conditions		Detection	LOD/LOQ	Ref.
			Column	Elution			
Tussilago farfara *Lithospermum erythrorhizon*	echimidine echimidine *N*-oxide erucifoline erucifoline *N*-oxide europine europine *N*-oxide heliotrine heliotrine *N*-oxide intermedine intermedine *N*-oxide jacobine jacobine *N*-oxide lasiocarpine lasiocarpine *N*-oxide lycopsamine lycopsamine *N*-oxide monocrotaline monocrotaline *N*-oxide retrorsine retrorsine *N*-oxide senecionine senecionine *N*-oxide seneciphylline	HPLC	Shim-pack GIST-C18 (Shimadzu Corporation, Kyoto, Japan) 150 mm × 2.1 mm, 2 μm	Mobile phase A: 0.1% formic acid in 5 mM ammonium formate Mobile phase B: 0.1% formic acid plus 5 mM ammonium formate in 100% methanol Gradient: 1.5 min, 1% B; 1.5–3.0 min, 1–15% B; 3.0–18.0 min, 15–30% B; 18.0–19.0 min, from 30 to 95% B 19.0–21.0 min, 95% B; 21.1 min, 1% B.	QqQ; MS/MS Mode: ESI + MRM	LOD: 0.5–1.7 ng/g LOQ: 1.7–6.4 ng/g	[36]
Sorghum oregano mixed herbal tea	echimidine echinatine erucifoline europine heliotrine indicine intermedine jacobine lasiocarpine lycopsamine monocrotaline retronecine retrorsine senecionine seneciphylline senecivernine senkirkine trichodesmine echimidine *N*-oxide echinatine *N*-oxide erucifoline *N*-oxide europine *N*-oxide heliotrine *N*-oxide indicine *N*-oxide intermedine *N*-oxide jacobine *N*-oxide	UHPLC	Waters Acquity UPLC® BEH Amide (Waters, Milford, MA, USA) 100 mm × 2.1 mm; 1.7 μm	Mobile phase A: water with ammonium formate 5 mM Mobile phase B: acetonitrile: water 95:5, *v*/*v*, with formic acid (0.1%, *v*/*v*). Gradient: 1.5 min, 1% B; 1.5–3.0 min, 1–15% B; 3.0–18.0 min, 15–30% B; 18.0–19.0 min, from 30 to 95% B; 19.0–21.0 min 95% B; 21.1 min, return to 1% B.	Q-TRAP; MS/MS Mode: ESI + MRM	LOD: - LOQ: 0.5–10 ng/g	[33]

Table 4. *Cont.*

Sample Type	PAs	Separation Technique	Chromatographic Conditions		Detection	LOD/LOQ	Ref.
			Column	Elution			
Sorghum oregano mixed herbal tea	intermedine *N*-oxide jacobine *N*-oxide lasiocarpine *N*-oxide lycopsamine *N*-oxide monocrotaline *N*-oxide retrorsine *N*-oxide senecionine *N*-oxide seneciphylline *N*-oxide senecivernine *N*-oxide	UHPLC	Waters Acquity UPLC® BEH Amide (Waters, Milford, MA, USA) 100 mm × 2.1 mm; 1.7 µm	Mobile phase A: water with ammonium formate 5 mM Mobile phase B: acetonitrile: water 95:5, *v*/*v*, with formic acid (0.1%, *v*/*v*). Gradient: 1.5 min, 1% B; 1.5–3.0 min, 1–15% B; 3.0–18.0 min, 15–30% B; 18.0–19.0 min, from 30 to 95% B; 19.0–21.0 min 95% B; 21.1 min, return to 1% B.	Q-TRAP; MS/MS Mode: ESI + MRM	LOD: - LOQ: 0.5–10 ng/g	[33]
tea	echimedine heliotrine lasiocarpine lycopsamine monocrotaline monocrotaline *N*-oxide retrorsine-*N*-oxide retrorsine senecionine-*N*-oxide senecionine seneciphylline *N*-oxide seneciphylline senkirkine trichodesmine europine-*N*-oxide intermedine jacobine europine jacobine *N*-oxide lasiocarpine *N*-oxide heliotrine *N*-oxide	UPLC	Waters X-Bridge (Waters, Milford, MA, USA) C18, 100 mm × 2.1 mm, 3.5 µm	Mobile Phase A: 5 mM ammonium formate and 0.1% formic acid Mobile Phase B: 95% methanol with 5 mM ammonium formate and 0.1% formic acid Gradient: 5% B for 0.5 min, increasing B from 5% to 30% for 6.5 min, from 30% to 95% for 4 min and then holding for 2 min, decreasing to 5% for 0.1 min, and finally holding for 1.9 min	QqQ; MS/MS Mode: ESI + MRM	LOD: 0.1–3.0 ng/g LOQ: 0.3–9.0 ng/g	[113]
plant material tea		SFC	CHIRALPAK®, IG-3/SFC, (Daicel Chiral Technologies, Shanghai, China) 100 mm × 3 mm, 3 µm,	Mobile Phase A: CO_2 Mobile Phase B: 50 mM Ammonium formate in methanol Mobile Phase C: Methanol Mobile Phase D: 0.1% Formic acid	QqQ; MS/MS Mode: ESI +	LOQ: 2–200 ng/g	[114]

Table 4. *Cont.*

Sample Type	PAs	Separation Technique	Chromatographic Conditions		Detection	LOD/LOQ	Ref.
			Column	Elution			
plant based food herbal tea	erucifoline *N*-oxide europine europine *N*-oxide, jacobine, retrorsine, retrorsine *N*-oxide, seneciphylline *N*-oxide, senecivernine *N*-oxide trichodesmine	UHPLC	Waters Acquity UPLC® BEH C18 (Waters, Milford, MA, USA) 100 mm × 2.1 mm, 1.7 µm	Mobile phase A: water with 0.1% ammonia Mobile phase B: acetonitrile. Gradient: Starting at 5% of phase B, kept for 1 min, rising to 15% till 2 min before a new isocratic separation for 1 min, increasing to 20% (from 3 to 5 min), 25% (from 5 to 6 min), 50% (from 6 to 9 min) and 95% (from 9 to 10 min).	QqQ; MS/MS Mode: ESI + MRM	LOD: - LOQ: 0.5–1 ng/g	[115]
maize	total	HPLC	Synergy Max-RP 80 Å (Phenomenex, Aschaffenburg, Germany) 150 mm × 2.1 mm, 4 µm,	Mobile phase A: 0.3% formic acid in water Mobile phase B: 0.3% formic acid in acetonitrile) Gradient: 2 min (95% A), 14 min (95–40% A), 15 min (40–0% A), 18 min (0% A), 19 min (95% A), 30 min (reequilibration 95% A).	Q-TRAP; MS/MS Mode: ESI + MRM	-	[81]
Gynura japonica milk	senecionine, seneciphylline, senkirkine, retrorsine	DART-MS HPLC-MS	Waters Acquity UHPLC BEH C18 (Waters, Milford, MA, USA) 2.1 mm × 100 mm, 1.7 µm	Mobile phase A: water with 0.1% formic acid Mobile phase B: acetonitrile Gradient: 0–3 min, B 3%; 3–6 min, B 3–10%; 6–8 min, B 10–100%; 8–10 min, B 100–3%; 10–15 min, Re-equilibration, B 3%.	IT; MS/MS Mode: ESI +	LOD: 0.55–0.85 ng/mL LOQ: 1.83–2.82 ng/mL	[116]

Table 4. *Cont.*

Sample Type	PAs	Separation Technique	Chromatographic Conditions		Detection	LOD/LOQ	Ref.
			Column	Elution			
herbal food supplements	monocrotaline, intermedine, monocrotaline *N*-oxide, indicine, lycopsamine, europine, europine *N*-oxide, indicine *N*-oxide, riddelliine, junction, riddelline *N*-oxide, trichodesmine, retrorsine, retrorsine *N*-oxide, heliotrine, seneciphylline, heliotrine *N*-oxide, seneciphylline *N*-oxide, integerrimine, senecionine, senecionine *N*-oxide, senkirkine, echimidine, lasiocarpine, lasiocarpine *N*-oxide	UHPLC	Agilent Poroshell 120 EC-C18 (Agilent Technologies, Palo Alto, CA, USA) 2.1 mm × 150 mm, 2.7 µm	Mobile phase A: water with 0.1% formic acid Mobile phase B: acetonitrile with 0.1% formic acid Gradient 0–23 min, 3–4% B; 23–45 min, 4–15% B; 45–55 min, 15–25% B 55–57 min 25–100% B. 3 min wash-100% B 5 min Re-equilibration 3% B	QToF-MS/MS Mode: ESI + TIC	LOD: 0.05–5 ng/mL LOQ: -	[78]
black tea, green tea mixed tea flavoured tea herbal tea (chamomile, sage linden, fennel, rosehips) culinary herb samples (thyme, peppermint)	29 pyrrolizidine alkaloids	UHPLC	Agilent Poroshell 120 EC-C18 (Agilent Technologies, Palo Alto, CA, USA) 2.1 mm × 150 mm, 2.7 µm	Mobile phase A: 0.1% formic acid in water Mobile phase B: 0.1% formic acid in acetonitrile Gradient: 0–23 min, 3–4% B; 23–45 min, 4–15% B; 45–55 min, 15–25% B 55–57 min to 100% B. 3 min wash with 100% B 5 min reequilibration with 3% B.	Q-TOF/MS Mode: ESI + Product Ion	LOD 0.105–0.867 ng/g LOQ 0.357–2.890 ng/g	[102]
milk	51 pyrrolizidine alkaloids	HPLC	Kinetex EVO C18 (Phenomenex, Torrance, CA, USA), 100 mm × 2.1 mm, 2.6 µm. Kinetex EVO C18 (Phenomenex, Torrance, CA, USA), 150 mm × 2.1 mm, 5 µm.	acidic conditions: Mobile phase A: water with ammonium formate and formic acid 5 mmol/L Mobile phase B: acetonitrile/water (95/5, *v*/*v*), 26.5 mmol/L alkaline conditions: Mobile phase A: ammonium carbonate in water 10 mmoL/L Mobile phase B: acetonitrile	QqQ; MS/MS Mode: ESI + MRM	LOD: 0.005–0.054 ng/g LOQ: 0.009–0.123 ng/g	[117]

Table 4. *Cont.*

Sample Type	PAs	Separation Technique	Chromatographic Conditions		Detection	LOD/LOQ	Ref.
			Column	Elution			
black tea, peppermint tea, mixed herbal tea, valerian herbal supplement, alfalfa, hay, sunflower expeller, bovine compound feed	43 pyrrolizidine alkaloids	UPLC	alkaline conditions: Waters Acquity UPLC BEH C18 (Waters, Milford, MA, USA) 2.1 mm × 150 mm, 1.7 µm acidic conditions: Waters Acquity UPLC CSH C18 (Waters, Milford, MA, USA) 2.1 mm × 150 mm, 1.7 µm	alkaline conditions: Mobile phase A: 10 mM ammonium carbonate in water, pH 9 Mobile phase B: acetonitrile acidic conditions: Mobile phase A: 0.1% formic acid in water Mobile phase B: acetonitrile	QqQ; MS/MS Mode: ESI + MRM	LOD: - LOQ: 10 ng/g	[118]
honey	retronecine	GC	Zebron ZB-5MS (Phenomenex, Torrance, CA, USA), 30 m × 0.25 mm; film 0.25 µm	-	Q; MS Mode: positive SIM	LOD: 2 ng/g LOQ: 6 ng/g	[108]
honey	echimidine heliotrine intermedine lasiocarpine lycopsamine retrorsine seneciphylline senecionine senkirkine	UHPLC	Supelco Analytical C8 (Supelco, Bellefonte, PA, USA), 150 mm × 3 mm, 2.7 µm	Mobile phase A: 0.5% formic acid in water Mobile phase B: acetonitrile	Q; MS Mode: ESI + SIM	LOD: - LOQ: 0.08–4.3 ng/g	[98]
honey	lycopsamine senecionine senecionine *N*-oxide heliosupine echimidine	HPLC	Phenomenex Synergi hydro-RP C18, (Phenomenex, Torrance, CA, USA), 100 mm × 30 mm, 2.5 µm	Mobile phase A: 0.1% formic acid in water Mobile phase B: 0.1% formic acid in acetonitrile	QqQ; MS/MS Mode: ESI + SIM	LOD: 0.45–0.67 ng/mL LOQ: 1.21–1.79 ng/mL	[99]
Tussilago farfara	senecionine senkirkine	HPLC-DAD	Waters Xterra C18 (Waters, Milford, MA, USA) 3.9 mm × 150 mm, 5 µm	Mobile phase A: 0.1% formic acid in 20 mM $NH_4CH_3CO_2$; Mobile phase B: 0.1% formic acid in acetonitrile	Q; MS Mode: ESI + SIM	LOD: 0.26/1.32 ng/g LOQ: 1.04/5.29 ng/g	[119]
Pardoglossum cheirifolium	9 pyrrolizidine alkaloids	GC	Restek Rxi-1 ms (Restek, Bellefonte, PA, USA), 30 m × 0.25 mm; film 0.25 µm	-	Q; MS TIC	LOD: - LOQ: -	[110]
tea, potato, beans.	15 pyrrolizidine alkaloids	UHPLC	Waters Acquity HSS T3 (Waters, Milford, MA, USA) 2.1 mm × 50 mm, 1.7 µm	Mobile phase A: water with 0.1 % formic acid and ammonium formate 4 mmol/L. Mobile phase B: methanol	Q-Orbitrap-MS/MS Mode: ESI + HRMS	LOD: 1.18–13.28 ng/g LOQ: -	[104]

Table 4. *Cont.*

Sample Type	PAs	Separation Technique	Chromatographic Conditions		Detection	LOD/LOQ	Ref.
			Column	Elution			
herbal infusions, rooibos, anise, lemon balm, chamomile, thyme, peppermint, lemon verbena, mixtures of teas of *Camellia sinensis*, flavoured teas, 73 plant-based food supplements (formulated as solid forms, infusions, and sirups).	118 pyrrolizidine alkaloids	UHPLC	Phenomenex Luna Omega Polar C_{18} (Phenomenex, Torrance, CA, USA), 2.1 mm × 100 mm, 1.6 µm	Mobile phase A: 0.1% formic acid in water Mobile phase B: 0.1% formic acid in acetonitrile	Q-Orbitrap-HRMS/MS Mode: HESI-II + Full MS/dd-MS^2	LOD: 0–1.5 ng/mL LOQ: 0.1–2.1 ng/g in solids; 1–12 ng/g in infusions	[105]
Common heliotrope (*Heliotropium europaeum*) *Heliotropium popovii* Chamomile (*Matricaria recutita*)	35 pyrrolizidine alkaloids	UHPLC	Waters Acquity UPLC BEH C18 (Waters, Milford, MA, USA) 150 mm × 2.1 mm, 1.7 µm	Mobile phase A: 10 mM ammonium carbonate in water, pH 9 Mobile phase B: acetonitrile	Q-Orbitrap-MS/MS Mode: HESI-II + Full MS Scan	LOD: - LOQ: -	[106]
rooibos, chamomile, red tea, black tea, green tea, white tea, linden, horsetail, mixture of herbs.	28 pyrrolizidine alkaloids	HPLC	C18	Mobile phase A: 0.1% formic acid in water Mobile phase B: 0.1% formic acid in acetonitrile	Q-Orbitrap-MS/MS Mode: ESI + HRMS	LOD: - LOQ: 5 ng/g	[107]
Crotalaria (*Fabaceae*) species	45 pyrrolizidine alkaloids	UHPLC	Hypersil GOLD aQ C18 (Thermo Scientific, Waltham, MA, USA) 100 mm × 2.1 mm, 1.9 µm	Mobile phase A: formic acid in water Mobile phase B: formic acid in acetonitrile (various formic acid concentrations: 0.05, 0.2, and 0.35% *v*/*v*)	Orbitrap-MS Mode: HESI-II + Full MS Scan	LOD: 0.35 ng/mL LOQ: -	[120]

6. Safety Ensuring, Prevention, and Market Control

The European Union (EU) has a range of tools to ensure food safety, including the Rapid Alert System for Food and Feed (RASFF). It has been set up for exchanging information between member countries and supports the rapid response of food safety authorities to public health. It is effective at every stage of the food chain. An interactive, searchable online database called RASFF Window provides public access to summary information about the most recently transmitted RASFF notifications and allows searching for information on any notification issued in the past (currently limited to 2020 and later) [121]. The database was searched on 28 April 2024, using the keyword "pyrrolizidine", and 136 records were collected in Table 5. The permissible PA level was exceeded in the products mentioned.

Table 5. Notifications from the RASFF Window database regarding exceedances of PA content in food products on the EU market (accessed on 31 May 2024).

No.	Product	Country of Origin	Notifying Country	Determined Level of PAs (μg/kg—ppb)	Maximum Level (μg/kg—ppb)	Notification Date
1	Psyllium Fibre Food Supplement	UK	Ireland	1177.0 ± 111.7 1113.5 ± 109.3	400	31 May 2024
2	Cumin	Turkey	Germany	8374 ± 3685	400	22 May 2024
3	Dill tops rubbed	Poland	Germany	1300 2000	400	16 May 2024
4	Ground oregano	Romania, Turkey	France	2563 ± 560	1000	7 May 2024
5	Dill	Poland, Spain	The Netherlands	840	400	3 May 2024
6	Cumin powder	Turkey	Bulgaria	3248.5 ± 1299.4 3232.5 ± 1293	400	29 April 2024
7	Herbes de Provence	France	France	2800 ± 700	600	19 April 2024
8	Ground cumin	Belgium	Belgium	773	400	18 April 2024
9	Pollen	France	Switzerland	3300	500	18 April 2024
10	Black tea—naturally flavoured maple	India	Belgium	347	150	15 April 2024
11	Dried oregano	Turkey	France	7861 ± 3931	1000	11 April 2024
12	Cumin seeds	Turkey	France	34,149.4 ± 17,074.7	400	11 April 2024
13	Cumin powder	Germany	Belgium	8860	400	29 March 2024
14	Oregano	Turkey	Switzerland	8062	1000	28 March 2024
15	Oregano	Turkey	Switzerland	24,231	1000	26 March 2024
16	Cumin	India	Czech Republic	985	400	11 March 2024
17	Black Tea	Kenya	Poland	540 ± 291	150	5 March 2024
18	Dried oregano	Belgium	France	1781.5	1000	5 March 2024
19	Gokshura/Lifepower/Karela	The Netherlands	The Netherlands	3920	400	28 February 2024
20	Kmin rzymski mielony (Ground cumin)	Poland	Poland	3340 ± 1169	400	22 February 2024
21	Herbata Czarna Earl Grey (Earl Gray Tea black tea)	Kenya	Poland	525 ± 180, 540 ± 291	150	20 February 2024
22	Cumin	India	Poland	1914 ± 670	400	20 February 2024
23	Parsley leaves	Poland	Romania	1400	400	15 February 2024
24	Kmin rzymski (Cumin)	Austria	Poland	776 ± 273	400	14 February 2024
25	Dried parsley leaves	Poland	Poland	3249 ± 459	400	26 January 2024
26	Oregano	Turkey	The Netherlands	2400	1000	19 January 2024
27	Green tea	Germany	The Netherlands	165	150	17 January 2024
28	Food Supplement	Norway	Sweden	1100	400	11 January 2024
29	Chamomile herbal tea	Czech Republic	Czech Republic	1936	400	10 January 2024
30	Pollen	Spain	Belgium	1430	500	9 January 2024
31	Oregano	Greece	The Netherlands	2600 ± 1300	1000	3 January 2024
32	Oregano	Turkey	The Netherlands	21,000	1000	28 December 2023
33	Ground cumin	Belgium	Belgium	752	400	22 December 2023
34	Oregano	Turkey	The Netherlands	1245	1000	21 December 2023
35	Oregano	Turkey	Poland	7941 ± 1571	1000	21 December 2023
36	Oregano	Jordan	Ireland	49,432.8 ± 5776.1	1000	21 December 2023
37	Cumin	Turkey	Germany	711	400	20 December 2023
38	Cumin, ground	Turkey	Germany	6080	400	13 December 2023
39	Black cumin seeds	Turkey	France	1054.6 ± 527.3	400	12 December 2023
40	Blackberry leaves	Albania	Germany	5170 ± 1293	200	8 December 2023
41	Mint tea (Mentha bruh, *Mentha piperita*)	Serbia	Croatia	>8550.5	400	5 December 2023
42	Dried oregano	Turkey	France	3626.4 ± 1813.2	1000	27 November 2023
43	Ground cumin	Spain	Belgium	2790	400	24 November 2023
44	Ground oregano	Greece	Germany	23,350	1000	24 November 2023
45	Chili powder	India, The Netherlands, Spain, Turkey	Belgium	2790	0	20 November 2023
46	Pollen	Spain	Belgium	1070	500	13 November 2023
47	Spiskummin (Cumin)	Lebanon	Sweden	1060, 1850, 2160	400	10 November 2023

Table 5. *Cont.*

No.	Product	Country of Origin	Notifying Country	Determined Level of PAs (μg/kg—ppb)	Maximum Level (μg/kg—ppb)	Notification Date
48	Dried oregano	Turkey	Italy	3910 ± 773	1000	9 November 2023
49	Whole lovage leaf	Germany	The Netherlands	1310	1000	6 November 2023
50	Chives, grinded	Germany	The Netherlands	553	0	2 November 2023
51	Rosemary	France	The Netherlands	967	400	30 October 2023
52	Cumin and Organic Cumin	Egypt, India	Denmark	16,000, 1600	400	27 October 2023
53	Dried oregano	Turkey	Poland	3640 ± 1274	1000	26 October 2023
54	Cumin	Lebanon	Denmark	12,000	400	24 October 2023
55	Herbal tea	Morocco	Germany	594 ± 148	200	19 October 2023
56	Cumin seed	Turkey	Belgium	1306	400	9 October 2023
57	Cumin powder	Turkey	Bulgaria	>16,221	400	6 October 2023
58	Dried oregano	Turkey	Bulgaria	8640.7	1000	21 September 2023
59	Peppermint herbal tea	Poland	Czech Republic	657	400	20 September 2023
60	Oregano	Turkey	Luxembourg	3292 ± 745	1000	14 September 2023
61	Herbal infusion	China	Belgium	786	200	21 August 2023
62	Tarragon	France	Belgium	1120	400	11 August 2023
63	Cumin seeds crushed or ground	Turkey	Greece	2074 ± 415	400	7 August 2023
64	Dried oregano	Turkey	Greece	4285 ± 857	1000	7 August 2023
65	Kmin rzymski mielony (Cumin)	India, Poland	Poland	1217	400	11 July 2023
66	Ground cumin	Turkey	Belgium	23,813	400	26 June 2023
67	Ground cumin	Turkey	Greece	8281	400	26 June 2023
68	Cumin	Turkey	Germany	13,600	400	21 June 2023
69	Cumin	Turkey	Belgium	2259 ± 890	400	14 June 2023
70	Cumin seeds	Spain	Luxembourg	717 ± 108	400	14 June 2023
71	Herbata Loyd Earl grey	Poland	Poland	240 ± 10	150	12 May 2023
72	Ground cumin	Turkey	Bulgaria	1553.4	400	2 May 2023
73	Dried oregano	Turkey	Sweden	2263	1000	12 April 2023
74	Organic oregano, rubbed	Germany, Greece	Germany	24,000	1000	28 March 2023
75	Dried oregano	Poland	Czech Republic	1448	1000	28 March 2023
76	Oregano rubbed	Greece	Germany	17,000	1000	22 March 2023
77	Cumin grain	Belgium, France	France	10,000	400	7 March 2023
78	Cumin Whole	India	Ireland	527.1 ± 87.9	400	17 February 2023
79	Borage	Italy	Germany	>59,999	1000	7 February 2023
80	Ground cumin	Belgium, Syria	Belgium	16,596, 13,551.4	400	3 February 2023
81	*Ginkgo biloba* extract	France	Belgium	702	400	30 January 2023
82	Camomille tea	France	Belgium	2470	400	27 January 2023
83	Cumin seeds	Turkey	France	1148.9 ± 574.4, 660.9 ± 330.5, 563.7 ± 281.9	400	27 January 2023
84	Licorice root ground Zoethoutwortel gemalen	France	The Netherlands	1558	400	23 January 2023
85	Herbal tea mix	Morocco	Norway	11,608.3	200	12 January 2023
86	Black Tea (ceai negru)	Poland	Romania	700	150	10 January 2023
87	Pollen	Poland	Poland	1187 ± 301	500	3 January 2023
88	Cumin fines	Turkey	Spain	7290 ± 3650	400	29 December 2022
89	Ground cumin	-	Belgium	5298, 2926	400	16 December 2022
90	Dried oregano	Turkey	Poland	13,921 ± 2735	1000	13 December 2022
91	Cumin	-	Greece	17,512	400	13 December 2022
92	Ground cumin	India	Germany	4040 ± 1620	400	1 December 2022
93	Ground cumin	Afghanistan, France	Belgium	23,899, 14,249	400	22 November 2022
94	Oregano (dried)	Turkey	Belgium	1983.5	1000	21 November 2022
95	Dried oregano	Turkey	France	5174 ± 2587	1000	17 November 2022
96	Dried oregano	Turkey	Poland	8236 ± 1564	1000	15 November 2022
97	Ground cumin	-	Belgium	3697 ± 1395, 10,118 ± 3915	400	3 November 2022
98	Oregano	Greece	The Netherlands	30,313	1000	2 November 2022
99	Origano secco	Turkey	Italy	5591 ± 1177	1000	19 October 2022
100	Comino	Turkey	Spain	8170 ± 4090	400	11 October 2022
101	Ground cumin	Turkey	Switzerland	4436	400	10 October 2022
102	Dried oregano	Turkey	Bulgaria	>2500	400	10 October 2022
103	Ground cumin	Turkey	Ireland	1191.4 ± 197.8	400	25 August 2022
104	Cumin seeds	India	Switzerland	154,000, 2780, 14,100	400	15 June 2022
105	Cumin	Turkey	Sweden	12,350, 10,560	0	10 June 2022
106	Ground cumin	Turkey	Bulgaria	>2500	0	12 May 2022
107	Dried oregano	Turkey	Bulgaria	2154	400	10 May 2022
108	Dried oregano	Turkey	Bulgaria	2644.1	400	7 May 2022
109	Ground cumin	Turkey	Bulgaria	1505.4	400	24 April 2022
110	Ground Cumin	Turkey	Ireland	1723.8, 4810.6 ± 801.4	400	22 April 2022
111	Dried oregano	Turkey	Finland	6970	0	30 March 2022

Table 5. *Cont.*

No.	Product	Country of Origin	Notifying Country	Determined Level of PAs (μg/kg—ppb)	Maximum Level (μg/kg—ppb)	Notification Date
112	Semillas de comino (Cumin seeds)	Turkey	Spain	50,000	400	7 March 2022
113	Ground cumin	Turkey	Czech Republic	11,907.7	0	1 March 2022
114	Organic bee feed	Spain	The Netherlands	97, 42, 880	500	5 January 2022
115	Oregano	Spain	Denmark	14,000 ± 5000	0	23 December 2021
116	Chamomile Tea	Uzbekistan	Denmark	5400	0	22 December 2021
117	Oregano	Turkey	Germany	2785, 2568	0	28 October 2021
118	Cumin seeds	Turkey	Germany	9474	0	19 October 2021
119	Oregano	Turkey	Switzerland	4879	0	2 June 2021
120	Oregano	Turkey	Germany	2079	0	20 May 2021
121	Organic cumin	Turkey	Germany	10,483.39	5000	14 May 2021
122	Cumin	Turkey	Germany	10,906.77	0	7 May 2021
123	Cumin	Turkey	Germany	10,406.94	5000	5 May 2021
124	Kräutertee (Herbal tea)	Czech Republic	Germany	2928.10	0	1 April 2021
125	Oregano	Turkey	Switzerland	8895	0	26 March 2021
126	Kreuzkümmel, gemahlen (Ground cummin)	Turkey	Germany	27,500 ± 970	0	12 February 2021
127	Kreuzkümmel, gemahlen (Ground cummin)	The Netherlands	Germany	21,200 ± 5300	0	21 January 2021
128	Ground cumin	Turkey	Switzerland	9948	0	24 December 2020
129	Ground cumin	Turkey	Switzerland	20,377, 5786	0	23 December 2020
130	Ground cumin	Turkey	Switzerland	5522	0	23 December 2020
131	Kreuzkümmel, gemahlen (Ground cummin)	Turkey	Germany	11,700 ± 2900	0	4 December 2020
132	Ground cumin	The Netherlands	Germany	55,176	0	18 November 2020
133	Cumin (Kreuzkümmel)	Lebanon	Germany	22,000, 18,900	0	18 November 2020
134	Anissamen (Anise seeds)	Egypt	Germany	12,184, 15,114, 1206 ± 188	0	20 August 2020
135	Cumin ganz	Syria	Germany	57,827	0	19 August 2020
136	Cumin, Organic	Turkey	Switzerland	29,120	0	30 June 2020
137	Bio Cumin	Turkey	Germany	56.100	0	30 April 2020
138	Ground cumin and dry oregano	Turkey	Denmark	15,000, 7200	0	24 April 2020
139	Oregano	Turkey	Germany	6620	0	30 March 2020
140	Dried camomile tea	Poland	Belgium	530	0	11 February 2020
141	Oregano getrocknet	Turkey	Germany	16,962 ± 8481	0	5 February 2020
142	Rubbed oregano	Turkey	Germany	8836	0	4 February 2020

7. Conclusions

The growing interest in plant-based medicinal products and dietary supplements should not be associated with the misconception that these products are inherently safe and free of side effects. PAs are a class of natural toxins that draw significant attention due to their presence in honey and medicinal plants. PAs are proven to be carcinogenic, genotoxic, and hepatotoxic. Metabolic activation of PAs leads to the formation of adducts with DNA, which is considered to be the main cause of the carcinogenic effects of PAs. The toxic nature of PAs poses a potential risk to human health. Human consumption of PAs from food and herbal medicinal products has likely remained stable over recent years, but new, sensitive analytical methods can now detect very low levels of PAs. However, once a problem is identified, regulatory action to mitigate it should be considered. Therefore, the development and validation of sensitive analytical methods, especially those based on LC-MS, are of great importance to ensure consumer safety and improve public health. In this review, we introduced sensitive and selective analytical methods for the determination of PAs in various materials. Reporting new cases of contamination is important to ensure the safety of herbal products.

8. Future Directions

Despite significant progress in the determination of PAs in various matrices, the structural diversity and chemical properties of PAs present unique challenges to analysts. The continuous development of analytical methods is essential to improving the detection and quantification of PAs. From a future perspective, the determination of PA in plant and bee products should be a mandatory preventive step during production. The development of convenient, fast, and sensitive electrochemical sensors can be an alternative to complex mass spectrometry-based methods.

Author Contributions: Conceptualization, supervision, formal analysis, A.L.-C. and E.U.S.; visualization, A.L.-C.; writing—original draft preparation, A.L.-C. wrote part of the paper, K.T. wrote part of the paper, K.S. wrote part of the paper, and E.U.S. wrote part of the paper; writing—review and editing, A.L.-C. and E.U.S. All authors have read and agreed to the published version of the manuscript.

Funding: The project was financially supported by the Polish Ministry of Science and Higher Education: subvention No. 10/2024.

Institutional Review Board Statement: Not applicable.

Informed Consent Statement: Not applicable.

Data Availability Statement: Not applicable.

Conflicts of Interest: The authors declare no conflicts of interest.

References

1. Sarker, S. Pharmacognosy in modern pharmacy curricula. *Pharmacogn. Mag.* **2012**, *8*, 91. [CrossRef] [PubMed]
2. Joint FAO/WHO Expert Committee on Food Additives (JECFA). *Supplement 2: Pyrrolizidine Alkaloids*; WHO Food Additives Series 71-S2; World Health Organization: Geneva, Switzerland; Food and Agriculture Organization of the United Nations: Rome, Italy, 2020; pp. 1–405.
3. FAO; WHO. *Code of Practice for Weed Control to Prevent and Reduce Pyrrolizidine Alkaloid Contamination in Food and Feed CAC/RCP 74-2014*; World Health Organization: Geneva, Switzerland; Food and Agriculture Organization of the United Nations: Rome, Italy, 2014.
4. COT—Committee on Toxicity of Chemicals in Food, Consumer Products and the Environment. COT Statement on Pyrrolizidine Alkaloids in Food. 2008. Available online: https://cot.food.gov.uk/sites/default/files/cot/cotstatementpa200806.pdf (accessed on 27 April 2024).
5. ANZFA—Australia New Zealand Food Authority. Pyrrolizidine Alkaloids in Food. 2001. Available online: https://www.foodstandards.gov.au/sites/default/files/publications/Documents/TR2.pdf (accessed on 27 April 2024).
6. Centre for Food Safety Food and Environmental Hygiene Department; The Government of the Hong Kong Special Administrative Region. Pyrrolizidine Alkaloids in Food. 2017. Available online: https://www.cfs.gov.hk/sc_chi/programme/programme_rafs/files/PA_Executive_Report_c.pdf (accessed on 27 April 2024).
7. PAFF Committee. Summary Report of the Standing Committee on Plants, Animals, Food and Feed Held in Brussels on 1 July 2014. 2014. Available online: https://food.ec.europa.eu/system/files/2016-10/cs_contaminants_catalogue_mycotoxins_monitoring_recommendations_en.pdf (accessed on 27 April 2024).
8. European Food Safety Authority; EFSA Panel on Contaminants in the Food Chain (CONTAM). Opinion of the Panel on contaminants in the food chain [CONTAM] related to pyrrolizidine alkaloids as undesirable substances in animal feed. *EFSA J.* **2007**, *5*, 447. [CrossRef]
9. European Food Safety Authority; EFSA Panel on Contaminants in the Food Chain (CONTAM). Scientific Opinion on Pyrrolizidine alkaloids in food and feed. *EFSA J.* **2011**, *9*, 2406. [CrossRef]
10. Committee on Herbal Medicinal Products. *Public Statement on Contamination of Herbal Medicinal Products/Traditional Herbal Medicinal Products with Pyrrolizidine Alkaloids*; Committee on Herbal Medicinal Products: Amsterdam, The Netherlands, 2016.
11. European Food Safety Authority (EFSA). Dietary exposure assessment to pyrrolizidine alkaloids in the European population. *EFSA J.* **2016**, *14*, e04572. [CrossRef]
12. Knutsen, H.K.; Alexander, J.; Barregård, L.; Bignami, M.; Brüschweiler, B.; Ceccatelli, S.; Cottrill, B.; Dinovi, M.; Edler, L.; Grasl-Kraupp, B.; et al. Risks for human health related to the presence of pyrrolizidine alkaloids in honey, tea, herbal infusions and food supplements. *EFSA J.* **2017**, *15*, e04908. [CrossRef] [PubMed]
13. PAFF Committee. Summary Report of The Standing Committee on Plants, Animals, Food and Feed Held in Brussels on 17 September 2018. 2018. Available online: https://food.ec.europa.eu/system/files/2018-10/reg-com_toxic_20180917_sum.pdf (accessed on 28 April 2024).
14. European Commission. Commission Regulation (EU) 2020/2040 of 11 December 2020 Amending Regulation (EC) No 1881/2006 as Regards Maximum Levels of Pyrrolizidine Alkaloids in Certain Foodstuffs (Text with EEA Relevance). *Off. J. Eur. Union* **2020**, *63*, 1–5. Available online: https://eur-lex.europa.eu/eli/reg/2020/2040/oj (accessed on 28 April 2024).
15. Casado, N.; Morante-Zarcero, S.; Sierra, I. The concerning food safety issue of pyrrolizidine alkaloids: An overview. *Trends Food Sci. Technol.* **2022**, *120*, 123–139. [CrossRef]
16. CCCF—Codex Committee on Contaminants in Foods. The 15th Session of the Codex Committee on Contaminants in Foods. 2022. Available online: https://www.fao.org/fao-who-codexalimentarius/meetings/detail/en/?meeting=CCCF&session=15 (accessed on 27 April 2024).

17. CCCF—Codex Committee on Contaminants in Foods. Request for Comments on the Recommendations for Follow-Up Risk Management Actions on Pyrrolizidine Alkaloids CL, 2.0.2.3./.4.0.-C.F. 2023. Available online: https://www.fao.org/fao-who-codexalimentarius/sh-proxy/es/?lnk=1&url=https%253A%252F%252Fworkspace.fao.org%252Fsites%252Fcodex%252FCircular%252520Letters%252FCL%2525202023-40%252Fcl23_40e.pdf (accessed on 27 April 2024).
18. Peloso, M.; Minkoumba Sonfack, G.; Paduano, S.; De Martino, M.; De Santis, B.; Caprai, E. Pyrrolizidine Alkaloids in Food on the Italian Market. *Molecules* **2023**, *28*, 5346. [CrossRef] [PubMed]
19. Moreira, R.; Pereira, D.; Valentão, P.; Andrade, P. Pyrrolizidine Alkaloids: Chemistry, Pharmacology, Toxicology and Food Safety. *Int. J. Mol. Sci.* **2018**, *19*, 1668. [CrossRef] [PubMed]
20. Wei, X.; Ruan, W.; Vrieling, K. Current Knowledge and Perspectives of Pyrrolizidine Alkaloids in Pharmacological Applications: A Mini-Review. *Molecules* **2021**, *26*, 1970. [CrossRef]
21. Aniszewski, T. Alkaloid chemistry. In *Alkaloids*; Elsevier: Amsterdam, The Netherlands, 2015; pp. 99–193.
22. Schramm, S.; Köhler, N.; Rozhon, W. Pyrrolizidine Alkaloids: Biosynthesis, Biological Activities and Occurrence in Crop Plants. *Molecules* **2019**, *24*, 498. [CrossRef]
23. Mroczek, T.; Glowniak, K. Pyrrolizidine Alkaloids. In *Natural Products in the New Millennium: Prospects and Industrial Application*; Springer: Dordrecht, The Netherlands, 2002; pp. 1–46.
24. Kopp, T.; Abdel-Tawab, M.; Mizaikoff, B. Extracting and Analyzing Pyrrolizidine Alkaloids in Medicinal Plants: A Review. *Toxins* **2020**, *12*, 320. [CrossRef] [PubMed]
25. Al-Subaie, S.F.; Alowaifeer, A.M.; Mohamed, M.E. Pyrrolizidine Alkaloid Extraction and Analysis: Recent Updates. *Foods* **2022**, *11*, 3873. [CrossRef] [PubMed]
26. Schrenk, D.; Gao, L.; Lin, G.; Mahony, C.; Mulder, P.P.J.; Peijnenburg, A.; Pfuhler, S.; Rietjens, I.M.C.M.; Rutz, L.; Steinhoff, B.; et al. Pyrrolizidine alkaloids in food and phytomedicine: Occurrence, exposure, toxicity, mechanisms, and risk assessment—A review. *Food Chem. Toxicol.* **2020**, *136*, 111107. [CrossRef] [PubMed]
27. Lu, Y.-S.; Qiu, J.; Mu, X.-Y.; Qian, Y.-Z.; Chen, L. Levels, Toxic Effects, and Risk Assessment of Pyrrolizidine Alkaloids in Foods: A Review. *Foods* **2024**, *13*, 536. [CrossRef] [PubMed]
28. Jayawickreme, K.; Świstak, D.; Ozimek, E.; Reszczyńska, E.; Rysiak, A.; Makuch-Kocka, A.; Hanaka, A. Pyrrolizidine Alkaloids—Pros and Cons for Pharmaceutical and Medical Applications. *Int. J. Mol. Sci.* **2023**, *24*, 16972. [CrossRef] [PubMed]
29. Letsyo, E.; Madilo, F.K.; Effah-Manu, L. Pyrrolizidine alkaloid contamination of food in Africa: A review of current trends and implications. *Heliyon* **2024**, *10*, e24055. [CrossRef] [PubMed]
30. Robertson, J.; Stevens, K. Pyrrolizidine alkaloids: Occurrence, biology, and chemical synthesis. *Nat. Prod. Rep.* **2017**, *34*, 62–89. [CrossRef] [PubMed]
31. Crews, C. Methods for Analysis of Pyrrolizidine Alkaloids. In *Natural Products*; Springer: Berlin/Heidelberg, Germany, 2013; pp. 1049–1068.
32. Crews, C.; Berthiller, F.; Krska, R. Update on analytical methods for toxic pyrrolizidine alkaloids. *Anal. Bioanal. Chem.* **2010**, *396*, 327–338. [CrossRef] [PubMed]
33. Dzuman, Z.; Jonatova, P.; Stranska-Zachariasova, M.; Prusova, N.; Brabenec, O.; Novakova, A.; Fenclova, M.; Hajslova, J. Development of a new LC-MS method for accurate and sensitive determination of 33 pyrrolizidine and 21 tropane alkaloids in plant-based food matrices. *Anal. Bioanal. Chem.* **2020**, *412*, 7155–7167. [CrossRef]
34. Valese, A.C.; Daguer, H.; Muller, C.M.O.; Molognoni, L.; Da Luz, C.F.P.; De Barcellos Falkenberg, D.; Gonzaga, L.V.; Brugnerotto, P.; Gorniak, S.L.; Barreto, F.; et al. Quantification of pyrrolizidine alkaloids in *Senecio brasiliensis*, beehive pollen, and honey by LC-MS/MS. *J. Environ. Sci. Health Part B* **2021**, *56*, 685–694. [CrossRef]
35. Bandini, T.B.; Spisso, B.F. Development and validation of an LC-HRMS method for the determination of pyrrolizidine alkaloids and quinolones in honey employing a simple alkaline sample dilution. *J. Food Meas. Charact.* **2021**, *15*, 4758–4770. [CrossRef]
36. Ko, K.Y.; Jeong, S.H.; Choi, E.Y.; Lee, K.; Hong, Y.; Kang, I.; Cho, S.; Lee, C. A LC–ESI–MS/MS analysis procedure coupled with solid phase extraction and MeOH extraction method for determination of pyrrolizidine alkaloids in Tussilago farfara and Lithospermi erythrorhzion. *Appl. Biol. Chem.* **2021**, *64*, 53. [CrossRef]
37. Jiao, W.; Zhu, L.; Shen, T.; Wang, L.; Li, Q.X.; Wang, C.; Wu, X.; Chen, H.; Hua, R. Simultaneous determination of 15 pyrrolizidine alkaloids and their N-oxides in weeds, soil, fresh tea leaves, and tea: Exploring the pollution source of pyrrolizidine alkaloids in tea. *Food Chem.* **2024**, *434*, 137305. [CrossRef] [PubMed]
38. Mukherjee, P.K. LC–MS: A Rapid Technique for Understanding the Plant Metabolite Analysis. In *Quality Control and Evaluation of Herbal Drugs*; Elsevier: Amsterdam, The Netherlands, 2019; pp. 459–479.
39. Ganzera, M.; Sturm, S. Recent advances on HPLC/MS in medicinal plant analysis—An update covering 2011–2016. *J. Pharm. Biomed. Anal.* **2018**, *147*, 211–233. [CrossRef] [PubMed]
40. Schenk, A.; Siewert, B.; Toff, S.; Drewe, J. UPLC TOF MS for sensitive quantification of naturally occurring pyrrolizidine alkaloids in Petasites hybridus extract (Ze 339). *J. Chromatogr. B* **2015**, *997*, 23–29. [CrossRef] [PubMed]
41. Chen, Y.; Li, L.; Xu, J.; Liu, Y.; Xie, Y.; Xiong, A.; Wang, Z.; Yang, L. Mass spectrometric analysis strategies for pyrrolizidine alkaloids. *Food Chem.* **2024**, *445*, 138748. [CrossRef] [PubMed]
42. Tábuas, B.; Cruz Barros, S.; Diogo, C.; Cavaleiro, C.; Sanches Silva, A. Pyrrolizidine Alkaloids in Foods, Herbal Drugs, and Food Supplements: Chemistry, Metabolism, Toxicological Significance, Analytical Methods, Occurrence, and Challenges for Future. *Toxins* **2024**, *16*, 79. [CrossRef]

43. Sharma, B.; Bhatia, R.; Ganti, S.S.; Rangra, N.K. Recent Trends in the Detection of Alkaloids through Analytical, Bioanalytical, and Electrochemical Techniques: Analytical Techniques Used in Detection of Alkaloids. *Curr. Pharm. Anal.* **2024**, *20*, 241–263. [CrossRef]
44. Dey, P.; Kundu, A.; Kumar, A.; Gupta, M.; Lee, B.M.; Bhakta, T.; Dash, S.; Kim, H.S. Analysis of alkaloids (indole alkaloids, isoquinoline alkaloids, tropane alkaloids). In *Recent Advances in Natural Products Analysis*; Elsevier: Amsterdam, The Netherlands, 2020; pp. 505–567.
45. Gutiérrez-Grijalva, E.P.; López-Martínez, L.X.; Contreras-Angulo, L.A.; Elizalde-Romero, C.A.; Heredia, J.B. Plant Alkaloids: Structures and Bioactive Properties. In *Plant-Derived Bioactives*; Springer Singapore: Singapore, 2020; pp. 85–117.
46. Bhambhani, S.; Kondhare, K.R.; Giri, A.P. Diversity in Chemical Structures and Biological Properties of Plant Alkaloids. *Molecules* **2021**, *26*, 3374. [CrossRef]
47. Heinrich, M.; Mah, J.; Amirkia, V. Alkaloids Used as Medicines: Structural Phytochemistry Meets Biodiversity—An Update and Forward Look. *Molecules* **2021**, *26*, 1836. [CrossRef]
48. Stegelmeier, B.L.; Edgar, J.A.; Colegate, S.M.; Gardner, D.R.; Schoch, T.K.; Coulombe, R.A.; Molyneux, R.J. Pyrrolizidine alkaloid plants, metabolism and toxicity. *J. Nat. Toxins* **1999**, *8*, 95–116. [PubMed]
49. Koleva, I.I.; Van Beek, T.A.; Soffers, A.E.M.F.; Dusemund, B.; Rietjens, I.M.C.M. Alkaloids in the human food chain—Natural occurrence and possible adverse effects. *Mol. Nutr. Food Res.* **2012**, *56*, 30–52. [CrossRef] [PubMed]
50. Mulder, P.P.J.; Sánchez, P.L.; These, A.; Preiss-Weigert, A.; Castellari, M. Occurrence of Pyrrolizidine Alkaloids in food. *EFSA Support. Publ.* **2015**, *12*, 859E. [CrossRef]
51. Prakash, A.S.; Pereira, T.N.; Reilly, P.E.B.; Seawright, A.A. Pyrrolizidine alkaloids in human diet. *Mutat. Res./Genet. Toxicol. Environ. Mutagen.* **1999**, *443*, 53–67. [CrossRef] [PubMed]
52. Kowalczyk, E.; Sieradzki, Z.; Kwiatek, K. Determination of Pyrrolizidine Alkaloids in Honey with Sensitive Gas Chromatography-Mass Spectrometry Method. *Food Anal. Methods* **2018**, *11*, 1345–1355. [CrossRef]
53. Wiedenfeld, H. Pyrrolizidine Alkaloids. In *Encyclopedia of Toxicology*; Elsevier: Amsterdam, The Netherlands, 2014; pp. 1170–1174.
54. Ober, D.; Hartmann, T. Homospermidine synthase, the first pathway-specific enzyme of pyrrolizidine alkaloid biosynthesis, evolved from deoxyhypusine synthase. *Proc. Natl. Acad. Sci. USA* **1999**, *96*, 14777–14782. [CrossRef]
55. Wink, M. Quinolizidine and Pyrrolizidine Alkaloid Chemical Ecology—A Mini-Review on Their Similarities and Differences. *J. Chem. Ecol.* **2019**, *45*, 109–115. [CrossRef] [PubMed]
56. National Toxicology Program. Toxicology and carcinogenesis studies of riddelliine (CAS No. 23246-96-0) in F344/N rats and B6C3F1 mice (gavage studies). *Natl. Toxicol. Program. Tech. Rep. Ser.* **2003**, *508*, 1–280.
57. European Commission. COMMISSION REGULATION (EU) 2023/915 of 25 April 2023 on Maximum Levels for Certain Contaminants in Food and Repealing Regulation (EC) No 1881/2006. *Off. J. Eur. Union* **2023**, *66*, 103–157. Available online: https://eur-lex.europa.eu/eli/reg/2023/915/oj (accessed on 28 April 2024).
58. Yang, M.; Ma, J.; Ruan, J.; Ye, Y.; Fu, P.P.-C.; Lin, G. Intestinal and hepatic biotransformation of pyrrolizidine alkaloid N-oxides to toxic pyrrolizidine alkaloids. *Arch. Toxicol.* **2019**, *93*, 2197–2209. [CrossRef]
59. Wang, Z.; Han, H.; Wang, C.; Zheng, Q.; Chen, H.; Zhang, X.; Hou, R. Hepatotoxicity of Pyrrolizidine Alkaloid Compound Intermedine: Comparison with Other Pyrrolizidine Alkaloids and Its Toxicological Mechanism. *Toxins* **2021**, *13*, 849. [CrossRef] [PubMed]
60. Chou, M.W.; Wang, Y.-P.; Yan, J.; Yang, Y.-C.; Beger, R.D.; Williams, L.D.; Doerge, D.R.; Fu, P.P. Riddelliine N-oxide is a phytochemical and mammalian metabolite with genotoxic activity that is comparable to the parent pyrrolizidine alkaloid riddelliine. *Toxicol. Lett.* **2003**, *145*, 239–247. [CrossRef] [PubMed]
61. He, Y.; Zhang, W.; Ma, J.; Xia, Q.; Song, Z.; Zhu, L.; Zhang, C.; Liu, J.; Ye, Y.; Fu, P.P.; et al. Blood Pyrrole–DNA Adducts Define the Early Tumorigenic Risk in Patients with Pyrrolizidine Alkaloid-Induced Liver Injury. *Environ. Sci. Technol. Lett.* **2021**, *8*, 551–557. [CrossRef]
62. He, Y.; Zhu, L.; Ma, J.; Lin, G. Metabolism-mediated cytotoxicity and genotoxicity of pyrrolizidine alkaloids. *Arch. Toxicol.* **2021**, *95*, 1917–1942. [CrossRef] [PubMed]
63. Ruan, J.; Yang, M.; Fu, P.; Ye, Y.; Lin, G. Metabolic Activation of Pyrrolizidine Alkaloids: Insights into the Structural and Enzymatic Basis. *Chem. Res. Toxicol.* **2014**, *27*, 1030–1039. [CrossRef] [PubMed]
64. Frei, H.; Lüthy, J.; Brauchli, J.; Zweifel, U.; Würgler, F.E.; Schlatter, C. Structure/activity relationships of the genotoxic potencies of sixteen pyrrolizidine alkaloids assayed for the induction of somatic mutation and recombination in wing cells of Drosophila melanogaster. *Chem. Biol. Interact.* **1992**, *83*, 1–22. [CrossRef]
65. Schoental, R.; Head, M.A.; Peacock, P.R. Senecio Alkaloids: Primary Liver Tumours in Rats as a Result of Treatment with (1) A Mixture of Alkaloids from S. Jacobaea Lin.; (2) Retrorsine; (3) Isatidine. *Br. J. Cancer* **1954**, *8*, 458–465. [CrossRef]
66. Edgar, J.A.; Molyneux, R.J.; Colegate, S.M. Pyrrolizidine Alkaloids: Potential Role in the Etiology of Cancers, Pulmonary Hypertension, Congenital Anomalies, and Liver Disease. *Chem. Res. Toxicol.* **2015**, *28*, 4–20. [CrossRef] [PubMed]
67. Hadi, N.S.A.; Bankoglu, E.E.; Stopper, H. Genotoxicity of pyrrolizidine alkaloids in metabolically inactive human cervical cancer HeLa cells co-cultured with human hepatoma HepG2 cells. *Arch. Toxicol.* **2023**, *97*, 295–306. [CrossRef]
68. Cheeke, P.R. Toxicity and Metabolism of Pyrrolizidine Alkaloids. *J. Anim. Sci.* **1988**, *66*, 2343. [CrossRef]
69. He, X.; Xia, Q.; Shi, Q.; Fu, P.P. Effects of glutathione and cysteine on pyrrolizidine alkaloid-induced hepatotoxicity and DNA adduct formation in rat primary hepatocytes. *J. Environ. Sci. Health Part C* **2020**, *38*, 109–123. [CrossRef]

70. Zheng, P.; Xu, Y.; Ren, Z.; Wang, Z.; Wang, S.; Xiong, J.; Zhang, H.; Jiang, H. Toxic Prediction of Pyrrolizidine Alkaloids and Structure-Dependent Induction of Apoptosis in HepaRG Cells. *Oxid. Med. Cell Longev.* **2021**, *2021*, 8822304. [CrossRef] [PubMed]
71. Steinhoff, B. Pyrrolizidine alkaloid contamination in herbal medicinal products: Limits and occurrence. *Food Chem. Toxicol.* **2019**, *130*, 262–266. [CrossRef] [PubMed]
72. Yang, M.; Ma, J.; Ruan, J.; Zhang, C.; Ye, Y.; Pi-Cheng Fu, P.; Lin, G. Absorption difference between hepatotoxic pyrrolizidine alkaloids and their N-oxides—Mechanism and its potential toxic impact. *J. Ethnopharmacol.* **2020**, *249*, 112421. [CrossRef] [PubMed]
73. Chapter 2.8.26. Contaminant pyrrolizidine alkaloids. In *European Pharmacopoeia*, 11th ed.; European Medicines Agency: Amsterdam, The Netherlands, 2023.
74. European Commission. Commission Implementing Regulation (EU) 2023/2783 of 14 December 2023 Laying Down the Methods of Sampling and Analysis for the Control of the Levels of Plant Toxins in Food and Repealing Regulation (EU) 2015/705. *Off. J. Eur. Union* **2023**, 1–13. Available online: https://eur-lex.europa.eu/legal-content/EN/TXT/PDF/?uri=OJ:L_202302783 (accessed on 28 April 2024).
75. EURL EU Reference Laboratory WU&, R. Guidance Document on Performance Criteria for Methods of Analysis for Mycotoxins and Plant Toxins in Food and Feed (Draft Version). 2024. Available online: https://www.wur.nl/en/show/eurlmp-guidance-document-performance-criteria-draft-17.05.2024.htm (accessed on 27 June 2024).
76. ECA Academy (ECA). USP Adopts New Chapter on Pyrrolizidine Alkaloids. 2023. Available online: https://www.gmp-compliance.org/gmp-news/usp-adopts-new-chapter-on-pyrrolizidine-alkaloids (accessed on 28 April 2024).
77. European Tea Committee (ETC). *Code of Practice to Prevent and Reduce Pyrrolizidine Alkaloid Contamination in Raw Materials for Tea and Herbal Infusions*; European Tea Committee (ETC): Hamburg, Germany, 2018.
78. Avula, B.; Sagi, S.; Wang, Y.-H.; Zweigenbaum, J.; Wang, M.; Khan, I.A. Characterization and screening of pyrrolizidine alkaloids and N-oxides from botanicals and dietary supplements using UHPLC-high resolution mass spectrometry. *Food Chem.* **2015**, *178*, 136–148. [CrossRef] [PubMed]
79. Wuilloud, J.C.A.; Gratz, S.R.; Gamble, B.M.; Wolnik, K.A. Simultaneous analysis of hepatotoxic pyrrolizidine alkaloids and N-oxides in comfrey root by LC-ion trap mass spectrometry. *Analyst* **2004**, *129*, 150. [CrossRef] [PubMed]
80. Oberlies, N.H.; Kim, N.-C.; Brine, D.R.; Collins, B.J.; Handy, R.W.; Sparacino, C.M.; Wani, M.C.; Wall, M.E. Analysis of herbal teas made from the leaves of comfrey (*Symphytum officinale*): Reduction of N-oxides results in order of magnitude increases in the measurable concentration of pyrrolizidine alkaloids. *Public. Health Nutr.* **2004**, *7*, 919–924. [CrossRef] [PubMed]
81. Letsyo, E.; Adams, Z.S.; Dzikunoo, J.; Asante-Donyinah, D. Uptake and accumulation of pyrrolizidine alkaloids in the tissues of maize (*Zea mays* L.) plants from the soil of a 4-year-old Chromolaena odorata dominated fallow farmland. *Chemosphere* **2021**, *270*, 128669. [CrossRef] [PubMed]
82. Wells, D.A. Chapter 11 Solid-phase extraction: High throughput techniques. *Prog. Pharm. Biomed. Anal.* **2003**, *5*, 361–432.
83. Izcara, S.; Casado, N.; Morante-Zarcero, S.; Sierra, I. A Miniaturized QuEChERS Method Combined with Ultrahigh Liquid Chromatography Coupled to Tandem Mass Spectrometry for the Analysis of Pyrrolizidine Alkaloids in Oregano Samples. *Foods* **2020**, *9*, 1319. [CrossRef]
84. Izcara, S.; Casado, N.; Morante-Zarcero, S.; Pérez-Quintanilla, D.; Sierra, I. Miniaturized and modified QuEChERS method with mesostructured silica as clean-up sorbent for pyrrolizidine alkaloids determination in aromatic herbs. *Food Chem.* **2022**, *380*, 132189. [CrossRef] [PubMed]
85. Villanueva-Cañongo, C.; Pérez-Hernández, N.; Hernández-Carlos, B.; Cedillo-Portugal, E.; Joseph-Nathan, P.; Burgueño-Tapia, E. Complete ^{1}H NMR assignments of pyrrolizidine alkaloids and a new eudesmanoid from *Senecio polypodioides*. *Magn. Reson. Chem.* **2014**, *52*, 251–257. [CrossRef] [PubMed]
86. Oplatowska, M.; Elliott, C.T.; Huet, A.-C.; McCarthy, M.; Mulder, P.P.J.; Von Holst, C.; Delahaut, P.; Van Egmond, H.P.; Campbell, K. Development and validation of a rapid multiplex ELISA for pyrrolizidine alkaloids and their N-oxides in honey and feed. *Anal. Bioanal. Chem.* **2014**, *406*, 757–770. [CrossRef] [PubMed]
87. Ronkainen, N.J.; Halsall, H.B.; Heineman, W.R. Electrochemical biosensors. *Chem. Soc. Rev.* **2010**, *39*, 1747. [CrossRef] [PubMed]
88. Grieshaber, D.; MacKenzie, R.; Vörös, J.; Reimhult, E. Electrochemical Biosensors—Sensor Principles and Architectures. *Sensors* **2008**, *8*, 1400–1458. [CrossRef] [PubMed]
89. Sumitha, M.S.; Xavier, T.S. Recent advances in electrochemical biosensors—A brief review. *Hybrid. Adv.* **2023**, *2*, 100023. [CrossRef]
90. Gañán, J.; Martínez-García, G.; Morante-Zarcero, S.; Pérez-Quintanilla, D.; Sierra, I. Nanomaterials-modified electrochemical sensors for sensitive determination of alkaloids: Recent trends in the application to biological, pharmaceutical and agri-food samples. *Microchem. J.* **2023**, *184*, 108136. [CrossRef]
91. Haroon, N.; Stine, K.J. Electrochemical Detection of Hormones Using Nanostructured Electrodes. *Coatings* **2023**, *13*, 2040. [CrossRef]
92. Câmpean, A.; Tertiş, M.; Săndulescu, R. Voltammetric determination of some alkaloids and other compounds in pharmaceuticals and urine using an electrochemically activated glassy carbon electrode. *Open Chem.* **2011**, *9*, 688–700. [CrossRef]
93. Gao, Y.; Wang, H.; Guo, L. Simultaneous determination of theophylline and caffeine by large mesoporous carbon/Nafion modified electrode. *J. Electroanal. Chem.* **2013**, *706*, 7–12. [CrossRef]

94. Shu, X.; Bian, F.; Wang, Q.; Qin, X.; Wang, Y. Electrochemical Sensor for Simultaneous Determination of Theophylline and Caffeine Based on a Novel poly(folic acid)/graphene Composite Film Modified Electrode. *Int. J. Electrochem. Sci.* **2017**, *12*, 4251–4264. [CrossRef]
95. Kesavan, S.; Abraham John, S. Fabrication of aminotriazole grafted gold nanoparticles films on glassy carbon electrode and its application towards the simultaneous determination of theophylline and uric acid. *Sens. Actuators B Chem.* **2014**, *205*, 352–362. [CrossRef]
96. Senturk, H.; Eksin, E.; Zeybek, U.; Erdem, A. Detection of Senecionine in Dietary Sources by Single-Use Electrochemical Sensor. *Micromachines* **2021**, *12*, 1585. [CrossRef] [PubMed]
97. Xiao, Y.; Yi, H.; Wang, G.; Chen, S.; Li, X.; Wu, Q.; Zhang, S.; Deng, K.; He, Y.; Yang, X. Electrochemiluminescence sensor for point-of-care detection of pyrrolizidine alkaloids. *Talanta* **2022**, *249*, 123645. [CrossRef] [PubMed]
98. Martinello, M.; Cristofoli, C.; Gallina, A.; Mutinelli, F. Easy and rapid method for the quantitative determination of pyrrolizidine alkaloids in honey by ultra performance liquid chromatography-mass spectrometry: An evaluation in commercial honey. *Food Control* **2014**, *37*, 146–152. [CrossRef]
99. Mudge, E.M.; Jones, A.M.P.; Brown, P.N. Quantification of pyrrolizidine alkaloids in North American plants and honey by LC-MS: Single laboratory validation. *Food Addit.Contam. Part A* **2015**, *32*, 2068–2074. [CrossRef] [PubMed]
100. Sleeman, R.; Carter, J.F. MASS SPECTROMETRY | Selected Ion Monitoring. In *Encyclopedia of Analytical Science*; Elsevier: Amsterdam, The Netherlands, 2005; pp. 423–430.
101. Kaufmann, A. High Mass Resolution Versus MS/MS. *Compr. Anal. Chem.* **2012**, *58*, 169–215. [CrossRef]
102. Gumus, Z.P. Assessment of Toxic Pyrrolizidine and Tropane Alkaloids in Herbal Teas and Culinary Herbs Using LC-Q-ToF/MS. *Foods* **2023**, *12*, 3572. [CrossRef]
103. Yi, Y.; Lu, Y.; Liu, H.; Wang, Z.; Li, S.; Huang, X.; Chai, Y.; Zhang, X.; Li, Z.; Chen, H. Insight into pyrrolizidine alkaloids degradation and the chemical structures of their degradation products using ultra high performance liquid chromatography and Q-Exactive Orbitrap mass spectrometry. *J. Hazard. Mater.* **2024**, *471*, 134260. [CrossRef]
104. Wang, H.; Xu, X.; Wang, X.; Guo, W.; Jia, W.; Zhang, F. An analytical strategy for discovering structural analogues of alkaloids in plant food using characteristic structural fragments extraction by high resolution orbitrap mass spectrometry. *LWT* **2022**, *154*, 112329. [CrossRef]
105. Rizzo, S.; Celano, R.; Piccinelli, A.L.; Russo, M.; Rastrelli, L. Target screening method for the quantitative determination of 118 pyrrolizidine alkaloids in food supplements, herbal infusions, honey and teas by liquid chromatography coupled to quadrupole orbitrap mass spectrometry. *Food Chem.* **2023**, *423*, 136306. [CrossRef] [PubMed]
106. Louisse, J.; Mulder, P.P.J.; Gerssen, A.; Stoopen, G.; Rijkers, D.; Van de Schans, M.G.M.; Peijnenburg, A.A.C.M. Bioassay-directed analysis-based identification of relevant pyrrolizidine alkaloids. *Arch. Toxicol.* **2022**, *96*, 2299–2317. [CrossRef] [PubMed]
107. León, N.; Miralles, P.; Yusà, V.; Coscollà, C. A green analytical method for the simultaneous determination of 30 tropane and pyrrolizidine alkaloids and their N-oxides in teas and herbs for infusions by LC-Q-Orbitrap HRMS. *J. Chromatogr. A* **2022**, *1666*, 462835. [CrossRef]
108. Cramer, L.; Beuerle, T. Detection and Quantification of Pyrrolizidine Alkaloids in Antibacterial Medical Honeys. *Planta Med.* **2012**, *78*, 1976–1982. [CrossRef] [PubMed]
109. Conradie, J.; Stewart, M.J.; Steenkamp, V. GC/MS identification of toxic pyrrolizidine alkaloids in traditional remedies given to two sets of twins. *Ann. Clin.Biochem. Int. J. Lab. Med.* **2005**, *42*, 141–144. [CrossRef] [PubMed]
110. Benamar, H.; Tomassini, L.; Frezza, C.; Marouf, A.; Bennaceur, M.; Nicoletti, M. First study on the pyrrolizidine alkaloids of *Pardoglossum cheirifolium* (L.) E.Barbier & Mathez.: GC-MS analysis of their volatile components in the whole plant. *Nat. Prod. Res.* **2021**, *35*, 4098–4103. [CrossRef] [PubMed]
111. Martinello, M.; Manzinello, C.; Gallina, A.; Mutinelli, F. In-house validation and application of UHPLC-MS/MS method for the quantification of pyrrolizidine and tropane alkaloids in commercial honey bee-collected pollen, teas and herbal infusions purchased on Italian market in 2019–2020 referring to recent European Union regulations. *Int. J. Food Sci. Technol.* **2022**, *57*, 7505–7516. [CrossRef]
112. Han, H.; Jiang, C.; Wang, C.; Wang, Z.; Chai, Y.; Zhang, X.; Liu, X.; Lu, C.; Chen, H. Development, optimization, validation and application of ultra high performance liquid chromatography tandem mass spectrometry for the analysis of pyrrolizidine alkaloids and pyrrolizidine alkaloid N-oxides in teas and weeds. *Food Control* **2022**, *132*, 108518. [CrossRef]
113. Kwon, Y.; Koo, Y.; Jeong, Y. Determination of Pyrrolizidine Alkaloids in Teas Using Liquid Chromatography–Tandem Mass Spectrometry Combined with Rapid-Easy Extraction. *Foods* **2021**, *10*, 2250. [CrossRef]
114. Grüning, A.; Schad, G.J.; Stenzler, J. Determination of Pyrrolizidine Alkaloids in Plant Material Using SFC-MS/MS. 2019. Available online: https://www.shimadzu.com/an/literature/lcms/ego119034.html (accessed on 16 April 2024).
115. Picron, J.-F.; Herman, M.; Van Hoeck, E.; Goscinny, S. Analytical strategies for the determination of pyrrolizidine alkaloids in plant based food and examination of the transfer rate during the infusion process. *Food Chem.* **2018**, *266*, 514–523. [CrossRef]
116. Chen, Y.; Li, L.; Xiong, F.; Xiong, A.; Wang, Z.; Yang, L. Rapid identification and determination of pyrrolizidine alkaloids in herbal and food samples via direct analysis in real-time mass spectrometry. *Food Chem.* **2021**, *334*, 127472. [CrossRef] [PubMed]
117. Klein, L.M.; Gabler, A.M.; Rychlik, M.; Gottschalk, C.; Kaltner, F. A sensitive LC–MS/MS method for isomer separation and quantitative determination of 51 pyrrolizidine alkaloids and two tropane alkaloids in cow's milk. *Anal. Bioanal. Chem.* **2022**, *414*, 8107–8124. [CrossRef] [PubMed]

118. EURL EU Reference Laboratory WU&, R. Determination of Pyrrolizidine Alkaloids in Plant-Based Food and FEED Materials, Including (Herbal) Teas, Herbal Food Supplements, Fodder and Feedstuffs by LC-MS/MS. 2019. Available online: https://www.wur.nl/nl/show/eurl-mp-method_002-pyrrolizidine-alkaloids-by-lc-msms-v2.htm (accessed on 27 June 2024).
119. Jiang, Z.; Liu, F.; Goh, J.J.L.; Yu, L.; Li, S.F.Y.; Ong, E.S.; Ong, C.N. Determination of senkirkine and senecionine in Tussilago farfara using microwave-assisted extraction and pressurized hot water extraction with liquid chromatography tandem mass spectrometry. *Talanta* **2009**, *79*, 539–546. [CrossRef] [PubMed]
120. Prada, F.; Stashenko, E.E.; Martínez, J.R. LC/MS study of the diversity and distribution of pyrrolizidine alkaloids in *Crotalaria* species growing in Colombia. *J. Sep. Sci.* **2020**, *43*, 4322–4337. [CrossRef]
121. European Commission. RASFF Window. 2024. Available online: https://food.ec.europa.eu/safety/rasff_en (accessed on 28 April 2024).

molecules

Review

Research Progress of *Ferula ferulaeoides*: A Review

Zhengqiong Chen [1,2], Gang Zhou [2,*] and Shengjun Ma [1,*]

1 College of Food Science and Pharmacy, Xinjiang Agricultural University, Urumqi 830052, China; chen18208822026@163.com
2 Key Laboratory of Ethnic Medicine and Traditional Chines, Xinjiang Uygur Autonomous Region Institute for Drug Control, Urumqi 830054, China
* Correspondence: zhougang740414@126.com (G.Z.); wlmqmsj@sina.com (S.M.)

Abstract: *Ferula ferulaeoides* (Steud.) Korov is one of the traditional ethnic medicines in Xinjiang Uygur and Kazakh of China, which mainly contains volatile oils, terpenoids, coumarins and other chemical components. Previous work has shown that *F. ferulaeoides* exhibited insecticide, antibacterial, antitumor properties, and so on. In this paper, the chemical composition, pharmacological activity, and quality control of *F. ferulaeoides* were reviewed, and the application of *F. ferulaeoides* in the food industry was explored, so as to provide some reference for the quality evaluation of *F. ferulaeoides* and its further development and utilization.

Keywords: *Ferula ferulaeoides*; chemical composition; pharmacological activity; quality control

1. Introduction

Ferula is a perennial herb belonging to the genus *Ferula* L. in the family Apiaceae, which is an important source of resin used in folk medicine [1]. There are about 180 species in the world, mainly distributed in the Mediterranean and Central Asia [2]. Among them, there are 27 species in China, 7 of which are unique [3], mainly distributed in Xinjiang, Tibet, Qinghai, Yunnan, and other provinces [4]. *Ferula* plants have a long history as a medicine. *Ferula* was utilized to treat many diseases as early as ancient Persia, so it was called "God's food". Nowadays, *Ferula* is still an indispensable member of Chinese herbal medicine in the international market. The medicinal material of *Ferula* is an oil–glue–resin air-dried lump, which is isolated from the genus *Ferula*. It is bitter in taste and warm in nature, and has benefits for the spleen and stomach. *Ferula* comprises three major fractions, including resin (40–64%), gum (25%) and essential oil (10–17%). The resin fraction includes ferulic acid and its esters, coumarins, sesquiterpene coumarins, and other terpenoids. The gum contains glucose, galactose, l-arabinose, rhamnose, glucuronic acid, polysaccharides, and glycoproteins, and the volatile fraction includes sulfur-containing compounds, monoterpenes, and other volatile terpenoids [5]. *Ferula ferulaeoides* (Steud.) Korov is one of the genera *Ferula*, which is often distributed in sand dunes or gravel Artemisia deserts in Mongolia, Russia, and Kazakhstan, and mainly grows in the Gobi desert in the marginal area of Junggar, Xinjiang in China [6]. *F. ferulaeoides* is widely used as a substitute for medicinal *Ferula* in Xinjiang. Its active ingredients are mainly volatile components. It has significant pharmacological activities in the treatment of insect accumulation, meat accumulation, lumps, abdominal pain, malaria, dysentery, etc. It is mostly used clinically to treat cold pain in the heart and abdomen, chronic gastroenteritis, gastric ulcers, rheumatoid arthritis, and other diseases [7].

In recent years, *F. ferulaeoides* has attracted the attention of researchers at home and abroad. It has been discovered that the main chemical components of *F. ferulaeoides* include volatile oil, resin, gum, etc. Sesquiterpenes, coumarins, and other chemical components may be the main pharmacological active components. What is more, modern pharmacological studies have shown that *F. ferulaeoides* has significant pharmacological activities such as insecticidal, bacteriostatic, anti-tumor, etc. In order to further develop and utilize this national medicinal material and fully understand its research status, this paper expounds on the research status of the chemical composition, pharmacological activity, and quality control of *F. ferulaeoides*, and looks forward to the application of *F. ferulaeoides* in the food industry, with the objective of providing some reference for the quality evaluation and in-depth development and utilization of resources.

2. Chemical Constituents of *F. ferulaeoides*

The chemical constituents of *F. ferulaeoides* mainly include coumarins, sesquiterpenes, and volatile oils [8]. The coumarins are one of the characteristic components of *Ferula*, and they are also the first compounds isolated from *Ferula*. The coumarins in *F. ferulaeoides* are mainly in the form of sesquiterpene derivatives, and simple coumarins are relatively rare. Most of the sesquiterpene coumarins in other *Ferula* plants are 7-hydroxy coumarins, while *F. ferulaeoides* contains mainly furan coumarins. As another characteristic component of *Ferulae*, sesquiterpenes mainly exist in the form of esters and lactone, and a few exist in the form of ketone derivatives (Table 1, Structures of sesquiterpenes are shown Figures 1–3). The constituents of the volatile oil obtained from *F. ferulaeoides* have been studied more over recent years. More than 100 compounds have been separated from the volatile oil, including monoterpenes, sesquiterpenes, fatty family compounds, aromatic family compounds, and alcohol ester compounds, etc., (Table 2) [6,8,9]. At the same time, it also contains phenols or phenolic acids, phenylpropanoids, steroids and other compounds (Table 3). In addition, polysaccharides, flavonoids, and ferulic acid are also effective components of *F. ferulaeoides*, but at present, there are few studies on polysaccharides and flavonoids, and they still remain in the extraction process and content determination. Currently, ultrasonic extraction and Soxhlet extraction are the most commonly used methods to extract polysaccharides and flavonoids of *F. ferulaeoides* [10,11]. Ferulic acid is a phenolic acid, rarely exists in free form, mainly binds with oligosaccharides, polyamines, lipids, and polysaccharides, and has antibacterial and anti-inflammatory, antioxidant, antithrombotic, and other pharmacological effects [12].

Table 1. Terpenoid Derivatives from Fructus *F. ferulaeoides*.

Type	No.	Compound	Formula	Molecular Weight	References
Sesquiterpene–coumarins	1	Badrakemin	$C_{24}H_{30}O_4$	382	[13]
	2	Badrakemin acetate	$C_{26}H_{32}O_5$	424	[13]
	3	Conferol acetates	$C_{26}H_{32}O_5$	424	[13]
	4	Ferulin A	$C_{25}H_{33}O_5$	413	[14]
	5	Ferulin B	$C_{24}H_{32}O_5$	400	[15]
	6	Ferulin C	$C_{25}H_{34}O_5$	414	[14]
	7	2,3-dihydro-7-hydroxy-2S*,3R*-dimethyl-3-[4,8-dimethyl-3(E),7-nonadienyl]-furo [3,2-c] coumarin	$C_{24}H_{30}O_4$	382	[14]
	8	2,3-dihydro-7-hydroxy-2R*,3R*-dimethyl-3-[4,8-dimethyl-3(E),7-nonadienyl]-furo [3,2-c] coumarin	$C_{24}H_{30}O_4$	382	[14,16]
	9	2,3-dihydro-7-hydroxy-2S*,3R*-dimethyl-3-[4-methyl-5-(4-methyl-2-furyl)-3(E)-pentenyl]-furo [3,2-c] coumarin	$C_{24}H_{26}O_5$	394	[14]
	10	2,3-dihydro-7-methoxy-2S*,3R*-dimethyl-3-[4,8-dimethyl-3(E),7-nonadienyl]-furo [3,2-c] coumarin	$C_{25}H_{32}O_4$	396	[15]
	11	2,3-dihydro-7-hydroxy-2S*,3R*-dimethyl-2-[4,8-dimethyl-3(E),7-nonadienyl]-furo [3,2-c] coumarin	$C_{24}H_{30}O_4$	382	[15]
	12	2,3-dihydro-7-hydroxy-2R*,3R*-dimethyl-2-[4,8-dimethyl-3(E),7-nonadienyl]-furo [3,2-c] coumarin	$C_{24}H_{30}O_4$	382	[15]
	13	2,3-dihydro-7-hydroxy-2S*,3R*-dimethyl-2-[4,8-dimethyl-3(E),7-nonadien-6-onyl]-furo [3,2-c] coumarin	$C_{25}H_{32}O_4$	396	[15]
	14	2,3-dihydro-7-hydroxy-2S*,3R*-dimethyl-2-[4-methyl-5-(4-methyl-2-furyl)-3(E)-pentenyl]-furo [3,2-c] coumarin	$C_{24}H_{26}O_5$	394	[14]

Table 1. *Cont.*

Type	No.	Compound	Formula	Molecular Weight	References
	15	2,3-dihydro-7-hydroxy-2R*,3R*-dimethyl-2-[4-methyl-5-(4-methyl-2-furyl)-3(E)-pentenyl]-furo [3,2-c] coumarin	$C_{24}H_{26}O_5$	394	[14]
	16	2,3-dihydro-7-methoxy-2S*,3R*-dimethyl-2-[4,8-dimethyl-3(E),7-nonadienyl]-furo [3,2-c] coumarin	$C_{25}H_{32}O_4$	396	[15]
	17	2,3-dihydro-7-methoxy-2R*,3R*-dimethyl-2-[4,8-dimethyl-3(E),7-nonadienyl]-furo [3,2-c] coumarin	$C_{25}H_{32}O_4$	396	[15]
	18	2,3-dihydro-7-methoxy-2S*,3R*-dimethyl-2-[4,8-dimethyl-3(E),7-nonadien-6-onyl]-furo [3,2-c] coumarin	$C_{25}H_{30}O_5$	410	[15]
	19	2,3-dihydro-7-methoxy-2S*,3R*-dimethyl-2-[4-methyl-5-(4-methyl-2-furyl)-3(E)-pentenyl]-furo [3,2-c] coumarin	$C_{25}H_{28}O_5$	408	[15]
	20	(trans)-2,3-dimethyl-3-[9-hydroxymethyl-4-methyl-3E,7Z-nonadienyl]-7-hydroxy-2(3H)-furo [3,2-c] coumarin	$C_{24}H_{31}O_5$	399	[15]
	21	7-hydroxy-2-[1-hydroxy-1,5,9-trimethyl-4E,8-decadienyl]-2(3H)-furo [3,2-c] coumarin	$C_{24}H_{31}O_5$	399	[15]
	22	2-[1-hydroxy-1,5,9-trimethyl-4E,8-decadienyl]-7-methoxy-2(3H)-furo [3,2-c] coumarin	$C_{25}H_{33}O_5$	408	[15]
	23	4,7-dihydroxy-3-[3,7,11-trimethyl-2(E),6(E),10-dodecatrienyl] coumarin	$C_{24}H_{30}O_4$	382	[14]
Sesquiterpene–heteroketone compounds	24	Ferulin D	$C_{25}H_{32}O_4$	396	[15]
	25	Ferulin E	$C_{24}H_{31}O_4$	383	[14]
	26	2,3-dihydro-7-hydroxy-2S*,3R*-dimethyl-2-[4,8-dimethyl-3(E),7-nonadienyl]-furo [3,2-b] chromone	$C_{24}H_{30}O_4$	382	[14]
	27	2,3-dihydro-7-hydroxy-2S*,3R*-dimethyl-2-[4-methyl-5-(4-methyl-2-furyl)-3(E),7-pentenyl]-furo [2,3-b] chromone	$C_{24}H_{26}O_5$	394	[15]
	28	2,3-dihydro-7-hydroxy-2R*,3R*-dimethyl-2-[4-methyl-5-(4-methyl-2-furyl)-3(E),7-pentenyl]-furo [2,3-b] chromone	$C_{24}H_{26}O_5$	394	[15]
	29	4-hydro-7-hydroxy-2-methyl-2-[4,8-dimethyl-3E,7-nonadienyl]-2(3H)-pyro [2,3-b] chromone	$C_{24}H_{31}O_4$	383	[14]
	30	2,3-dihydro-7-hydroxy-2S*,3R*-dimethyl-2-[4,8-dimethyl-3(E),7-nonadienyl]-furo [3,2-b] chromone	$C_{24}H_{30}O_4$	382	[15]
	31	2,3-dihydro-7-hydroxy-2R*,3R*-dimethyl-2-[4,8-dimethyl-3(E),7-nonadienyl]-furo[3,2-b] chromone	$C_{24}H_{30}O_4$	382	[15]
Sesquiterpene–phenolic compounds	32	Dshamirone (secoammoresinol)	$C_{23}H_{33}O_3$	356	[15]
	33	(4E,8E)-1-(2,4-dihydroxyphenyl)-5,9,13-trimethyl-tetradeca-4,8,12-trien-1-one	$C_{23}H_{32}O_3$	356	[14]
	34	(4E,8E)-1-(2,4-dihydroxyphenyl)-2-methoxycarbonyl-5,9,13-trimethyltetradeca-4,8,12-trien-1-one	$C_{25}H_{34}O_5$	414	[14]
	35	1-(2,4-dihydroxyphenyl)-3,7,11-trimethyl-3-vinyl-6(E),10-dodecadiene-1-one	$C_{23}H_{32}O_3$	356	[15]
	36	8,9-oxoisopropanyldshami-rone	$C_{22}H_{30}O_5$	374	[17]
	37	Ferulaeolactone A	$C_{24}H_{34}O_6$	418	[17]
	38	8-acetoxy-9-hydroxydshmirone	$C_{24}H_{34}O_5$	402	[17]
	39	Ferulaeone A	$C_{25}H_{33}O_5$	413	[15]
	40	Ferulaeone B	$C_{25}H_{37}O_6$	433	[15]
	41	Ferulaeone C	$C_{23}H_{31}O_3$	355	[15]
	42	Ferulaeone D	$C_{23}H_{28}O_4$	369	[15]
	43	Ferulaeone E	$C_{33}H_{49}O_4$	509	[15]
	44	Ferulaeone F	$C_{33}H_{49}O_4$	509	[15]
	45	Ferulaeone G	$C_{33}H_{43}O_5$	519	[15]
	46	Ferulaeone H	$C_{38}H_{55}O_5$	591	[15]
	47	3-(2,4-dihydroxybenzoyl)-4S*,5R*-dimethyl-5-[4,8-dimethyl-3(E),7(E)-nonadien-1-yl] tetrahydro-2-furanone	$C_{24}H_{32}O_5$	400	[15]
	48	3-(2-hydroxyl-4-methoxybenzoyl)-4S*,5R*-dimethyl-5-[4,8-dimethyl-3(E),7(E)-nonadien-1-yl] tetrahydro-2-furanone	$C_{25}H_{34}O_5$	414	[15]

Table 1. *Cont.*

Type	No.	Compound	Formula	Molecular Weight	References
	49	3-(2,4-dihydroxybenzoyl)-4R*,5R*-dimethyl-5-[4,8-dimethyl-3(E),7(E)-nonadien-1-yl] tetra-hydro-2-furanone	$C_{24}H_{32}O_5$	400	[15]
	50	Ferulactone A	$C_{24}H_{32}O_9$	463	[18]
	51	Ferulactone B	$C_{24}H_{30}O_7$	429	[18]
	52	3S*-(2,4-dihydroxybenzoyl)-4R*,5R*-dimethyl-5-(4,8-dimethyl-3(E),7(E)-nonadien-1-yl) tetrahydro-2-furanone	$C_{24}H_{32}O_5$	400	[19]
	53	1-(2,4-dihydroxyphenyl)-2-hydroxy-5,9,13-trimethy1-4(E),8(E),12-tetradecatrien-1-one	$C_{23}H_{32}O_4$	372	[19]
	54	1-(2,4-dihydroxyphenyl)-3,7,11-trimethyl-3-vinyl-6(E),10-dodecadiene-1,9-dione	$C_{23}H_{30}O_4$	370	[19]
	55	1-(2,4-dihydroxyphenyl)-3,7-dimethyl-3-vinyl-8-(4-methyl-2-furyl)-6(E)-octen-l-one	$C_{23}H_{28}O_4$	368	[15]
	56	3S*-(2,4-dihydroxybenzoyl)-4R*,5R*-dimethyl-5-[4-methyl-5-(4-methyl-2-furyl)-3(E)-penten-1-yl] tetrahydro-2-furanone	$C_{24}H_{28}O_6$	412	[15]
	57	3S*-(2,4-dihydroxybenzoyl)-4R*,5S*-dimethyl-5-[4-methyl-5-(4-methyl-2-furyl)-3(E)-penten-l-yl] tetrahydro-2-furanone	$C_{24}H_{28}O_6$	421	[15]

Table 2. Main chemical compositions in volatile oil of *F. ferulaeoides*.

Type	Compound	References
Monoterpenoids	α-pinene, β-pinene, camphene, δ-3-carene, limonene, D-limonene, L-limonene, α-phellandrene, β-thujene, γ-terpinene, ocimene, β-ocimene, terpinolene, α-terpinolene, myrcene, β-myrcene, sabinene, fenchene, fenchol, borneol, (R)-camphor, citronellol, P-cymen-8-ol, α-terpined, 1,7,7-trimethyl-exo-bicyclo[2.2.1]heptan-2-ol, tanacetone, Z-3-pnen-2-ol, isogeraniol), 3-methoxy-p-cymene, dihydrocarvyl aetate, thymylether methyl, 2-camphanol acetate, exobornyl acetate, α-fenchyl acetate, [1,3,6-octatriene,3,7-dimethyl-,(Z)-], 3-isopropylidene-5-methyl-hex-4-en-2-one)	[8]
Sesquiterpene	Farnesene, α-guaiene, α-farnesene, β-farnesene, α-curcumene, β-curcumene, α-elemene, β-elemene, valencene, (−)-alloaromadendrene, trans-caryophyllene, aristolene, α-gurjunene, α-cedrene, β-cedrene, hujopsenet, α-himachalene, β-himachalene, α-bulnesene, isoledene, (Z,E)-α-farnesene, 1.2,3,4,4A,5,6,8A-octa-hydro-naphthalene, bergamotene, γ-selinene, α-bergamotene, zingiberene, isocaryophyllene, ledene, β-bisabolene, τ-guaiene, guaiol, α-guaiol, δ-guaiol, bulnesol, 10-O-γ-eudesmol, α-eudesmol, τ-eudesmol, β-eudesmol, elemiol, cedrol, hinesol, agarospirol, ginsenol, calareneexoide, eudesm-7(11)-en-4-ol, β-bisabolol, nerolidol, trans-nerolidol, 1,2- propanediol,3-methoxy-, 2,6,6-trimethyl-1- methylen-cyclohex-2-ene, 5,9-undecadien-2-one,6,10-dimethyl-,(Z)-, 1,1-dimethyl-2,4-di(1-propenyl)cyclohexane, (Z)-2,6,10- trimethyl-1,5,9-undecatriee, 5-β-H,7-β, 10-α-selina-4(19),11-diethy, 4-(1E)-1,3-butadien-1-yl-3,5,5-trimethyl]	[8]
Aromatic	Toluene, resorcinol, O-cymene, benzene, 2-methoxy-4-vinylphenol, 1-methyl-4-(1-methylethyl) benzene, 4-(1-methylethyl)-benzene-methanol, 1-1,5-dimethyl-4-hexenyl-4-methyl benzene	[8]
Alcohol esters	cyclopentanol,2-methyl acetate, neryl acetate, butane-2,3-diol, (2S,3S)-(+)-2,3-butane diol, 4,8-dimethyl-3,7-nonadien-2-ol, 4-terpinyl acetate, α-terpinyl acetate, ethyl palmitate, ethyl linoleate	[8]

Table 3. Other chemical constituents of *F. ferulaeoides*.

Type	No.	Compound	Formula	Molecular Weight	References
Phenylpropanoids	1	myristicin	$C_{11}H_{12}O_3$	192	[19]
Phenols, phenolic acids	2	2,4-dihydroxyacetophenone	$C_8H_8O_3$	152	[19]
	3	2-hydroxy-4-methoxyacetophenone	$C_9H_{10}O_3$	166	[19]
	4	2,4-dihydroxybenzoic acid	$C_7H_6O_4$	154	[19]
	5	2,4-dihydroxy-α-oxobenzeneacetic acid	$C_8H_8O_4$	168	[19]
	6	β-resorcylic acid	$C_7H_6O_4$	154	[19]
	7	Methoxyresorcylic acid	$C_8H_8O_4$	168	[19]
	8	Umbelliferone	$C_9H_6O_3$	162	[6]
	9	Lehmannmlone	$C_{24}H_{30}O_4$	382	[6]
	10	methyl 2,4-dihydroxybenzoate	$C_8H_8O_4$	168	[6]
	11	ethyl 2,4-dihydroxybenzoate	$C_9H_{10}O_4$	182	[6]
Steroids	12	β-sitosterol	$C_{29}H_{50}O$	414	[19]
	13	Daucossterol	$C_{35}H_{60}O_6$	576	[6]

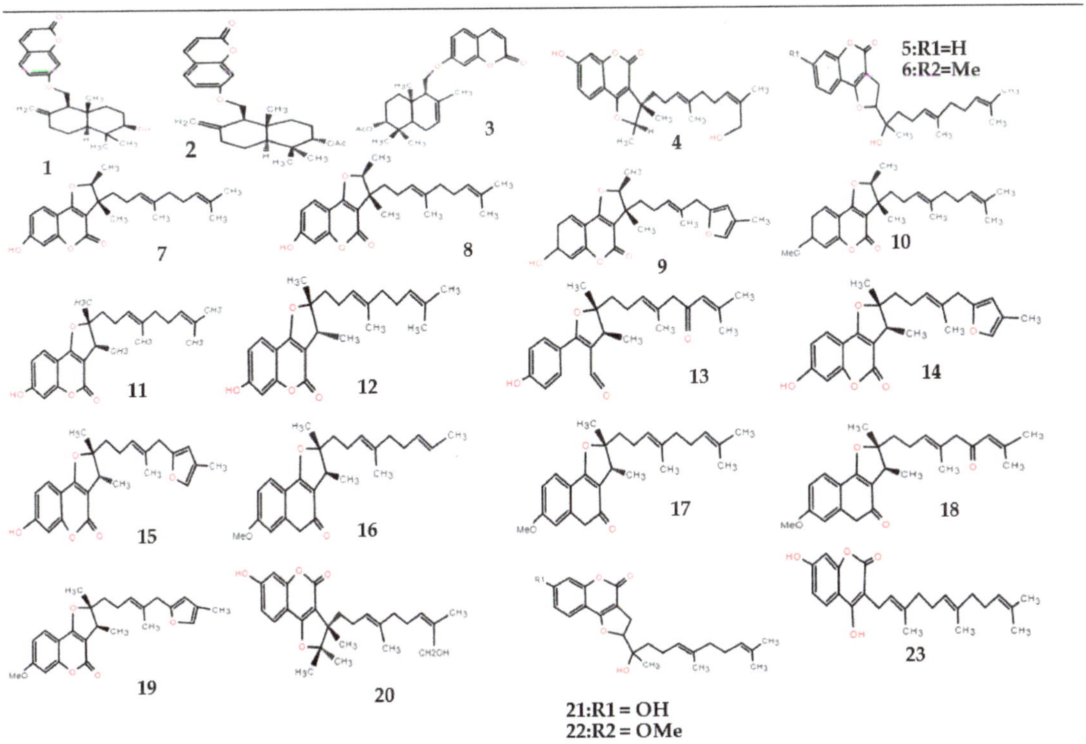

Figure 1. Structural diagram of sesquiterpene–coumarins of *F. ferulaeoides*.

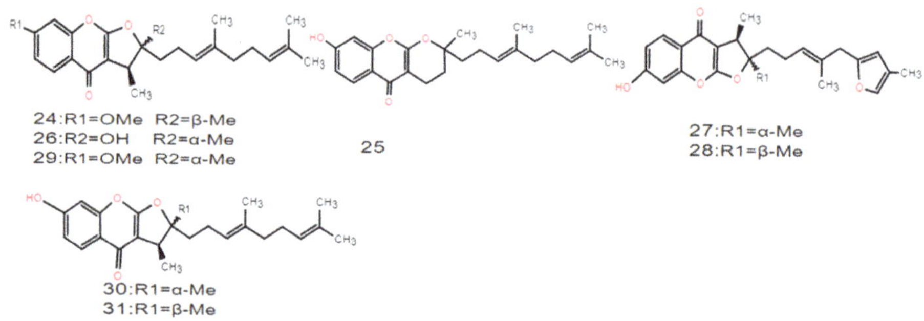

Figure 2. Structural diagram of sesquiterpene–heteroketone compounds of *F. ferulaeoides*.

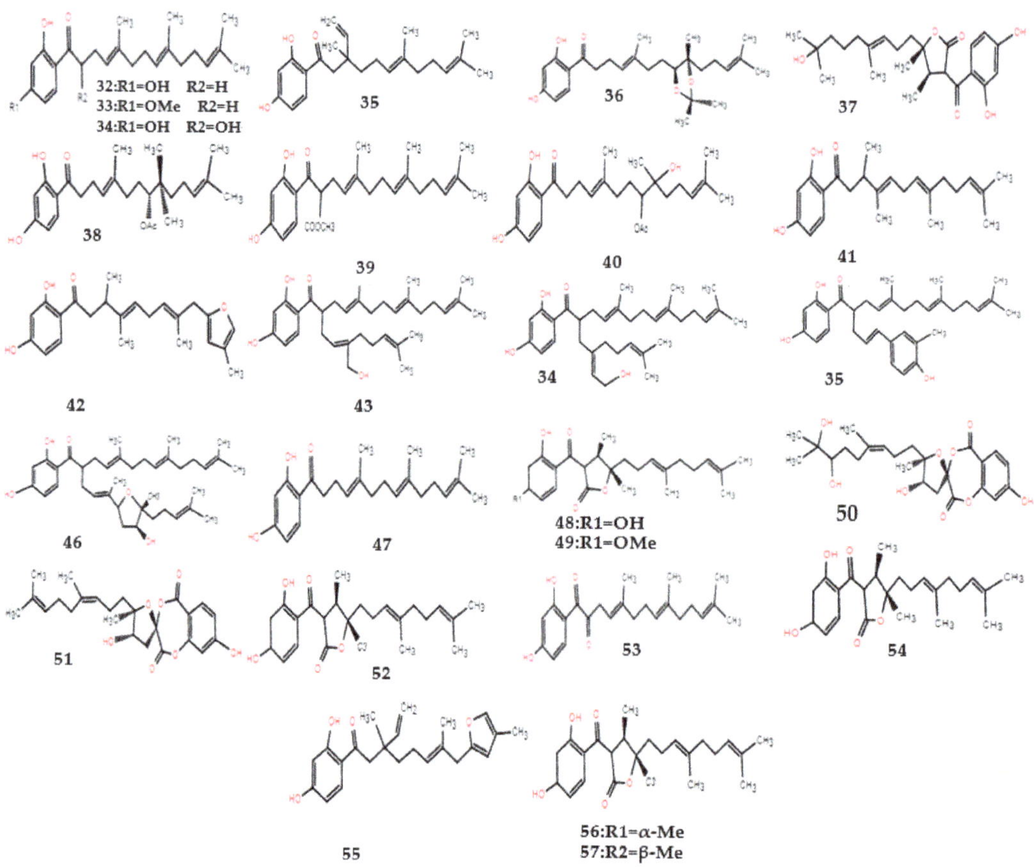

Figure 3. Structural diagram of sesquiterpene–phenolic compounds of *F. ferulaeoides*.

3. Pharmacological Effects of *F. ferulaeoides*

3.1. Antimicrobial Activity

The emergence of bacterial resistance to different classes of antibacterial agents such as β-lactams, quinolones, and macrolides is a major problem that seriously affects human health. Therefore, during the past two decades, researchers have paid great attention to the development of antimicrobial agents, especially those of a natural origin. In addition to their efficacy, most of the natural products are non-toxic, so they can be used as a safe treatment strategy [20]. Among medicinal plants, the *Ferula* species has been identified as a rich source of antimicrobial compounds, with a variety of Ferulic plants having antibacterial effects (Table 4). The inhibitory effect of *F. ferulaeoides* on microorganisms was mainly manifested in *Staphylococcus aureus, Bacillus subtilis*, and *Sarcina*. Liu et al. isolated some terpenoid derivatives from *F. ferulioides* and used them in the antibacterial experiments of drug-resistant *S. aureus* strains including SA1199B (resistant to fluoroquinolones), XU212 (resistant to both tetracycline and methicillin), ATCC25923 (non-resistant), RN4220 (resistant to erythromycin), EMRSA15 (epidemic hospital MRSA), and EMRSA16 (epidemic hospital MRSA). These terpenoid derivatives showed obvious antibacterial activity [17,21]. Gao et al. carried out bacteriostatic tests on the extract and alcohol extract of *Ferula sinkiangensis, Ferula ferulaeoides*, and *Ferula sinkiangensis* leaves by the disc diffusion method. The results showed that the alcohol extracts of three kinds of *Ferula* leaves had good bacteriostatic effects on *S. aureus, B. subtilis*, and *Eight fold aureus*, among which *F. ferulaeoides* had the strongest inhibitory effect [22].

Table 4. Antibacterial Effect of *Ferula* Plants.

Ferula Plants	Antibacterial Ingredients	Microorganism	References
F. lycia	essential oil	*Haemophilus influenza T*	[23]
F. glauca	essential oil	*Streptococcus mutans, Enterococcus faecalis, Escherichia coli*	[23]
F. heuffelii	essential oil	*Micrococcus luteus, Staphylococcus epidermidis, B. subtilis,Micrococcus flavus*	[23]
F. assafoetida	organic extractswater extracts	*coli, S. aureus, E. faecalis, Shigella flexneri, Klebsiella pneumonia*	[21,23]
F. gummosa	oleo-resin	E. *coli, P. aeruginosa, S. aureus, Salmonella enteritidis, Listeria monocytogenes*	[23]
	essential oil	*C. albicans, S. epidermidis, S. aureus, E. coli, B. cereus, E. faecalis, P. aeruginosa*	
F. communis	petroleum ether extract	*S. aureus, B. subtilis, Streptococcus durans,* E. faecalis	[23]
F. szovitsiana	essential oil	*B. subtilis*	[23]
F. hermonis	essential oil	*S. typhi, P. aeruginosa, E. coli, S. aureus, S. fecalis*	[23]
F. vesceritensis	essential oil	*E. coli, K. pneumonia, S. aureus, P. aeruginosa*	[23]
F. kuhistanica	essential oil	*S. aureus*	[23]
F. ferulaeoides	sesquiterpene derivatives alcohol extract	Drug-resistant *S. aureus* *S. aureus, B. subtilis, Sarcina*	[14][22]
F. elaeochytris	essential oil	*S. aureus*	[24]
F.pseudalliacea	sesquiterpene coumarins	*S. aureus, Enterococcus faecium, B. cereus, E. coli, P. aeruginosa*	[23]
F. tunetana	essential oil	*Salmonella typhimurium, S. epidermidis, Micrococcus luteus, B. Cereus, B. subtilis*	[25]
F. tingitana L.	essential oil	*Bacillus subtilis, Neisseria gonorrhoeae*	[26]

3.2. *Antitumor Effect*

In recent years, the treatment of various cancers with traditional Chinese medicine has attracted the extensive attention of scholars at home and abroad, and there is an urgent need to find safe and effective anti-tumor drugs from traditional Chinese medicine resources [27]. *F. ferulaeoides* is a traditional Chinese medicine in Xinjiang, and its anti-tumor research has also attracted the attention of domestic scholars. Yang et al. detected the inhibition effects of different extracts (with volatile oil, 95% ethanol extract, petroleum ether, chloroform, ethylacetate, n-butanol, and water fraction) from *F. ferulaeoides* on five types of gastric cancer cell lines (AGS, MKN-45, BGC-823, MGC-803 and SGC-7901). As a result, volatile oil, 95% ethanol extract and its petroleum ether, chloroform, and ethyl acetate fraction on five types of gastric cancer cells had different proliferation inhibition effects. Among them, the chloroform fraction had a good sensitivity to the five types of gastric cancer cell lines, with the highest sensitivity in the gastric cancer cell lines SGS-7901, and the volatile oil had a strong inhibitory effect on gastric cancer cell AGS [28]. Malignant peripheral-nerve sheath tumors (MPNSTs) are the sixth most common invasive soft tissue sarcoma, originating from the Schwann cell lineage or its precursors, with highly invasive properties in relation to surrounding peripheral nerves and there is currently no effective treatment. DAW22, a natural sesquiterpene coumarin isolated from *F. ferulaeoides* by Li et al., was found to inhibit cell proliferation and colony formation in five established human MPNST cancer cell lines, which provided strong evidence for DAW22 as a potential new alternative therapy for MPNST patients [29]. Other than that, DAW22 can induce apoptosis in C6 glioma cells occurred via the mitochondria-mediated and death-receptor pathways. It inhibited C6 glioma cell growth in a time- and concentration-dependent manner with an IC_{50} value (at 24 h) of 18.92 μM [30]. *F. ferulaeoides* has also been studied in inhibiting the activity of cervical cancer cells. Ma et al. [31] studied the anti-cervical cancer activity of four kinds of ethanol extracts from Xinjiang *Ferula* including *F. ferulaeoides*. The results showed that *F. syreitschikowii*, *F. feurlaeoides*, *F. akitschkensis*, and *F. soongarica* could inhibit and promote the apoptosis of human cervical cancer Hela cells, and the apoptosis rates were 54.82%, 48.99%, 51.83% and 69.75%, respectively.

3.3. Anti-Inflammatory Effect

Researchers at home and abroad have found that *Ferula* has a clear anti-inflammatory effect [32]. The anti-inflammatory activity of *F. ferulaeoides* was studied as early as 1993. For instance, Ye et al. found that three species of *Ferula* including *F. ferulaeoides* could inhibit carrageenan-induced rat voix pedis swelling [33]. A patent in 2013 proved that ferulin A, B, C, D, and E in *F. ferulaeoides* have anti-inflammatory activity with effective doses in the range of 5–15 μ mol kg^{-1} [34]. In a study completed in 2021, the researchers established the inflammation model of lipopolysaccharide (LPS)-stimulated mouse macrophages RAW264.7 and detected the content of NO by the Griess reagent method. Moreover, they conducted the correlation analysis between the NO content and HPLC spectra of different polar extracts of *Ferula sinkiangensis* K.M Shen, *Ferula fukanensis* K.M Shen, and *Ferula. Ferulaeoides* by bivariate Pearson correlation analysis, and screened the anti-inflammatory active extracts. The results showed that the 95% ethanol extract, methylene chloride extract, and ethyl acetate extract of three kinds of *Ferula* had significant anti-inflammatory activity, and *F. ferulaeoides* had significant anti-inflammatory activity at the retention time of 86 min [35].

3.4. Insecticidal Activity

Insecticidal active components in plants are secondary metabolites produced in the long-term co-evolution process of plants and insects, which is relatively safe to human, livestock, crops, and the ecological environment, and insects cannot easily produce resistance to them. Therefore, it meets the requirements of people for ideal pesticides [36]. The main insecticidal effects of *F. ferulaeoides* are sesquiterpenes guaiacol and volatile oil. Liu isolated a compound with insecticidal activity, guaiacol, from the methylene chloride extract of *F. ferulaeoides* root, and tested its insecticidal activity against aphids. The results verify that the killing rate of guaiacol to aphid was almost 100%. Furthermore, guaiacol has obvious killing activity on the fourth instar armyworm, third instar cabbage moth, and the housefly [37]. *Oncomelania hupensis* (*O. hupensis*) is the unique intermediate host of Schistosoma japonicum. Studies have shown that guaiacol in *Ferula* has a killing effect. In 2007, Li et al. initially observed that *ferula* had the effect of killing *O. hupensis*, and in 2015, Fu and Zhao further proved that the mechanism of *O. hupensis* was killed mainly through guaiacol affecting its esterase and glycogen [38,39]. It was detected that the essential oil of *F.ferulaeoides.* had a strong repellent activity against the adults of the stored pest, Tribolium castaneum, and the 2nd instar larvae of Plutella maculipennis Curtis and the 10th instar larvae of yellow mealworm [40].

3.5. Toxic Effect

The acute toxicity of volatile oil of *F. ferulaeoides* is low. The toxicity and death of mice were observed after the mice were given an emulsion of the volatile oil of *F. ferulaeoides*. The mortality of the mice was used as the index, and the median lethal dose of the volatile oil to mice was determined by the Bliss method; the LD_{50} was 10240 mg·kg^{-1} g [41].

4. Research on Quality Control

4.1. Traits and Microstructure

F. ferulaeoides is characterized by granular, teardrop-like or irregular lumps, with a yellow-white to dark brown surface, light internal color, soft texture, and sticky teeth.

It smells similar to celery, but is stronger than celery, with a slightly bitter and pungent taste [42]. Ding et al. [43] observed spiral or reticulate ducts, non-glandular hairs, stone cells, and sub-prism of calcium oxalate in the microstructure of *F. ferulaeoides* powder, and found a large number of resin channels of different sizes distributed in the cross-sectional microstructure of the roots, stems, and leaves of *F. ferulaeoides*, with the maximum diameter up to 10 μm. The resin canal, as its name implies, refers to the secretory tissue that secretes and synthesizes resin, mainly the secretory tube. In another study, Liu et al. investigated the microstructure and ultrastructure of the secretory ducts from the perspective of development. Their research determined that the formation model of SDs in *F. ferulaeoides* was schizogenous and pectinase contributed to SDs' formation, while resin production was due to the activity of organelles and cytoplasm of secretory cells [44].

4.2. Studies on Fingerprint

Chinese medicine fingerprinting technology is an effective method for evaluating the merits of Chinese medicines, identifying authenticity, distinguishing species and ensuring their consistency and stability. In addition, it is currently the most effective means for identifying drug varieties and evaluating drug quality at home and abroad [45]. At the moment, the fingerprint applied to *F. ferulaeoides* mainly focuses on the DNA fingerprint and GC-MS fingerprint. By comparing the results of different primer increases, Miao [46] selected 15 primers of ISSR and 15 primers of RAPD to establish a DNA fingerprint profile for *F. ferulaeoides*, which provides a reliable method for scientific evaluation, effective control of herb quality, and rapid and accurate identification of similar species of the same genus. Sheng et al. [47] established GC-MS fingerprints of 44 samples of essential oils of *F. ferulaeoides* from 8 producing areas. Twelve common peaks were established by analyzing forty-four essential oil samples of *F. ferulaeoides*. The GC-MS fingerprint of in vitro anti-gastric cancer active parts from *F. ferulaeoides* was investigated and analyzed by GC-MS and principal component analysis by Wang and colleagues. The method can characterize the whole information of the chloroform extraction part of the *F. ferulaeoides* with 11 common peaks to a greater extent, and the corresponding compounds are, respectively, 3-methoxy-1,2-propanediol, D-limonene, L-borneol acetate, terpinyl acetate, 1,5,9-undecatriene, 2,6,10-trimethyl, α-cedrene, and a-bergsmotene, β-cedrene, 8-epi-γ-eudesmol, γ-eudesmol, and hinesol, which are aliphatic, aromatic, monoterpene, sesquiterpene, and their oxygenated derivatives [48]. The GC fingerprint of the volatile oil of *F. ferulaeoides* and *Ferula multicolor* was established by Luo. Two kinds of GC fingerprints for volatile oil from *ferula* have twenty common peaks in which the fingerprint similarity was more than 0.9. In addition, Luo also established two kinds of HPLC fingerprints for water extract from *ferula*, and there were thirteen common peaks in the HPLC fingerprint of *F. ferulaeoides* while there were twenty-one common peaks in the HPLC fingerprint of *Ferula* multicolor [49]. In summary, the common peaks of *Ferula ferulaeoides* are almost identified as a volatile oil, and it is necessary to strengthen the identification of other chemical constituents.

4.3. Content Determination

The determination of the content of *Ferulae* mainly focused on flavonoids, polysaccharides, and ferulic acid. The best extraction technology for total flavonoids is as follows: 80% ethanol, reflux extraction at 70 °C for 60 min, which could extract 29.45 mg/g of total flavonoids from 1 g of Uygur medicine *F. ferulaeoides* [10]. Ultrasonic extraction of *F. ferulaeoides* polysaccharide is better than Soxhlet extraction. The content of polysaccharide was determined by phenol-sulfuric acid colorimetry. The results show that the content of polysaccharide extracted by Soxhlet extraction and ultrasonic extraction is 2.87% and 3.22%, respectively [11]. The best extraction process of ferulic acid was the ratio of material to liquid ratio of 1:40, sodium hydroxide solution of 3%, and an extraction time of 30 min. Under these conditions, the content of ferulic acid from *F. ferulaeoides* reached 0.2998 mg/g [12]. The determination of coumarin content has also been involved; Zhu et al. established the UPLC method for the determination of DAW22 from *F. ferulaeoides* of different growth periods in Xinjiang. That indicated that the linear range of DAW22 was 6.21–124.2 ng (r = 10,000) with an average recovery of 99.81% (RSD 2.0%). The content of DAW22 in *F. ferulaeoides* growing on May 9 was the highest [50]. In general, there are few studies on the content determination of *F. ferulaeoides* and the research methods are relatively old. Therefore, modern technologies such as liquid chromatography–mass spectrometry should be used to strengthen the quantitative analysis of other chemical components, especially its characteristic components.

5. Prospect: Application of *F. ferulaeoides* in Food Industry

Firstly, *F. ferulaeoides* can be used as raw materials for food or food additives for food processing. There have been studies on the processing of ferulic acid into food and as a food additive. For example, some nomads of central Iran use the dried aerial parts of *F. assafoetida* L. in the preparation of their delicious local food, "Loghri", which also contains barley, tomato or tomato paste, beans, and other vegetables. In America, different organs of *F. assafoetida* L., either in fresh or dried form, are used for cooking as even small parts of this plant can give a pungent smell to foodstuffs. It has also found many applications as a condiment and flavoring agent in chocolates, seasonings, and soft drinks [2]. Moreover, it is proved by research that felllloylphenethylami and femloyltymmine could effectively inhibit the increase in acid values, polar substances' content, and the oxidative degradation of linoleic acid and 1inolenic acid in frying oil [51]. Secondly, the volatile oil and alcohol extract of *F. ferulaeoides* have a strong bacteriostatic effect, which can be used to make slow-release capsules or preservative agents for plastic wrap applied to food to have a bacteriostatic and preservative effect. Niazmand et al. [52] developed antimicrobial films by incorporating the hydroalcoholic extract of *Ferula asafetida* leaf and gum in the polymer matrix of LDP and investigated the antimicrobial effect of the films on different microorganisms and the capability of the produced films for extending the shelf life of the dough. The results show that as a bioactive packaging material, LDPE film containing *Ferulae* leaves and gum extract is promising, which not only improves food safety and quality, but also has good mechanical, thermal, and barrier properties. Valinezhad et al. [53] prepared a kind of composite material which called chitosan *Ferula gummosa* EO-nanocomposite (CS-FEO). Further, a transparent and flexible CS-FEO biopolymer film was prepared and characterized. The obtained results demonstrate that the prepared CS-FEO nanocomposite could be a potential candidate for food and biomedical applications as it holds the promising capability of proper interactions and advanced features. Finally, *F. ferulaeoides* can be used as a raw material for functional food development. With the improvement in people's quality of life and the enhancement in their awareness of health care, people's consumption of food with health care function has improved significantly. Functional food has gradually opened the market and become the future development trend of the food industry [54]. *F. ferulaeoides* contains a variety of bioactive ingredients, which is an ideal material for the development of functional food.

6. Conclusions

All in all, as a national medicine, the quantity of research on *F. ferulaeoides* is relatively small and not deep enough. First, the modern research of *F. ferulaeoides* mostly focuses on the studies of active ingredients such as sesquiterpenes, volatile oils, coumarins, etc., and less on active ingredients such as phenylpropanoids, phenolic acids, and other steroids. Second, the research on active ingredients mainly stays in the activity screening, mostly focusing on antibacterial, anti-tumor, anti-inflammatory, and insecticidal aspects, and there is a lack of research into its mechanisms. Third, there is a lack of drug safety evaluation in quality control research, and the identification of the fingerprint is insufficient in its depth. The existing reports are almost focused on the volatile oil, there are relatively few content determinations, and the method is more traditional. Fourth, as a kind of medicine and food homologous substance, there is almost no application research in food compared with other *Ferula* plants. In the future work, we should definitely strengthen the study of other active components of *F. ferulaeoides*, and the in-depth study of the material basis and mechanism of action. The identification of other chemical components should be added, and the quantitative analysis of other chemical components should be strengthened by using modern technologies such as liquid chromatography–mass spectrometry. In addition, a safety evaluation method for *F. ferulaeoides* should be established to provide a scientific basis for a comprehensive evaluation of the quality of medicinal materials. Eventually, we should speed up the development and utilization of *F. ferulaeoides* products, and deeply explore its development and utilization value.

Author Contributions: The authors (Z.C., G.Z. and S.M.) have equally contributed to this work. All authors have read and agreed to the published version of the manuscript.

Funding: This research was funded by the Medical Products Administration of Xinjiang Uygur Autonomous Region, Special project on the formulation of quality standards for local medicinal materials of *ferula* root (2021003).

Institutional Review Board Statement: Not applicable.

Informed Consent Statement: Not applicable.

Data Availability Statement: Data will be provided upon request.

Conflicts of Interest: The authors declare no conflict of interest.

References

1. Amalraj, A.; Gopi, S. Biological Activities and Medicinal Properties of Asafoetida: A Review. *J. Tradit. Complement. Med.* **2017**, *7*, 347–359. [CrossRef] [PubMed]
2. Mohammadhosseini, M.; Venditti, A.; Sarker, S.D.; Nahar, L.; Akbarzadeh, A. The Genus *Ferula*: Ethnobotany, Phytochemistry and Bioactivities—A Review. *Ind. Crops Prod.* **2019**, *129*, 350–394. [CrossRef]
3. Yaqoob, U.; Nawchoo, I.A. Distribution and Taxonomy of *Ferula* L.: A Review. *Res. Rev. J. Bot. Sci.* **2016**, *5*, 15–23.
4. Kan, M.M.; Wang, H.; Kang, D.J.; Yu, F.H. Study on chemical constituents of *Ferula syreitschikowii* in Xinjiang. *J. Shihezi Univ.* **2017**, *35*, 132. [CrossRef]
5. Salehi, M.; Naghavi, M.R.; Bahmankar, M. A Review of *Ferula* Species: Biochemical Characteristics, Pharmaceutical and Industrial Applications, and Suggestions for Biotechnologists. *Ind. Crops Prod.* **2019**, *139*, 111511. [CrossRef]
6. Liu, J. Study on Chemical Constituents and Quality Control Method of *Ferula ferulaeoides*. Master's Thesis, Inner Mongolia Medical University, Hohhot, China, 2019.
7. Wang, S. Study on Fingerprint of Active Site of Gastric Cancer In Vitro of *Ferula sinkiangensis*. Master's Thesis, Xinjiang Medical University, Ürümqi, China, 2015.
8. Yang, M.H.; Tang, D.P.; Sheng, P. Research Progress of *Ferula ferulaeoides*. *China Mod. Appl. Pharmacol.* **2020**, *37*, 2031–2041. [CrossRef]
9. Wang, L.; Sun, R.; Xu, M.; Shi, Y.Y. Research on the Development of Chemical Components, Pharmacological Activities and Toxicity of Resina *Ferulae*. *World Tradit. Chin. Med.* **2020**, *15*, 3887–3894.
10. Liu, H.D.; Li, G.F.; Niu, L.Z.; Zhuang, L.; Shi, Y.X. Research of the extraction process and determination of total flavonoids in Uygur medicine *Ferula ferulaeoides*. *Chin. J. Tradit. Chin. Med.* **2016**, *31*, 310–312.
11. Tan, Y.; Gao, T.; Kan, M.M.; Wang, H.; Yu, F.H. Determination of polysaccharide in five kinds of *Ferula* in Xinjiang from different parts. *Food Sci. Technol.* **2016**, *41*, 262–265. [CrossRef]
12. Feng, X.; Wang, L.; Liu, H.D.; Niu, L.Z.; Li, Y.X. Study on optimized extraction technology of ferulic acid from *Ferula ferulaeoides* (Steud.) Korov by an orthogonal experiment method. *Heilongjiang Anim. Husb. Vet. Med.* **2016**, *5*, 196–199. [CrossRef]
13. Eshbakova, K.A.; Saidkhodzhaev, A.I.; Vdovin, A.D.; Abdullaev, N.D. Terpenoid Coumarins from *Ferula ferulaeoides*. *Chem. Nat. Compd.* **2009**, *45*, 708. [CrossRef]
14. Liu, T.; Wang, S.; Xu, L.; Fu, W.; Gibbons, S.; Mu, Q. Sesquiterpenoids with Anti-MDR Staphylococcus Aureus Activities from *Ferula Ferulioides*. *Chem. Biodivers.* **2015**, *12*, 599–614. [CrossRef]
15. Meng, H.; Li, G.; Huang, J.; Zhang, K.; Wei, X.; Ma, Y.; Zhang, C.; Wang, J. Sesquiterpenoid Derivatives from *Ferula ferulaeoides* (Steud.) Korov. *Phytochemistry* **2013**, *86*, 151–158. [CrossRef]
16. Isaka, K.; Nagatsu, A.; Ondognii, P.; Zevgeegiin, O.; Gombosurengyin, P.; Davgiin, K.; Kojima, K.; Ogihara, Y. Sesquiterpenoid Derivatives from *Ferula Ferulaeioides*. V. *Chem. Pharm. Bull.* **2001**, *49*, 1072–1076. [CrossRef] [PubMed]
17. Liu, T.; Osman, K.; Kaatz, G.W.; Gibbons, S.; Mu, Q. Antibacterial Sesquiterpenoid Derivatives from *Ferula ferulaeoides*. *Planta Med.* **2013**, *79*, 701–706. [CrossRef]
18. Hu, Y.; Li, X.-D.; Li, G.-Y.; Li, N.; Zuo, W.-J.; Zeng, Y.-M.; Meng, H.; Li, X.; Wang, J.-H. Two Novel Sesquiterpenoids from the Roots of *Ferula ferulaeoides* (Steud.) Korov. *Helv. Chim. Acta* **2010**, *93*, 1019–1024. [CrossRef]

19. Yang, X.W. Bioactive Material Basis of Medicinal Plants in Genus *Ferula*. *Chin. Mod. Tradit. Chin. Med.* **2018**, *20*, 123–144. [CrossRef]
20. Erdem, S.A.; Nabavi, S.F.; Orhan, I.E.; Daglia, M.; Izadi, M.; Nabavi, S.M. Blessings in Disguise: A Review of Phytochemical Composition and Antimicrobial Activity of Plants Belonging to the Genus Eryngium. *DARU J. Pharm. Sci.* **2015**, *23*, 53. [CrossRef]
21. Bhatnager, R.; Rani, R.; Dang, A.S. Antibacterial Activity of *Ferula Asafoetida*: A Comparison of Red and White Type. *J. Appl. Biol. Biotechnol.* **2015**, *3*, 018–021. [CrossRef]
22. Gao, T.T.; Yu, F.H.; Tan, Y. Study on Antimicrobial Activity in vitro of Extracts from Three Species of *Ferula* Root. *Beifang Yuanyi* **2013**, *24*, 156–158.
23. Boghrati, Z.; Iranshahi, M. *Ferula* Species: A Rich Source of Antimicrobial Compounds. *J. Herb. Med.* **2019**, *16*, 100244. [CrossRef]
24. Khoury, M.; El Beyrouthy, M.; Eparvier, V.; Ouaini, N.; Stien, D. Chemical Diversity and Antimicrobial Activity of the Essential Oils of Four Apiaceae Species Growing Wild in Lebanon. *J. Essent. Oil Res.* **2018**, *30*, 25–31. [CrossRef]
25. Baccari, W.; Znati, M.; Zardi-Bergaoui, A.; Chaieb, I.; Flamini, G.; Ascrizzi, R.; Ben Jannet, H. Composition and Insecticide Potential against Tribolium Castaneum of the Fractionated Essential Oil from the Flowers of the Tunisian Endemic Plant *Ferula Tunetana* Pomel Ex Batt. *Ind. Crops Prod.* **2020**, *143*, 111888. [CrossRef]
26. Elghwaji, W.; El-Sayed, A.M.; El-Deeb, K.S.; ElSayed, A.M. Chemical Composition, Antimicrobial and Antitumor Potentiality of Essential Oil of *Ferula Tingitana* L. Apiaceae Grow in Libya. *Pharmacogn. Mag.* **2017**, *13*, S446–S451. [CrossRef]
27. Jin, B.; Liu, H.; Feng, J.; Yang, X.G.; Zhou, J.H. Application and Prospect of Traditional Chinese Medicine in Cancer Treatment. *Wisdom Health* **2019**, *5*, 50–51. [CrossRef]
28. Yang, M.; Luo, J.Y.; Qiao, M.L.; Sheng, P.; Yang, M.H. Mechanism of Uygur Medicine Root of Ferula ferulaeoides in Resisting Gastric Cancer Activity and Inducing Cell Apoptosis and Cell Cycle Arrest in vitro. *Chin. J. Exp. Formul.* **2018**, *24*, 112–122.
29. Li, X.X.; Zhang, S.J.; Chiu, A.P.; Lo, L.H.; Huang, J.; Rowlands, D.K.; Wang, J.; Keng, V.W. Targeting of AKT/ERK/CTNNB1 by DAW22 as a Potential Therapeutic Compound for Malignant Peripheral Nerve Sheath Tumor. *Cancer Med.* **2018**, *7*, 4791–4800. [CrossRef]
30. Yang, Y.; He, P.Y.; Zhang, Y.; Li, N. Natural Products Targeting the Mitochondria in Cancers. *Molecules* **2020**, *26*, 92. [CrossRef]
31. Ma, R.T.; Sun, J.N.; Zhao, S.J.; Zhang, H.Y. Study on anti-cervical cancer cell activity of 4 *Ferula* ethanol extracts from Xinjiang. *J. Xinjiang Med. Univ.* **2022**, *45*, 1341–1347.
32. Motai, T.; Daikonya, A.; Kitanaka, S. Sesquiterpene Coumarins from *Ferula Fukanensis* and Their Pro-Inflammatory Cytokine Gene Expression Inhibitory Effects. *Chem. Pharm. Bull.* **2013**, *61*, 618–623. [CrossRef]
33. Ye, E.B.; Liu, F.; Xiong, Y.J.; Chen, G.; Yang, X.Z.; Gao, Y. Anti-inflammatory and immunosuppressive effects of three kinds of Xinjiang *Ferula*. *Northwest J. Pharm.* **1993**, *8*, 72–75.
34. Li, G.Y. A Group of Sesquiterpene Coumarins, Sesquiterpene Chromogenic Ketones. CN201310023821.2, 16 September 2015.
35. Sun, J.N.; Ma, R.T.; Zhao, S.J.; Zhang, H.Y. Study on the spectrum-effect relationship of anti-inflammatory activity of different polar parts of three kinds of *Ferula*. *Chem. Biol. Eng.* **2022**, *39*, 24–34.
36. Ahmed, N.; Alam, M.; Saeed, M.; Ullah, H.; Iqbal, T.; Al-Mutairi, K.A.; Shahjeer, K.; Ullah, R.; Ahmed, S.; Ahmed, N. Botanical Insecticides Are a Non-Toxic Alternative to Conventional Pesticides in the Control of Insects and Pests. In *Global Decline of Insects*; Books on Demand: Paris, France, 2021; pp. 1–19.
37. Liu, T. Chemical Constituents from Two Plants of Apiaceae and Their Bioligocal Activities. Ph.D. Thesis, Fudan University, Shanghai, China, 2013.
38. Li, J.W.; Zhao, H.M. Study on killing snails and schistosome miracidia by *Ferula sinkiangensis*. *Tradit. Chin. Vet. Med.* **2007**, *4*, 17–19. [CrossRef]
39. Fu, K.; Zhao, H.M. Study on the molluscicidal mechanism of Ferulic guaiol against Oncomelania. *Tradit. Chin. Vet. Med.* **2015**, *34*, 56–58. [CrossRef]
40. Lin, N. Insecticidal and Antifungal Activities of Volatile oil from *Ferula ferulaeoidis*. Master's Thesis, Shihezi University, Shihezi, China, 2008.
41. Luo, X.; Ma, G.Z.; Shan, M.; Wang, J.; Yu, F.S.; Teng, L.; Wang, C.H. Comparative study on acute toxicity and chemical constituents of volatile oil from two kinds of *Ferulae*. *Chin. Pat. Med.* **2015**, *37*, 1130–1135.
42. Wang, C.X.; Xu, F.; Chen, Y.; Tang, W.; He, J.; Zhao, J.; Zhang, Y.F. Identification study of three commonly used medicinal herbs. *Xinjiang Tradit. Chin. Med.* **2011**, *29*, 50–52.
43. Ding, X.; Tan, Y.; Yan, L.; Cheng, Y.H.; Zhao, W.B. Study on the identification of Ferula ferulaeoides. *Chin. Med. Mater.* **2011**, *34*, 879–881.
44. Liu, M.; Zhao, Y.; Ma, Y.; Liu, S.; Yao, J.; Chi, Y.; Li, H.; Liao, K.; Zhu, Y. The Study of Schizogenous Formation of Secretory Ducts in *Ferula ferulaeoides* (Steud.) Korov. *Protoplasma* **2022**, *259*, 679–689. [CrossRef]
45. Xu, S.; Xu, W.F.; Kuang, Y.M.; Jin, P.F. Research progressin quality evaluation of Chinese herbal medicine *Scutellaria barbata* Herbal. *Northwest J. Pharm.* **2023**, *38*, 222–226.
46. Miao, L.J. Fingerprints Study on *Ferula ferulaeoides* (Steud.) Korov. from National Medicine in Xinjiang. Master's Thesis, Xinjiang Medical University, Ürümqi, China, 2014.
47. Sheng, P.; Tang, D.P.; Miao, L.J.; Zhan, W.J.; Sun, Y.; Wang, Y.J.; Wang, G.P. Fingerprints study on *Ferula ferulaeoides* (Steud.) Korov. from nztional medicine in Xinjiang. *China Mod. Appl. Pharmacol.* **2015**, *32*, 30–37. [CrossRef]

48. Wang, S.; Sheng, P.; Yao, L.; Du, B.J. GC-MS fingerprint of in vitro anti-gastric cancer active parts from roots of Uygur medicine *Ferula ferulaeoides*. *Chin. Herb. Med.* **2015**, *46*, 2874–2879.
49. Luo, X. Preliminary Study on Chemical Fingerprint and Antiulcer Activity of Two Species of *ferula*. Master's Thesis, Xinjiang Medical University, Ürümqi, China, 2015.
50. Zhu, Y.; Zhang, K.; Liang, X.; Tang, Y.; Huang, J.; Wang, J.H. Content Determination of Sesquiterpene Coumarin (DAW22) in *Ferula ferulaeoides* by UPLC. *Chin. J. Exp. Prescr.* **2016**, *22*, 63–65. [CrossRef]
51. Jin, W.H.; Zhang, J.L.; Gao, L.; Wang, X.G.; Wang, X.S. Evaluation of antioxidative properties of ferulic acid amides with different dosages in flying system of soybean oil. *China Fats Oils* **2021**, *46*, 89–92+116. [CrossRef]
52. Niazmand, R.; Razavizadeh, B.M.; Sabbagh, F. Simulating Release Model and Antimicrobial Efficiency of LDPE Film Carrying Ferula Asafetida Leaf and Gum Extracts. *Polym. Bull.* **2022**, *79*, 1151–1174. [CrossRef]
53. Valinezhad, N.; Talebi, A.F.; Alamdari, S. Preparation, Characterization, and Antibacterial Effects of Ferula Gummosa Essential Oil–Chitosan (CS-FEO) Nanocomposite. *Nanochem. Res.* **2022**, *7*, 85–92.
54. Li, T.T.; Zhu, Y.F.; Li, H.J.; Shi, S.L.; Yu, D. Research progress of functional foods. *Mod. Food* **2022**, *28*, 79–81. [CrossRef]

MDPI AG
Grosspeteranlage 5
4052 Basel
Switzerland
Tel.: +41 61 683 77 34

Molecules Editorial Office
E-mail: molecules@mdpi.com
www.mdpi.com/journal/molecules

www.ingramcontent.com/pod-product-compliance
Lightning Source LLC
Chambersburg PA
CBHW040753250926
51850CB00007BB/1485
9783725848652